PectinTechnological and Physiological Properties

Vassilis Kontogiorgos
Editor

Pectin: Technological and Physiological Properties

Springer

Editor
Vassilis Kontogiorgos
School of Agriculture and Food Sciences
The University of Queensland
Brisbane, QLD, Australia

ISBN 978-3-030-53420-2 ISBN 978-3-030-53421-9 (eBook)
https://doi.org/10.1007/978-3-030-53421-9

This Springer imprint is published by the registered company Springer Nature Switzerland AG
The registered company address is: Gewerbestrasse 11, 6330 Cham, Switzerland

Preface

Polysaccharides are routinely used in food and pharmaceutical industries as they show a broad range of functional properties that are employed to create structures with reproducible physical properties. In recent years, there is a need to create advanced formulations beyond traditional food applications that, for example, delay lipid digestion or deliver bioactives in the gastrointestinal tract at the site of interest. In addition, another area of focus is the identification of bioactive polysaccharides that exhibit physiological characteristics that may improve health. The need to understand how to form and stabilise structures for these applications has boosted research on the fundamental chemical, physical and physiological properties of polysaccharides. Pectin is an ideal macromolecule for such applications, as it is highly tunable because of its complex chain architecture, type and amount of substituent groups, charge and charge density and, in certain cases, the presence of protein. Achieving these research objectives, however, is not trivial, and strong foundations are always needed.

The present book was written with the goal to form the springboard for new research ventures using pectin as the primary biological material, as it covers the entire spectrum of its technological and physiological properties. The book aims to have an educational approach that can be used as a supplementary textbook in relevant university courses that deal, for example, with food chemistry, food physical chemistry, digestion and metabolism, functional biomaterials or biodegradable polymers. The chapters provide all the classic information with a measured, in-depth approach instigating curiosity for further reading. The book also deals with current topics such as emulsification and film-forming properties of pectin as well as its gastrointestinal fate and bioactivity.

The targeted audience of the book are undergraduate and postgraduate students, and researchers in academia or industry with no previous experience on pectin who would like a short but at the same time a thorough introduction in all aspects of pectin properties. The ambition is that this should be the first book that someone would reach to learn about pectin and I hope that readers find it a useful addition to their reference book collection.

Brisbane, QLD, Australia Vassilis Kontogiorgos

Contents

Contributors

K. Alba Department of Biological Sciences, University of Huddersfield, Huddersfield, UK

C. T. Anderson Department of Biology, The Pennsylvania State University, State College, PA, USA

A. Assifaoui School of Pharmacy, Université de Bourgogne Franche Comté, UMR PAM, PCAV team, Dijon, France

C. G. Biliaderis Department of Food Science and Technology, Faculty of Agriculture, Aristotle University of Thessaloniki, Thessaloniki, Greece

H. A. S. Binhamad Department of Chemical Sciences, University of Huddersfield, Huddersfield, UK

E. Bonnin INRAE, UR BIA Biopolymères Interactions Assemblages, Nantes, France

O. Chambin School of Pharmacy, Université de Bourgogne Franche Comté, UMR PAM, PCAV team, Dijon, France

S. W. Cui Guelph Research and Development Centre, Agriculture and Agri-Food Canada, Guelph, ON, Canada

H. Ding Guelph Research and Development Centre, Agriculture and Agri-Food Canada, Guelph, ON, Canada

K. Haider Quadram Institute Bioscience, Norwich Research Park, Norwich, UK

V. Kontogiorgos School of Agriculture and Food Sciences, The University of Queensland, Brisbane, QLD, Australia

A. Lazaridou Department of Food Science and Technology, Faculty of Agriculture, Aristotle University of Thessaloniki, Thessaloniki, Greece

G. A. Morris Department of Chemical Sciences, University of Huddersfield, Huddersfield, UK

J. Pelloux BIOPI Biologie des Plantes et Innovation SFR Condorcet, Université de Picardie, Amiens, France

M. C. Ralet INRAE, Biopolymères Interactions Assemblages, Nantes, France

D. Ropartz INRAE, Biopolymères Interactions Assemblages, Nantes, France

P. Wilde Quadram Institute Bioscience, Norwich Research Park, Norwich, UK

M. A. K. Williams School of Fundamental Sciences, Massey University, Palmerston North, New Zealand

The MacDiarmid Institute for Advanced Materials and Nanotechnology, Wellington, New Zealand

The Riddet Institute, Massey University, Palmerston North, New Zealand

Y. Yang Laboratory of Cell and Molecular Biology, Institute of Vegetable Science, Zhejiang University, Hangzhou, China

About the Editor

Vassilis Kontogiorgos is a food scientist by training having received all his degrees in Food Science from the Aristotle University of Thessaloniki, Greece (B.Sc. and M.Sc.) and University of Guelph, Canada (Ph.D.). He has worked as an NSERC research fellow at the Agriculture and Agri-Food Canada (Canada) and as faculty member at the Department of Biological Sciences, University of Huddersfield (UK) before joining the School of Agriculture and Food Sciences at the University of Queensland (Australia).

His research interests are focused on the area of polysaccharide isolation and characterisation with particular interest on pectin extraction and functionality. He is also working on the physical chemistry of colloidal systems that are fabricated with pectin and other polysaccharides. He is an associate editor of *Food Hydrocolloids* and *Food Biophysics*.

Chapter 1
Biosynthesis, Localisation, and Function of Pectins in Plants

Yang Yang and Charles T. Anderson

1.1 Pectin Biosynthesis

Pectins are a class of matrix polysaccharides in the cell wall that contains three major domains: homogalacturonan (HG), rhamnogalacturonan I (RG-I), and rhamnogalacturonan II (RG-II) (see Chap. 2). Galacturonic acid (GalA) monomers form the backbones of HG and RG-II domains, whereas rhamnose-GalA dimers form the backbone of RG-I. Both types of RG can be linked to HG, and HG and RG-I can also be covalently attached to an arabinogalactan protein (Fig. 1.1) (Atmodjo et al. 2013; Tan et al. 2013).

Pectic polysaccharides are synthesised in the Golgi apparatus, with related glycoproteins simultaneously synthesised in different Golgi cisternae (Moore et al. 1991) (Fig. 1.1). Newly synthesised pectins are packaged into vesicles and delivered to the cell wall (Moore et al. 1991). Approximately 67 transferases, including glycosyltransferases, methyltransferases, and acetyltransferases, are predicted to function in pectin biosynthesis. Glycosyltransferases function in synthesising pectin backbones and adding side chains, whereas methyltransferases and acetyltransferases catalyse pectin methyl-esterification and acetylation, respectively (Mohnen 2008). In addition, the synthesis and correct localisation of nucleotide sugars, which are the substrates for pectin synthesis, require the action of other enzymes, including interconversion enzymes, sugar kinases, UDP-sugar pyrophosphorylases, and nucleotide sugar transporters (Atmodjo et al. 2013).

HG, the most abundant domain of pectins, is a linear polymer of galacturonic acid (GalA), and is partially modified with methyl groups and acetyl groups at the

Y. Yang
Laboratory of Cell and Molecular Biology, Institute of Vegetable Science, Zhejiang University, Hangzhou, China

C. T. Anderson (✉)
Department of Biology, The Pennsylvania State University, University Park, PA, USA
e-mail: cta3@psu.edu

V. Kontogiorgos (ed.), *Pectin: Technological and Physiological Properties*,
https://doi.org/10.1007/978-3-030-53421-9_1

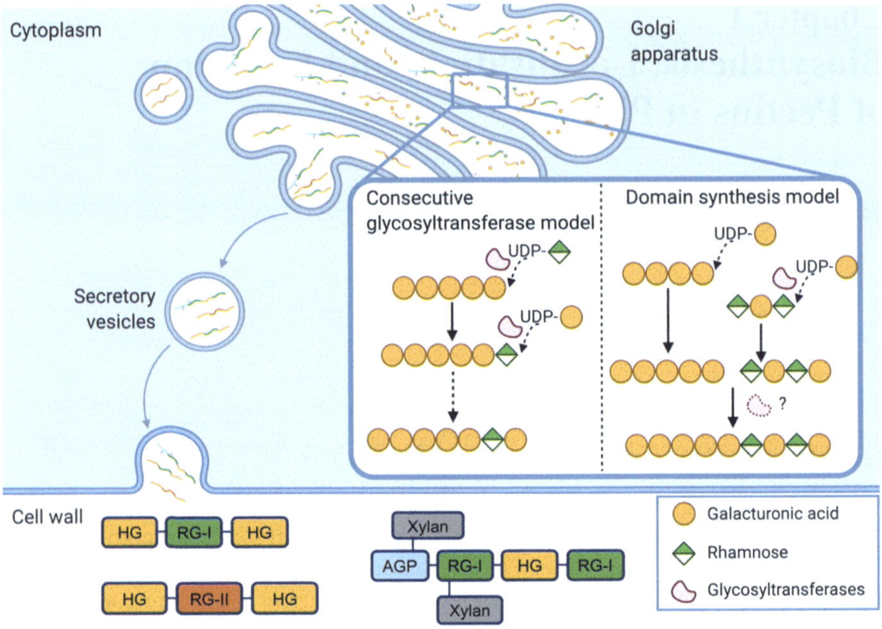

Fig. 1.1 Pectin synthesis and domain arrangement. Pectins are synthesised in the Golgi apparatus and secreted to the cell wall. The inset box shows current hypotheses about pectin synthesis: in the consecutive glycosyltransferase model, UDP-GalA or UDP-Rha are consecutively added to the non-reducing end of pectin chains by glycosyltransferases; in the domain synthesis model, pectin domains are elongated separately by glycosyltransferases, then linked with other domains. In the cell wall, pectin domains are arranged in at least three different covalently linked configurations. Figure made with BioRender

C-6 and O-2/O-3 positions, respectively. Galacturonosyltransferases (GAUTs) transfer UDP-GalA onto the non-reducing end of oligogalacturonide, resulting in HG chain elongation (Scheller et al. 1999). In *Arabidopsis thaliana*, the *GAUT* and *GAUT-LIKE* gene families contain 15 and 10 genes, respectively, some of which have been demonstrated to produce Golgi-localised proteins (Kong et al. 2011; Sterling et al. 2001). GAUT1 has UDP-GalA transferase activity, and GAUT7 is required for anchoring GAUT1 to the Golgi by forming a GAUT1:GAUT7 complex (Sterling et al. 2006; Atmodjo et al. 2011). GAUT11 also shows transferase activity for HG elongation, which is essential for synthesising the HG regions in RG-I domain (Voiniciuc et al. 2018), and GAUT4 has UDP-GalA transferase activity when transiently expressed (Biswal et al. 2018). HG undergoes methylesterification and acetylation before delivering to the wall. In *A. thaliana*, increased methyltransferase activities are detected from the isolated microsomal fractions of *COTTON GOLGI-RELATED2* (*CGR2*) and *CGR3* overexpression lines, respectively, indicating that these two genes encode putative methyltransferases (Kim et al. 2015). Mutations in another putative pectin methyltransferase gene, *QUASIMODO2/TUMOROUS SHOOT DEVELOPMENT2*, causes dramatic reduction in HG content (Mouille et al. 2007). In addition, the gene product of *POWDERY MILDEW*

RESISTANT 5, which encodes an acetyltransferase, transfers acetyl-CoA to pectin fragments as an acetyltransferase, and contributes to plant resistance to fungal disease (Chiniquy et al. 2019).

In contrast to HG, the RG-I backbone is formed by the repeating disaccharide of rhamnose-GalA (Mohnen 2008). Rhamnosyltransferases add rhamnose residues to the non-reducing end of RG-I oligosaccharides, and presumably, a galacturonosyltransferase alternates with the rhamnosyltransferase adds GalA residues to the RG-I backbone (Takenaka et al. 2018). A RG-I:rhamnosyltransferase (RRT), RRT1, has been characterised from *A. thaliana* with enzyme activity determined in vitro (Takenaka et al. 2018). The rhamnose residues of RG-I can be modified with side chains containing galactans and/or arabinans. RG-I galactosyltransferase and arabinosyltransferase activities are detected in microsomal fractions from suspension cultured cells and plant tissues (Geshi et al. 2002; Nunan and Scheller 2003). *ARABINAN DEFICIENT 1* and *2* encode putative arabinosyltransferases in *A. thaliana*, and mutation of these genes leads to decreased arabinan content in the cell wall (Harholt et al. 2012).

RG-II shares the same backbone structure with HG, but has complex side chains. Xylosyltransferase activity has been determined for *Arabidopsis* rhamnogalacturonan specific xylosyltransferas E (RGXT) 1, 2, and 3, which link xylose onto the fucose residues in the side chains of RG-II (Egelund et al. 2006, 2008). Up to now, several genes encoding putative RG-II synthesis enzymes have been functionally characterised in terms of their effects on pectin content and plant development (Funakawa and Miwa 2015). However, it remains unclear whether those genes encode *bona fide* enzymes and how most RG-II backbone and side chain residues are linked to the polymer.

Based on current studies, two models are proposed for pectin synthesis (Fig. 1.1): the consecutive glycosyltransferase model and the domain synthesis model (Atmodjo et al. 2013). In the consecutive glycosyltransferase model, nucleotide-sugar residues are consecutively added to the non-reducing end of growing backbones and side chains of pectin. Based on the mechanism of the GAUT1:GAUT7 complex in HG synthesis, this model has been updated into a two-phase model whose synthesis efficiency depends on chain length (Amos et al. 2018). During de novo synthesis and short-chain acceptor elongation, pectin synthesis undergoes a slow phase, which converts to a rapid phase when the chain acceptor reaches 11 residues in length. In the domain synthesis model, HG, RG-I, and RG-II domains are elongated separately by adding nucleotide sugars onto their growing chains, and oligosaccharide domains are then linked to other domains to form the final pectin structure (Atmodjo et al. 2013). However, the identity of any domain-linking proteins, or "pectin ligases," is unknown.

1.2 Localisation of Pectins in Plants

Pectins are deeply conserved from the Charophyte algal relatives of plants to angiosperms, and their abundance varies among different plant species. Although cell wall structure in non-vascular plants differs from that in vascular plants, pectins can still be detected in cell wall residues of Bryophytes and Charophytes (Popper and Fry 2003). For instance, in *Penium margaritaceum*, a unicellular Charophyte alga, HG is a major wall component and localises to the outer layer of the cell wall (Domozych 2014). Pectins are enriched in the primary walls of eudicots and non-grass monocots, making up 20–35% of the dry mass of the cell wall. However, pectins are reported to constitute only 5% and 0.1% of the primary cell walls of grasses and the secondary cell walls of angiosperms, respectively (Vogel 2008). The compositions of the primary cell walls in gymnosperms are similar to those of eudicots, in which GalA is the most abundant uronic acid (Popper and Fry 2004). In early-diverging vascular plants such as lycophytes, the amount of RG-II in the primary cell wall is comparable with that in angiosperms, but higher than in bryophytes. This increase in RG-II content corresponds with the evolution of lignified secondary walls in vascular plants (Matsunaga et al. 2004).

The plant cell wall is a unique structure, and about 90% of its dry mass consists of pectins, cellulose, hemicellulose (Albersheim et al. 1996). Prevailing cell wall models indicate that pectins form a gel-like matrix (Fig. 1.2a), with cellulose microfibrils and hemicellulose embedded in this matrix (Lampugnani et al. 2018). However, pectins exist in close proximity with both hemicelluloses and cellulose, and covalent crosslinks between pectins and hemicellulose have been detected (Broxterman and Schols 2018; Fry 2018).

Pectins show dynamic patterns in the walls during cell growth. HG has recently been reported to assemble into nanofilament structures, facilitating anisotropic cell growth (Haas et al. 2020). Putative metabolic pulse-labelling of RG-I also shows reorientation from a diffuse pattern into a fibrillar pattern along with cell elongation (Anderson et al. 2012) (Fig. 1.2a, b). During cytokinesis, pectins are one of the earliest wall components delivered to the growing cell plate (Samuels et al. 1995). Pectins are enriched in the middle lamella that lies between the primary walls of adjacent cells and function in cell adhesion (Daher and Braybrook 2015) (Fig. 1.2c). In wood cell walls, pectins are irregularly located in the middle lamella, which corresponds to the location where lignin first appears during secondary xylem development (Wi et al. 2005). They also localise to the boundaries of tip-growing and anisotropically growing cells (Fig. 1.2d, e). Pectin content varies in different tissues, and pectins are enriched in fruits, seed coats, and root caps (Fig. 1.2f–h). In kiwifruit, tomato, and strawberry fruits, pectins comprise over 45% of the dry mass of the cell wall, and relative pectin content increases along with fruit ripening (Redgwell et al. 1997). Therefore, fruits are major sources for extractable pectins (see Chap. 4). In addition, pectins are the dominant component of mucilage, where mainly consist of RG-I and a small proportion of HG (Francoz et al. 2015). During

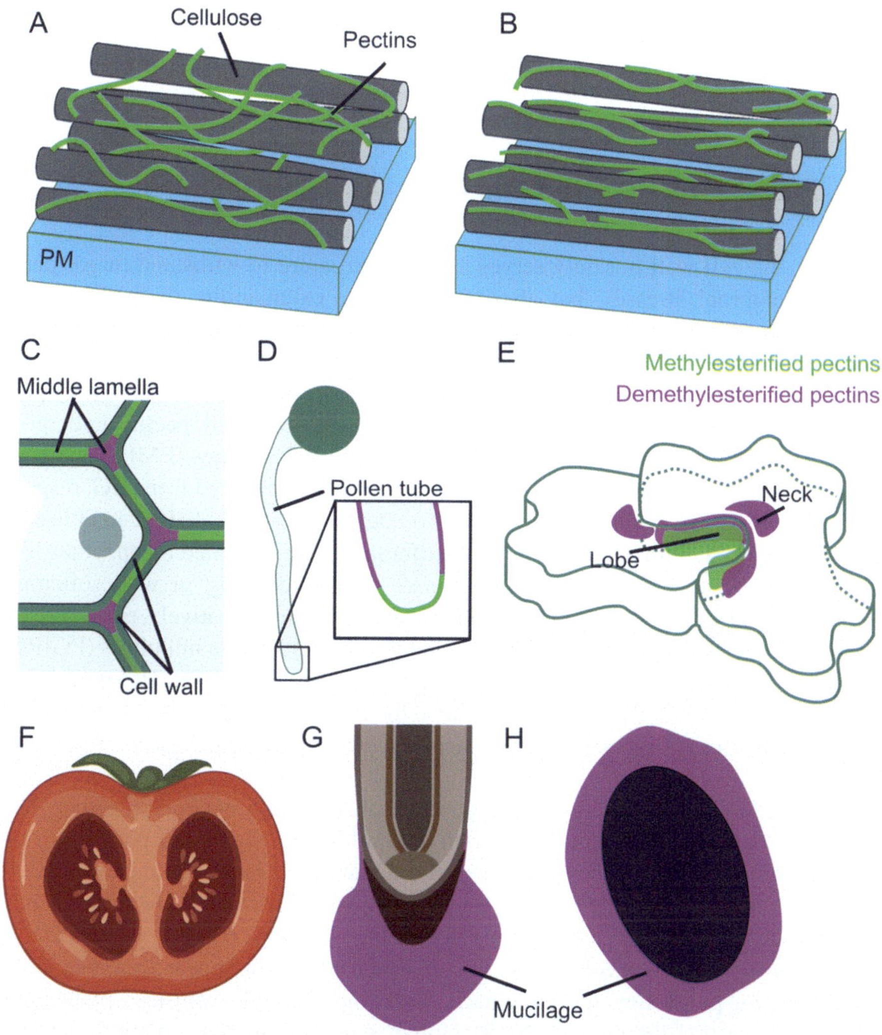

Fig. 1.2 Pectin localisation in the cell wall and distribution in cells and tissues. Pectin arrangements in the cell wall are shown in (**a**) and (**b**): a diffuse pattern (**a**) and a fibrillar pattern (**b**). Pectin distributions at the cellular level are shown in (**c–e**). In the middle lamella (**c**), methylesterified pectins (green) and demethylesterified pectins (magenta) are enriched at boundaries between adjacent cells and corner junctions, respectively. In the pollen tube wall (**d**), methylesterified pectins and demethylesterified pectins are asymmetrically located at apical and longitudinal walls, respectively. During lobe formation in pavement cells (**e**), demethylesterified pectins are enriched at neck regions of periclinal walls, and are less abundant at lobe regions of periclinal walls. Coupled with wall expansion, pectins within anticlinal walls undergo demethylesterification at walls adjacent to new lobe regions. At the tissue level, pectins are enriched in softening fruits (**f**). Mucilage released at the surface of root cap (**g**) and seed coat (**h**) is high in demethylesterifed pectins, which shows a red colour with Ruthenium red staining. PM, plasma membrane. Figure made with BioRender

seed germination and root growth, mucilage is extruded at the surface of seed coat and root cap, facilitating seed imbibition and root protection (Iijima et al. 2004; Western et al. 2000).

1.3 Pectin Functions in Plants

In plants, the cell wall not only serves as a stiff structure to withstand turgor pressure and external stresses, but also possesses the extensibility to facilitate cell growth and constrain cell shape (Cosgrove 2018). Pectin structure, status and localisation are important influencers of wall structure and mechanics, which modulate cell dynamics and further regulate plant development. Pectin modifications in particular are central in modulating wall mechanics. In the cell wall, pectins undergo a series of modifications. With the action of pectin methylesterases (PMEs) and pectin acetylesterases (PAEs), methyl and acetyl groups are removed from HG, respectively (Bonnin et al. 2008; Bordenave 1996). Demethylesterified HG can either be crosslinked by calcium, resulting in wall stiffening, or be degraded by enzymes like polygalacturonases (PGs) and pectate lyases (PLs), resulting in wall softening (Hocq et al. 2017). The activities of PMEs and PGs are negatively regulated by pectin methylesterase inhibitors (PMEIs) and polygalacturonase inhibitors (PGIPs), respectively (Sénéchal et al. 2014). For more information about HG-modifying enzymes, see Chap. 3. Below, we summarise the functions of pectins in cell dynamics and how these processes influence plant development and stress responses (Fig. 1.3).

The middle lamella functions as a glue for adhering adjacent cells together. The absence of cellulose and hemicellulose in the middle lamella makes pectins strong candidates for mediating cell adhesion and separation (Jarvis et al. 2003). Cell adhesion is mainly controlled at thickened corner junctions, which are strengthened by demethylesterified pectin with calcium-crosslinking (Parker et al. 2001) (Fig. 1.2c). Decreased levels of calcium-crosslinked pectin appear to reduce cell-cell adhesion, promoting the enzymatic isolation of single wall-less plant cells called protoplasts (Lionetti et al. 2015). Unlike at corner junctions, methylesterified pectins accumulate at regions between adjacent cells (Zamil and Geitmann 2017) (Fig. 1.2c). PMEs and PGs are required to degrade pectins and facilitate cell separation: for example, inhibiting PME activity prevents the separation of border cells from root tips (Wen et al. 1999). Downregulating PG activity also results in extended cell-cell adhesion in fruits (Posé et al. 2013). It is noteworthy that microspore separation requires the degradation of the primary cell walls of pollen mother cells, which envelop developing microspores. PMEs and PGs are required in this process, and mutations in PME and PG genes result in aberrantly connected tetrad pollen (Francis et al. 2006; Rhee and Somerville 1998). Cell adhesion defects are also reported in HG synthesis mutants: perturbing HG chain elongation and methylesterification causes cell adhesion defects in *Arabidopsis* hypocotyls (Bouton et al. 2002; Krupková et al. 2007; Mouille et al. 2007). However, in some of these mutants,

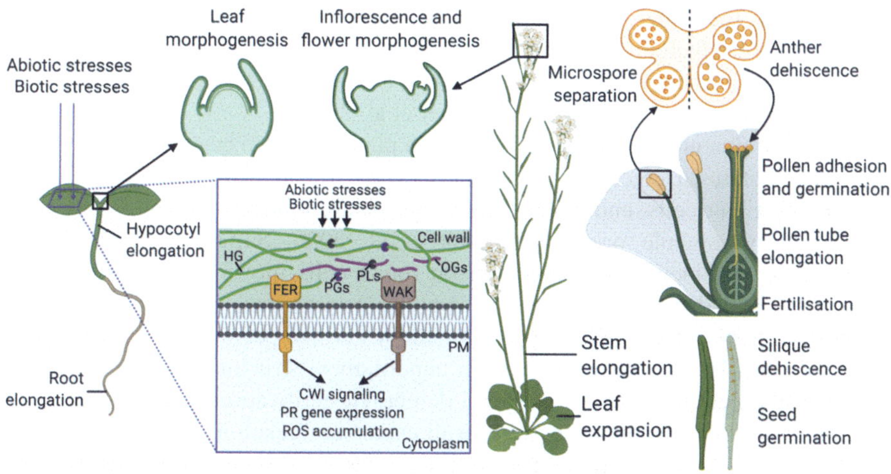

Fig. 1.3 Pectin functions in plant development and stress responses. Here, *Arabidopsis* is used as a model to show pectin functions in multiple developmental stages and stress responses. Pectins modulate organ elongation and expansion, like the elongation of hypoctyls, roots, and stems, and leaf expansion. Pectins function in organ morphogenesis by influencing organ patterning at apical meristems. During reproductive development, pectins mediate plant fertility by controlling pollen development, male-female interactions, fertilisation, and seed germination. The degradation of pectins also facilitates anther and silique dehicience, and organ abscission. For plant-environment interactions, pectins might triger cell wall intergity (CWI) signaling pathways via their interactions with FERONIA (FER) and other receptor-like kinases. Pectin degradation products, oligogalacturonides (OGs), might bind to Wall-associated kinase (WAK) receptors and trigger downstream immune responses (inset panel). *PG* polygalacturonase, *PL* pectate lyase, *PR gene* pathogen-related gene, *ROS* reactive oxygen species, *PM* plasma membrane. Figure made with BioRender

the cell adhesion defect does not directly result from the reduced HG content. Instead, lower HG levels appear to trigger pectin-related signalling pathways and abrogate cell adhesion (Verger et al. 2016).

Asymmetric wall mechanics based on pectin asymmetries and cellulose orientation underlie cell anisotropic growth in rapidly growing tissues like dark-grown hypocotyls. In the epidermis of the dark-grown hypocotyl, methylesterified pectins localise in fast-growing axial walls, whereas demethylesterified pectins with calcium crosslinking are present in slow-growing transverse walls. Correlated with this asymmetric pectin distribution, axial walls show higher elasticity than transverse walls (Daher et al. 2018). Pollen tubes are useful models for studying tip growth in plant cells. In the pollen tube wall, methylesterified pectins are enriched at the apex, whereas demethylesterified pectins are only present along longitudinal walls (Fig. 1.2d). This gradient of pectin distribution is coupled with the spatial profile of wall extensibility, controlling pollen tube elongation (Fayant et al. 2010). To maintain this asymmetric distribution, PMEIs are restricted to the pollen tube apex and inhibit PME activity in this region (Chebli et al. 2012; Röckel et al. 2008). Impaired pollen tube growth is reported in mutant lines lacking functional PME, PG, and PAE genes (Gou et al. 2012; Huang et al. 2009a; Jiang et al. 2005). Additionally, RG-II is enriched at the pollen tube apex, and its borate cross-linking maintains boron concentrations, which are important for pollen tube elongation (Iwai et al. 2006). RG-I is widely present in the pollen tube wall, and its side chain structure prevents the tight apposition of HG polymers (Lehner et al. 2010).

The development of complex cell shapes in plants can also be controlled by pectin level and status. Defective RG-I synthesis leads to helical growth of petal cells, distorted root hair cells, and twisted growth of petals and roots, which is independent from microtubule orientation (Diet et al. 2006; Saffer et al. 2017). In *Arabidopsis*, pavement cells develop from round cells into jigsaw puzzle piece-shaped cells with distinct neck (convex) and lobe (concave) regions. Lobe formation is closely related to the distribution of demethylesterified pectins (Fig. 1.2e). In the anticlinal wall, HG has been detected in nanofilament structures oriented perpendicular to the periclinal wall. Local HG demethylesterification is hypothesised to lead to radial swelling and asymmetric wall expansion, resulting in lobe formation (Haas et al. 2020). In the periclinal wall, the accumulation of demethylesterified pectins at neck regions leads to higher wall stiffness than in lobe regions. This uneven mechanics restricts periclinal wall expansion in neck regions, and influences the orientations of microtubules and newly deposited cellulose as well (Altartouri et al. 2019).

In addition, pectins facilitate dynamic cell processes, including stomatal opening and closing and mucilage secretion during seed imbibition. Stomatal dynamics are important for regulating gas exchange between plants and the environment. Wall flexibility in the guard cells that flank stomatal pores is mediated by pectic arabinan and HG (Amsbury et al. 2016; Jones et al. 2003; Rui et al. 2017): for example, highly methylesterified pectin levels cause reduced dynamic range in stomatal movements (Amsbury et al. 2016), and mutation of a PG gene leads to inhibited stomatal closing, whereas excess HG degradation catalysed by overexpression of the same PG accelerates stomatal opening (Rui et al. 2017). During seed imbibition, mucilage is released from the locally ruptured outer primary wall and forms a gel-like structure at the seed surface. Pectin demethylesterification guides the localisation of peroxidases in the outer cell wall, facilitating wall loosening in that region (Francoz et al. 2019). Mucilage is abundant in unbranched RG-I, which is essential for its hydration properties. Insufficient side chain trimming on RG-I results in incomplete and slow mucilage release, as well as failed mucilage expansion (Arsovski et al. 2009; Dean et al. 2007).

The multiple functions of pectins at the cellular level underlie their crucial roles in plant development (Fig. 1.3). Pectins function in anisotropic cell growth, contributing to organ elongation in hypocotyls, roots, and stems. Axial organ growth also shows a gradient in growth rate with the highest growth rate at the apical zone. In correlation with this growth gradient, pectin content, structure, and degree of methylesterification and acetylation also show apical-basal patterning in hypocotyls and inflorescence stems (Goldberg and Prat 1982; Phyo et al. 2017a). In inflorescence stems, the cell walls of the fastest growing segment contain more pectin and side chain structures, higher degrees of methylesterification and mobility, and less interaction with cellulose than in basal segments (Phyo et al. 2017a). In seedlings overexpressing a PG, the cell walls of hypocotyls show more methylesterified pectins and dispersed RG-I side chains, less calcium cross-linked HGs, and reduced interactions between pectin and cellulose (Phyo et al. 2017b). Pectins also facilitate organ expansion. Altered RG-II structure with reduced borate cross-linking results in dwarf rosettes (O'Neill et al. 2001). Altered expression of PGs affects rosette leaf expansion (Xiao et al. 2014, 2017; Rui et al. 2017).

Pectins modulate organ morphogenesis by mediating asymmetries in wall mechanics (Fig. 1.3). Leaf adaxial-abaxial polarity is essential for asymmetric leaf growth. In developing leaves, differential pectin methylesterification status among adaxial, middle, and abaxial domains is correlated with asymmetric wall mechanics, and further regulates leaf polarity. Ectopic expression of PME or PMEI genes in different leaf domains disturbs the asymmetry of wall mechanics, resulting in radially symmetrical leaves (Qi et al. 2017). In the inflorescence meristem, pectin demethylesterification status in subepidermal layers modulates floral primordium initiation and maintains phyllotaxis. Inhibiting PME activity results in pin-like meristems and cylindrical flower primordia, whereas overexpressing PME causes the ectopic formation of floral organs (Peaucelle et al. 2008, 2011).

During reproductive development, pectins are essential for the development of male and female gametophytes (Fig. 1.3). During early pollen development, methylesterified pectins are located on the outer layer of the microspore plasma membrane, as well as in the tapetum, which provides materials for pollen wall synthesis. In mature pollen walls, methylesterified and demethylesterified pectins are present at the inner layer (endintine) and outer layer (exintine) of the pollen intine, respectively (Aouali et al. 2001; Geitmann et al. 1995). Mutants for PME, PG, and PL genes show abnormal pollen intine development, reduced pollen activity, and delayed pollen germination (Huang et al. 2009a, b; Jiang et al. 2014; Leroux et al. 2015). To deliver sperm cells to the ovary, demethylesterified pectins enhance pollen tube adhesion to transmitting tracts (Jiang et al. 2005; Mollet et al. 2000). Along with the FERONIA receptor-like kinase, demethylesterified pectins at the filiform apparatus help to prevent multiple pollen tubes from penetrating a single female gametophyte (Duan et al. 2020). Apart from pectin modification, RG-II synthesis is necessary for male and female organ structure. Reduced RG-II content in a glucuronyltransferase mutant results in reduced pollen germination, branched pollen tube structure, and defective transmitting tracts, which affect the interaction between pollen tube and the ovary (Iwai et al. 2006).

Pectins and their modification are essential for fruit ripening, organ abscission, and organ dehiscence, which are often accompanied by pectin degradation and loss of cell-cell adhesion (Fig. 1.3). In the early stage of fruit ripening, pectins undergo demethylesterification upon dramatically increased PME activity, and galactan and arabinan side chains are trimmed from RG-I (Draye and Van Cutsem 2008; Orfila et al. 2001). In the following stages, PGs and PLs depolymerise pectin and lead to fruit softening (Jiménez-Bermúdez et al. 2002; Quesada et al. 2009). The process of fruit softening can be manipulated by changing the expression of pectin-degrading enzymes: downregulating the expression of PG and PL genes in strawberry helps to maintain fruit firmness and elongates the duration of postharvest softening without influencing fruit weight and shape (Jiménez-Bermúdez et al. 2002; Quesada et al. 2009). In fact, the first genetically engineered food, the Flavr Savr tomato, benefited from this idea. By knocking down the expression of a PG, these tomatoes show longer shelf life and increased resistance to rotting (Bruening and Lyons 2000). During organ abscission and dehiscence, pectin degradation occurs in the cell wall and middle lamella in abscission zones and dehiscence zone. Cell detachment in these regions helps to remove senescent organs from plants, and is also essential for releasing mature pollen garins and seeds from anthers and siliques, respectively (Ogawa et al. 2009; Yang et al. 2018).

In plant-environment interactions, pectins are involved in plant immune response to abiotic and biotic stresses (Fig. 1.3). During pathogen or parasitic plant invasion, the host cell wall acts as the first barrier to invasion and is degraded by exogenous pectin-modifying enzymes. Oligogalacturonides (OGs) are short fragments of degraded pectin, which can be released by endogenous, pathogen, or parasite PGs and PLs (Ferrari et al. 2013). During invasion, host PGIPs and PMEIs are upregulated to inhibit exogenous PG and PME activities, respectively. PGIPs can also regulate PG activity and trigger OG accumulation. OGs function as damage-associated molecular patterns, and OG signals are perceived by downstream receptors that activate defence responses, including the production of reactive oxygen species (ROS) and phytoalexins, callose accumulation, and the induction of expression of pathogen related (PR) genes (Benedetti et al. 2015; Ferrari et al. 2013; Gallego-Giraldo et al. 2020). Wall-associated kinase 1 (WAK1) is one well-characterised OG receptor, and can bind OGs in a calcium-dependent manner (Decreux and Messiaen 2005). As the most mobile wall components, pectins serve as cell-wall integrity sensors, and can be bound by other types of cell wall integrity receptors. One such receptor is FERONIA, which directly interacts with pectins via its extracellular domain and might trigger signalling pathways to maintain cell-wall integrity during abiotic and biotic stresses (Feng et al. 2018; Vaahtera et al. 2019). Pectins are complex and intriguing constituents of the cell wall with wide-ranging localisation and function across the plant kingdom. For more information on these fascinating biomolecules, see Anderson (2016, 2019), Atmodjo et al. (2013), Daher and Braybrook (2015), Palin and Geitmann (2012), Peaucelle et al. (2012), Sénéchal et al. (2014), and Wolf and Greiner (2012).

References

Albersheim P, Darvill AG, O'Neill MA et al (1996) An hypothesis: the same six polysaccharides are components of the primary cell walls of all higher plants. In: Visser J, Voragen AGJ (eds) Pectins and pectinases. Elsevier, London, pp 47–55

Altartouri B, Bidhendi AJ, Tani T et al (2019) Pectin chemistry and cellulose crystallinity govern pavement cell morphogenesis in a multi-step mechanism. Plant Physiol 181:127–141

Amos RA, Pattathil S, Yang JY et al (2018) A two-phase model for the non-processive biosynthesis of homogalacturonan polysaccharides by the GAUT1:GAUT7 complex. J Biol Chem 293:19047–19063

Amsbury S, Hunt L, Elhaddad N et al (2016) Stomatal function requires pectin de-methyl-esterification of the guard cell wall. Curr Biol 26:2899–2906

Anderson CT (2016) We be jammin: an update on pectin biosynthesis, trafficking and dynamics. J Exp Bot 67:495–502

Anderson CT (2019) Pectic polysaccharides in plants: structure, biosynthesis, functions, and applications. In: Cohen E, Merzendorfer H (eds) Extracellular sugar-based biopolymers matrices. Springer, Cham, pp 487–514

Anderson CT, Wallace IS, Somerville CR (2012) Metabolic click-labeling with a fucose analog reveals pectin delivery, architecture, and dynamics in *Arabidopsis* cell walls. Proc Natl Acad Sci U S A 109:1329–1334

Aouali N, Laporte P, Clément C (2001) Pectin secretion and distribution in the anther during pollen development in Lilium. Planta 213:71–79

Arsovski AA, Popma TM, Haughn GW et al (2009) AtBXL1 encodes a bifunctional beta-D-xylosidase/alpha-L-arabinofuranosidase required for pectic arabinan modification in *Arabidopsis* mucilage secretory cells. Plant Physiol 150:1219–1234

Atmodjo MA, Sakuragi Y, Zhu X et al (2011) Galacturonosyltransferase (GAUT)1 and GAUT7 are the core of a plant cell wall pectin biosynthetic homogalacturonan:galacturonosyltransfer ase complex. Proc Natl Acad Sci U S A 108:20225–20230

Atmodjo MA, Hao Z, Mohnen D (2013) Evolving views of pectin biosynthesis. Annu Rev Plant Biol 64(1):747–779

Benedetti M, Pontiggia D, Raggi S et al (2015) Plant immunity triggered by engineered in vivo release of oligogalacturonides, damage-associated molecular patterns. Proc Natl Acad Sci U S A 112:5533–5538

Biswal AK et al (2018) Sugar release and growth of biofuel crops are improved by downregulation of pectin biosynthesis. Nat Biotechnol 36:249–257

Bonnin E, Clavurier K, Daniel S et al (2008) Pectin acetylesterases from *Aspergillus* are able to deacetylate homogalacturonan as well as rhamnogalacturonan. Carbohydr Polym 74:411–418

Bordenave M (1996) Analysis of pectin methyl esterases. In: Linskens HF, Jackson JF (eds) Plant cell wall analysis. Modern methods of plant analysis. Springer, Cham, pp 165–180

Bouton S, Leboeuf E, Mouille G et al (2002) QUASIMODO1 encodes a putative membrane-bound glycosyltransferase required for normal pectin synthesis and cell adhesion in *Arabidopsis*. Plant Cell 14:2577–2590

Broxterman SE, Schols HA (2018) Interactions between pectin and cellulose in primary plant cell walls. Carbohydr Polym 192:263–272

Bruening G, Lyons JM (2000) The case of the FLAVR SAVR tomato. Calif Agric 54:6–7

Chebli Y, Kaneda M, Zerzour R, Geitmann A (2012) The cell wall of the *Arabidopsis* pollen tube—spatial distribution, recycling, and network formation of polysaccharides. Plant Physiol 160:1940–1955

Chiniquy D, Underwood W, Corwin J et al (2019) PMR5, an acetylation protein at the intersection of pectin biosynthesis and defense against fungal pathogens. Plant J 100:1022–1035

Cosgrove DJ (2018) Diffuse growth of plant cell walls. Plant Physiol 176:16–27

Daher FB, Braybrook SA (2015) How to let go: pectin and plant cell adhesion. Front Plant Sci 6:523

Daher FB, Chen Y, Bozorg B et al (2018) Anisotropic growth is achieved through the additive mechanical effect of material anisotropy and elastic asymmetry. elife 7:e38161

Dean GH, Zheng H, Tewari J et al (2007) The *Arabidopsis* MUM2 gene encodes a β-galactosidase required for the production of seed coat mucilage with correct hydration properties. Plant Cell 19:4007–4021

Decreux A, Messiaen J (2005) Wall-associated kinase WAK1 interacts with cell wall pectins in a calcium-induced conformation. Plant Cell Physiol 46:268–278

Diet A, Link B, Seifert GJ et al (2006) The *Arabidopsis* root hair cell wall formation mutant lrx1 is suppressed by mutations in the RHM1 gene encoding a UDP-l-rhamnose synthase. Plant Cell 18:1630–1641

Domozych DS, Sørensen I, Popper ZA, Ochs J, Andreas A, Fangel JU, Pielach A, Sacks C, Brechka H, Ruisi-Besares P, Willats WG, Rose JK (2014) Pectin metabolism and assembly in the cell wall of the charophyte green alga *Penium margaritaceum*. Plant Physiol 165(1):105–118
Draye M, Van Cutsem P (2008) Pectin methylesterases induce an abrupt increase of acidic pectin during strawberry fruit ripening. J Plant Physiol 165:1152–1160
Duan Q, Liu M-CJ, Kita D et al (2020) FERONIA controls pectin- and nitric oxide-mediated male–female interaction. Nature 2020:1–6
Egelund J, Petersen BL, Motawia MS et al (2006) *Arabidopsis thaliana* RGXT1 and RGXT2 encode Golgi-localized (1,3)-α-d-xylosyltransferases involved in the synthesis of pectic rhamnogalacturonan-II. Plant Cell 18:2593–2607
Egelund J, Damager I, Faber K et al (2008) Functional characterisation of a putative rhamnogalacturonan II specific xylosyltransferase. FEBS Lett 582:3217–3222
Fayant P, Girlanda O, Chebli Y et al (2010) Finite element model of polar growth in pollen tubes. Plant Cell 22:2579–2593
Feng W, Kita D, Peaucelle A et al (2018) The FERONIA receptor kinase maintains cell-wall integrity during salt stress through Ca^{2+} signaling. Curr Biol 28:666–675
Ferrari S, Savatin DV, Sicilia F et al (2013) Oligogalacturonides: plant damage-associated molecular patterns and regulators of growth and development. Front Plant Sci 4:49
Francis KE, Lam SY, Copenhaver GP (2006) Separation of *Arabidopsis* pollen tetrads is regulated by QUARTET1, a pectin methylesterase gene. Plant Physiol 142:1004–1013
Francoz E, Ranocha P, Burlat V, Dunand C (2015) *Arabidopsis* seed mucilage secretory cells: regulation and dynamics. Trends Plant Sci 20:515–524
Francoz E, Ranocha P, Le Ru A et al (2019) Pectin demethylesterification generates platforms that anchor peroxidases to remodel plant cell wall domains. Dev Cell 48:261–276
Fry SC (2018) Cell wall polysaccharide composition and covalent crosslinking. Annu Plant Rev Online 2018:1–42
Funakawa H, Miwa K (2015) Synthesis of borate cross-linked rhamnogalacturonan II. Front Plant Sci 6:223
Gallego-Giraldo L, Liu C, Pose-Albacete S et al (2020) Arabidopsis dehiscence zone polygalacturonase 1 (ADPG1) releases latent defense signals in stems with reduced lignin content. Proc Natl Acad Sci U S A 117:3281–3290
Geitmann A, Hudák J, Vennigerholz F, Walles B (1995) Immunogold localization of pectin and callose in pollen grains and pollen tubes of *Brugmansia suaveolens* — Implications for the self-incompatibility reaction. J Plant Physiol 147:225–235
Geshi N, Pauly M, Ulvskov P (2002) Solubilization of galactosyltransferase that synthesizes 1,4-beta-galactan side chains in pectic rhamnogalacturonan I. Physiol Plant 114:540–548
Goldberg R, Prat R (1982) Involvement of cell wall characteristics in growth processes along the mung bean hypocotyl. Plant Cell Physiol 23:1145–1154
Gou JY, Miller LM, Hou G et al (2012) Acetylesterase-mediated deacetylation of pectin impairs cell elongation, pollen germination, and plant reproduction. Plant Cell 24:50–65
Haas KT, Wightman R, Meyerowitz EM, Peaucelle A (2020) Pectin homogalacturonan nanofilament expansion drives morphogenesis in plant epidermal cells. Science 367:1003–1007
Harholt J, Jensen JK, Verhertbruggen Y et al (2012) ARAD proteins associated with pectic arabinan biosynthesis form complexes when transiently overexpressed in planta. Planta 236:115–128
Hocq L, Pelloux J, Lefebvre V (2017) Connecting homogalacturonan-type pectin remodeling to acid growth. Trends Plant Sci 22:20–29
Huang L, Cao J, Zhang A et al (2009a) The polygalacturonase gene BcMF2 from *Brassica campestris* is associated with intine development. J Exp Bot 60:301–313
Huang L, Ye Y, Zhang Y et al (2009b) BcMF9, a novel polygalacturonase gene, is required for both *Brassica campestris* intine and exine formation. Ann Bot 104:1339–1351
Iijima M, Higuchi T, Barlow PW (2004) Contribution of root cap mucilage and presence of an intact root cap in maize (*Zea mays*) to the reduction of soil mechanical impedance. Ann Bot 94:473–477

Iwai H, Hokura A, Oishi M et al (2006) The gene responsible for borate cross-linking of pectin rhamnogalacturonan-II is required for plant reproductive tissue development and fertilization. Proc Natl Acad Sci U S A 103:16592–16597

Jarvis MC, Briggs SPH, Knox JP (2003) Intercellular adhesion and cell separation in plants. Plant Cell Environ 26:977–989

Jiang L, Yang S-L, Xie L-F et al (2005) VANGUARD1 encodes a pectin methylesterase that enhances pollen tube growth in the *Arabidopsis* style and transmitting tract. Plant Cell 17:584–596

Jiang J, Yao L, Yu Y et al (2014) PECTATE LYASE-LIKE10 is associated with pollen wall development in *Brassica campestris*. J Integr Plant Biol 56:1095–1105

Jiménez-Bermúdez S, Redondo-Nevado J, Muñoz-Blanco J et al (2002) Manipulation of strawberry fruit softening by antisense expression of a pectate lyase gene. Plant Physiol 128:751–759

Jones L, Milne JL, Ashford D, McQueen-Mason SJ (2003) Cell wall arabinan is essential for guard cell function. Proc Natl Acad Sci U S A 100:11783–11788

Kim S-J, Held MA, Zemelis S et al (2015) CGR2 and CGR3 have critical overlapping roles in pectin methylesterification and plant growth in *Arabidopsis thaliana*. Plant J 82:208–220

Kong Y, Zhou G, Yin Y et al (2011) Molecular analysis of a family of *Arabidopsis* genes related to galacturonosyltransferases. Plant Physiol 155:1791–1805

Krupková E, Immerzeel P, Pauly M, Schmülling T (2007) The tumorous shoot development2 gene of *Arabidopsis* encoding a putative methyltransferase is required for cell adhesion and coordinated plant development. Plant J 50:735–750

Lampugnani ER, Khan GA, Somssich M, Persson S (2018) Building a plant cell wall at a glance. J Cell Sci 131:207373

Lehner A, Dardelle F, Soret-Morvan O et al (2010) Pectins in the cell wall of *Arabidopsis thaliana* pollen tube and pistil. Plant Signal Behav 5:1282–1285

Leroux C, Bouton S, Kiefer-Meyer M-C et al (2015) Pectin methylesterase48 is involved in *Arabidopsis* pollen grain germination. Plant Physiol 167:367–380

Lionetti V, Cervone F, De Lorenzo G (2015) A lower content of de-methylesterified homogalacturonan improves enzymatic cell separation and isolation of mesophyll protoplasts in *Arabidopsis*. Phytochemistry 112:188–194

Matsunaga T, Ishii T, Matsumoto S et al (2004) Occurrence of the primary cell wall polysaccharide rhamnogalacturonan ii in pteridophytes, lycophytes, and bryophytes. Implications for the evolution of vascular plants. Plant Physiol 134:339–351

Mohnen D (2008) Pectin structure and biosynthesis. Curr Opin Plant Biol 11:266–277

Mollet J-C, Park S-Y, Nothnagel EA, Lord EM (2000) A lily stylar pectin is necessary for pollen tube adhesion to an in vitro stylar matrix. Plant Cell 12:1737–1749

Moore PJ, Swords KMM, Lynch MA, Staehelin LA (1991) Spatial organization of the assembly pathways of glycoproteins and complex polysaccharides in the Golgi apparatus of plants. J Cell Biol 112:589–602

Mouille G, Ralet M-C, Cavelier C et al (2007) Homogalacturonan synthesis in *Arabidopsis thaliana* requires a Golgi-localized protein with a putative methyltransferase domain. Plant J 50:605–614

Nunan KJ, Scheller HV (2003) Solubilization of an arabinan arabinosyltransferase activity from mung bean hypocotyls. Plant Physiol 132:331–342

O'Neill MA, Eberhard S, Albersheim P, Darvill AG (2001) Requirement of borate cross-linking of cell wall rhamnogalacturonan II for Arabidopsis growth. Science 294:846–849

Ogawa M, Kay P, Wilson S, Swain SM (2009) Arabidopsis dehiscence zone polygalacturonase1 (ADPG1), ADPG2, and QUARTET2 are polygalacturonases required for cell separation during reproductive development in Arabidopsis. Plant Cell 21:216–233

Orfila C, Seymour GB, Willats WGT et al (2001) Altered middle lamella homogalacturonan and disrupted deposition of (1→5)-α-l-arabinan in the pericarp of cnr, a ripening mutant of tomato. Plant Physiol 126:210–221

Palin R, Geitmann A (2012) The role of pectin in plant morphogenesis. Biosystems 109:397–402

Parker CC, Parker ML, Smith AC, Waldron KW (2001) Pectin distribution at the surface of potato parenchyma cells in relation to cell-cell adhesion. J Agric Food Chem 49:4364–4371

Peaucelle A, Louvet R, Johansen JN et al (2008) *Arabidopsis* phyllotaxis is controlled by the methyl-esterification status of cell-wall pectins. Curr Biol 18:1943–1948

Peaucelle A, Braybrook SA, Le Guillou L et al (2011) Pectin-induced changes in cell wall mechanics underlie organ initiation in *Arabidopsis*. Curr Biol 21:1720–1726

Peaucelle A, Braybrook S, Höfte H (2012) Cell wall mechanics and growth control in plants: the role of pectins revisited. Front Plant Sci 3:121

Phyo P, Wang T, Kiemle SN et al (2017a) Gradients in wall mechanics and polysaccharides along growing inflorescence stems. Plant Physiol 175:1593–1607

Phyo P, Wang T, Xiao C et al (2017b) Effects of pectin molecular weight changes on the structure, dynamics, and polysaccharide interactions of primary cell walls of *Arabidopsis thaliana*: insights from solid-state NMR. Biomacromolecules 18:2937–2950

Popper ZA, Fry SC (2003) Primary cell wall composition of Bryophytes and Charophytes. Ann Bot 91:1–12

Popper ZA, Fry SC (2004) Primary cell wall composition of pteridophytes and spermatophytes. New Phytol 164:165–174

Posé S, Paniagua C, Cifuentes M et al (2013) Insights into the effects of polygalacturonase FaPG1 gene silencing on pectin matrix disassembly, enhanced tissue integrity, and firmness in ripe strawberry fruits. J Exp Bot 64:3803–3815

Qi J, Wu B, Feng S et al (2017) Mechanical regulation of organ asymmetry in leaves. Nat Plants 3:724–733

Quesada MA, Blanco-Portales R, Posé S et al (2009) Antisense down-regulation of the FaPG1 gene reveals an unexpected central role for polygalacturonase in strawberry fruit softening. Plant Physiol 150:1022–1032

Redgwell RJ, MacRae E, Hallett I et al (1997) In vivo and in vitro swelling of cell walls during fruit ripening. Planta 203:162–173

Rhee SY, Somerville CR (1998) Tetrad pollen formation in quartet mutants of *Arabidopsis thaliana* is associated with persistence of pectic polysaccharides of the pollen mother cell wall. Plant J 15:79–88

Röckel N, Wolf S, Kost B et al (2008) Elaborate spatial patterning of cell-wall PME and PMEI at the pollen tube tip involves PMEI endocytosis, and reflects the distribution of esterified and de-esterified pectins. Plant J 53:133–143

Rui Y, Xiao C, Yi H et al (2017) Polygalacturonase involved in expansion3 functions in seedling development, rosette growth, and stomatal dynamics in *Arabidopsis thaliana*. Plant Cell 29:2413–2432

Saffer AM, Carpita NC, Irish VF (2017) Rhamnose-containing cell wall polymers suppress helical plant growth independently of microtubule orientation. Curr Biol 27:2248–2259

Samuels AL, Giddings TH Jr, Staehelin LA (1995) Cytokinesis in tobacco BY-2 and root tip cells: a new model of cell plate formation in higher plants. J Cell Biol 130:1345–1357

Scheller HV, Lou DR, Ridley BL, Mohnen D (1999) Pectin biosynthesis: a solubilized α1,4-galacturonosyltransferase from tobacco, catalyzes the transfer of galacturonic acid from UDP-galacturonic acid onto the non-reducing end of homogalacturonan. Planta 207:512–517

Sénéchal F, Wattier C, Rustérucci C, Pelloux J (2014) Homogalacturonan-modifying enzymes: structure, expression, and roles in plants. J Exp Bot 65:5125–5160

Sterling JD, Quigley HF, Orellana A, Mohnen D (2001) The catalytic site of the pectin biosynthetic enzyme α-1,4-galacturonosyltransferase is located in the lumen of the Golgi. Plant Physiol 127:360–371

Sterling JD, Atmodjo MA, Inwood SE et al (2006) Functional identification of an *Arabidopsis* pectin biosynthetic homogalacturonan galacturonosyltransferase. Proc Natl Acad Sci U S A 103:5236–5241

Takenaka Y, Kato K, Ogawa-Ohnishi M et al (2018) Pectin RG-I rhamnosyltransferases represent a novel plant-specific glycosyltransferase family. Nat Plants 4:669–676

Tan L, Eberhard S, Pattathil S et al (2013) An Arabidopsis cell wall proteoglycan consists of pectin and arabinoxylan covalently linked to an arabinogalactan protein. Plant Cell 25:270–287
Vaahtera L, Schulz J, Hamann T (2019) Cell wall integrity maintenance during plant development and interaction with the environment. Nat Plants 5:924–932
Verger S, Chabout S, Gineau E, Mouille G (2016) Cell adhesion in plants is under the control of putative O-fucosyltransferases. Development 143:2536–2540
Vogel J (2008) Unique aspects of the grass cell wall. Curr Opin Plant Biol 11:301–307
Voiniciuc C, Engle KA, Günl M et al (2018) Identification of key enzymes for pectin synthesis in seed mucilage. Plant Physiol 178:1045–1064
Wen F, Zhu Y, Hawes MC (1999) Effect of pectin methylesterase gene expression on pea root development. Plant Cell 11:1129–1140
Western TL, Skinner DJ, Haughn GW (2000) Differentiation of mucilage secretory cells of the *Arabidopsis* seed coat. Plant Physiol 122:345–355
Wi SG, Singh AP, Lee KH, Kim YS (2005) The pattern of distribution of pectin, peroxidase and lignin in the middle lamella of secondary xylem fibres in alfalfa (*Medicago sativa*). Ann Bot 95:863–868
Wolf S, Greiner S (2012) Growth control by cell wall pectins. Protoplasma 249:169–175
Xiao C, Somerville C, Anderson CT (2014) polygalacturonase involved in expansion functions in cell elongation and flower development in *Arabidopsis*. Plant Cell 26:1018–1035
Xiao C, Barnes WJ, Zamil MS et al (2017) Activation tagging of *Arabidopsis* polygalacturonase involved in expansion promotes hypocotyl elongation, leaf expansion, stem lignification, mechanical stiffening, and lodging. Plant J 89:1159–1173
Yang Y, Yu Y, Liang Y et al (2018) A profusion of molecular scissors for pectins: classification, expression, and functions of plant polygalacturonases. Front Plant Sci 9:1208
Zamil MS, Geitmann A (2017) The middle lamella - More than a glue. Phys Biol 14:015004

Chapter 2
Pectin Structure

David Ropartz and Marie-Christine Ralet

2.1 Introduction

Pectin is a natural constituent of all terrestrial plants that is particularly abundant—together with hemicelluloses, cellulose, and low amounts of structural proteins—in the primary cell walls of eudicotyledons and non-graminaceous monocotyledons (Carpita and Gibeaut 1993). Once extracted from citrus peel or apple pomace, commercial pectin is widely used as gelling, thickening, stabilising and emulsifying agent in various food products such as jams, acidic milk drinks, ice creams, or salad dressings, and it is well agreed that the fine structure of pectin deeply affects its functionality and applicability (Willats et al. 2006). The existence of a "jelly" in tamarind extract was discovered more than two centuries ago by the French pharmacist and chemist Louis-Nicolas Vauquelin (1790). The word "pectin" from the Greek πηκτός ("pêktós", which means "thick") was first used in 1825 by Braconnot, who had resumed Vauquelin's work (Braconnot 1825a, b). Smolenski in 1923 was the first scientist to describe pectin as a polymer of galacturonic acid (GalA) and Kertesz (1951) defined pectin as a hetero-polysaccharide containing mainly partly methylesterified GalA together with some neutral sugars. In the 80s, the work of de Vries and co-workers was instrumental in showing that neutral sugars were present as side-chains arranged in blocks in so called "hairy regions" while >90% of the GalA residues could be isolated as chains comprising solely GalA (de Vries et al. 1981, 1982, 1983). Nowadays, it is widely accepted that pectin is a heterogonous macromolecule composed of interlinked distinct domains, the relative amount and structure of which vary according to the botanical origin, the organs and cell types considered, the stages of cellular development, and the precise location within the cell wall (Voragen et al. 2009). On top of this biological diversity, pectin structure varies broadly depending on the extraction

D. Ropartz · M.-C. Ralet (✉)
INRAE, Biopolymères Interactions Assemblages, Nantes, France
e-mail: david.ropartz@inrae.fr; marie.ralet@inrae.fr

V. Kontogiorgos (ed.), *Pectin: Technological and Physiological Properties*,
https://doi.org/10.1007/978-3-030-53421-9_2

method used. Different pectin populations that are more or less strongly anchored in the cell wall co-exist and a high number of extraction steps is needed to extract all pectin (Broxterman 2018).

Two main families of pectin structural elements are usually considered: galacturonans and rhamnogalacturonan I (RG-I). Galacturonans are made of a backbone of α-(1,4)-linked D-galacturonic acid (GalA) residues. This galacturonan backbone may be unbranched (homogalacturonan) or decorated with more or less complex side-chains. The backbone of RG-I is made of the diglycosyl repeating unit [2-α-L-Rha-(1,4)-α-D-GalA-(1] (Lau et al. 1985). Rhamnose (Rha) residues are ramified at *O*-4 (mainly) and *O*-3 (scarcely) positions with single to polymeric neutral sugar side chains that include arabinose (Ara) and galactose (Gal) residues in various combinations (Lau et al. 1985; Colquhoun et al. 1990). Four main types of polymeric side-chains are usually envisioned, arabinans, galactans, type I arabinogalactans (AG-I) and type II arabinogalactans (AG-II) (Yapo 2011).

As has been briefly described above, pectin is an extremely complex polysaccharide composed of as many as eighteen distinct monosaccharides connected to each other through twenty different linkages. In addition, several of these monosaccharides can be chemically modified by *O*-ether or *O*-ester groups. Because of polydispersity and polymolecularity, analyses on whole macromolecules are usually not sufficient to reveal all the structural details of pectin. For structural investigation, pectin is thereby commonly degraded into oligosaccharides by chemicals or enzymes. Enzymatic digestion is usually preferred since it offers additional information thanks to the specificity of cleaving. Degradation enzymes are, however, not available for all linkages. Degradation products are further fractionated usually by chromatographic means. The isolated structural elements thereby recovered are in the analytical range of a broad set of techniques (Schols and Voragen 2002) and very valuable structural information has been obtained by chemical analyses and NMR in the 90s. This information has however been restricted to major oligosaccharides that could be purified in quantities large enough to allow full structural characterisation. In the past fifteen years, significant developments have led to a re-emergence of the use of mass spectrometry (MS) for the structural characterisation of oligosaccharides (Dong et al. 2018). This review includes improvements that have been achieved in separation techniques (ultra-performance liquid chromatography (UPLC), capillary electrophoresis, ion mobility) and developments in activation approaches using electron, cation or photon interactions that are much more efficient than classical fragmentation methods (i.e., low energy collision induced (or activated) dissociation (LE-CI(A)D) for the determination of the fine structure of oligosaccharides. Very recent works have been conducted on instrument development for the structural characterisation of oligosaccharides based on ion activation (Ropartz et al. 2014, 2016, 2017), on the coupling of MS and infrared ion spectroscopy (Schindler et al. 2017; Mucha et al. 2018) and on high resolution ion mobility (Ujma et al. 2019; Ropartz et al. 2019). These approaches allow to shed new light on complex structures and mixtures of oligosaccharides, as can be found in pectin.

2.2 Galacturonans

2.2.1 Homogalacturonans

Homogalacturonan (HG) also known as the "smooth region" of pectin, is a linear homopolymer of α-(1,4)-linked D-galacturonic acid (GalA) residues that can be methyl and acetyl-esterified (Fig. 2.1). It usually accounts for approximately 60% of the total pectin amount (Mohnen 2008; Atmodjo et al. 2013) but there are some exceptions—sugar beet or soybean pectin for instance—in which the relative amount of HG is much lower (Voragen et al. 1995).

HG have been specifically isolated from several plant sources by exploiting the susceptibility to acid hydrolysis of the different glycosidic linkages (Thibault et al. 1993; Yapo et al. 2007; Ralet and Thibault 2009; Round et al. 2010) or by enzymatic means (Bonnin et al. 2002; Hellin et al. 2005). Depending on the method used for (1) pectin extraction, (2) HG isolation and (3) molar mass determination, the average degree of polymerisation (DP) ranges between 72 and 300. Ralet and Thibault (2009) isolated HG from pectin samples sequentially extracted from different plant sources: pineapple flesh (*Ananas comosus*, *Bromeliaceae*), leek (*Allium porrum*, *Alliaceae*), cucumber (*Cucumis sativus*, *Cucurbitaceae*), sugar beet root (*Beta vulgaris*, *Amaranthaceae*), fennel bulb (*Foeniculum vulgare*, *Apiaceae*), and lemon albedo (*Citrus medica*, *Rutaceae*) and showed that pectin samples encompassed various amounts of HG domains of very similar DP (70–100) and of low polydispersity, confirming the hypothesis of a HG length periodicity (Thibault et al. 1993).

2.2.1.1 Methyl-esterification

In HG, GalA units are usually partially methyl-esterified at *C*-6 (Fig. 2.1) and not only the degree of methyl esterification (i.e., the number of moles of methanol per 100 moles of GalA), but also the distribution of non-esterified GalA residues on HG segments are key features for pectin functional properties (Willats et al. 2006). Numerous investigations have thereby been devoted to understand the peculiar methyl-ester distribution patterns and their functional implications. The development of methods allowing characterising and quantifying degradation products

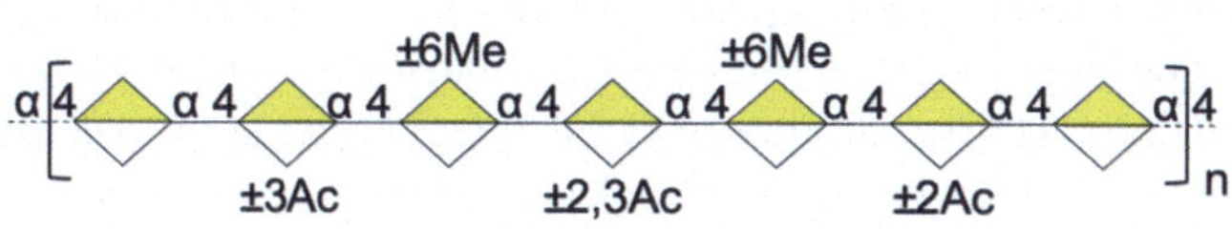

Fig. 2.1 Structure of homogalacturonans according to the symbol nomenclature for glycans (SNFG) introduced by Varki et al. (2015). Homogalacturonan is the dominant feature of pectin. It consists of partly methyl-esterified and, in some plant species, partly acetyl-esterified α-(1,4)-linked D-galacturonopyranosyl chains with a degree of polymerisation 100–300. The amount and distribution of methyl- and acetyl-esters onto homogalacturonans is a key factor for pectin functionality

obtained by treating well-defined pectin samples with HG-degrading enzymes was instrumental in providing such information. The concept of *degree of blockiness* calculated from the amount of oligogalacturonates released quantified by high-performance anion-exchange chromatography or by capillary electrophoresis, has been developed. This concept allowed differentiating pectin encompassing HG domains exhibiting subtle differences in methylesterification patterns (Guillotin et al. 2005; Daas et al. 2000; Limberg et al. 2000; Ström et al. 2007; Ngouemazong et al. 2011; Ralet et al. 2012). Parallel electrospray ionisation multistage mass spectrometry (ESI-MS^n) may be also used to provide information about the location of methyl-esterified GalA residues in oligogalacturonates of DP 3–10 (Körner et al. 1999; van Alebeek et al. 2000; Quéméner et al. 2003; Ralet et al. 2009, 2012). Recent advances in mass spectrometry and separation techniques allowed exploring complex mixtures and characterising oligosaccharides of higher DP (identification up to DP14, structural characterization up to DP9) (Ropartz et al. 2014). Model HGs were degraded by a pectin lyase and digestion products were characterised (Ropartz et al. 2014). The structure of each isomer was stated based on its methylation pattern. These approaches can bring different types of information: (1) the specificity of the enzyme and the tolerance of each subsite to the occurrence of methyl-esters can be stated, and (2) the structure of large oligosaccharides can be revealed, giving insights into parts of the polymers that are recalcitrant to enzymatic digestion.

2.2.1.2 Acetyl-Esterification

In few plant species, GalA residues in HG domains are partially acetyl-esterified at *O*-2 and/or *O*-3 (Rombouts and Thibault 1986; Ishii 1997; Needs et al. 1998; Perrone et al. 2002) (Fig. 2.1) and this has a strong negative impact on gelation (Pippen et al. 1950; Kohn and Furda 1968; Kohn and Malovikova 1978; Renard and Jarvis 1999; Oosterveld et al. 2000a; Ralet et al. 2003). The distribution of acetyl groups onto HG segments has been particularly studied in sugar beet. Keenan et al. (1985) showed by nuclear magnetic resonance (NMR) that acetyl groups could be attached on any of the available ring positions (*O*-2 and *O*-3) of GalA residues. Ralet et al. (2005, 2008) and Remoroza et al. (2014) used combinations of pectin-degrading enzymes to generate partly methyl esterified and acetyl esterified oligogalacturonates that were further separated by chromatographic means and analysed by ESI-MS^n. The large variety of oligogalacturonates that were identified and quantified revealed that (1) the occurrence of *O*-2 and *O*-3 acetyl esterification in roughly similar amounts, (2) the absence of 2,3-di-*O*-acetylation, and (3) the scarcity of GalA residues that are both methyl- and acetyl-esterified (Ralet et al. 2005). Pectins with high degree of methyl esterification (HM) having different patterns of ester distribution could be also discriminated (Remoroza et al. 2014). A blockwise distribution of acetyl groups was evidenced (Ralet et al. 2008) and, in commercially-extracted sugar beet pectin, blocks of (1) non-esterified, (2) partly methyl- and acetyl-esterified, and (3) highly methyl- and acetyl-esterified GalA residues were identified (Remoroza et al. 2014). The balance between intra-chain heterogeneity (i.e., the different blocks that are present within a single macromolecule) and inter-chain heterogeneity (some pectin molecules are very highly acetyl-esterified and others are not) is however very difficult to appraise.

2.2.2 *Galacturonans Substituted with More or Less Complex Side-Chains*

HG can be more or less heavily substituted at *O*-2 and/or *O*-3 by monomers or dimers of apiose or xylose leading to apiogalacturonan (AGA, Fig. 2.2a) or xylogalacturonan (XGA, Fig. 2.2b), respectively. It can also be substituted with complex side-chains to form rhamnogalacturonan II (RG-II, Fig. 2.2c).

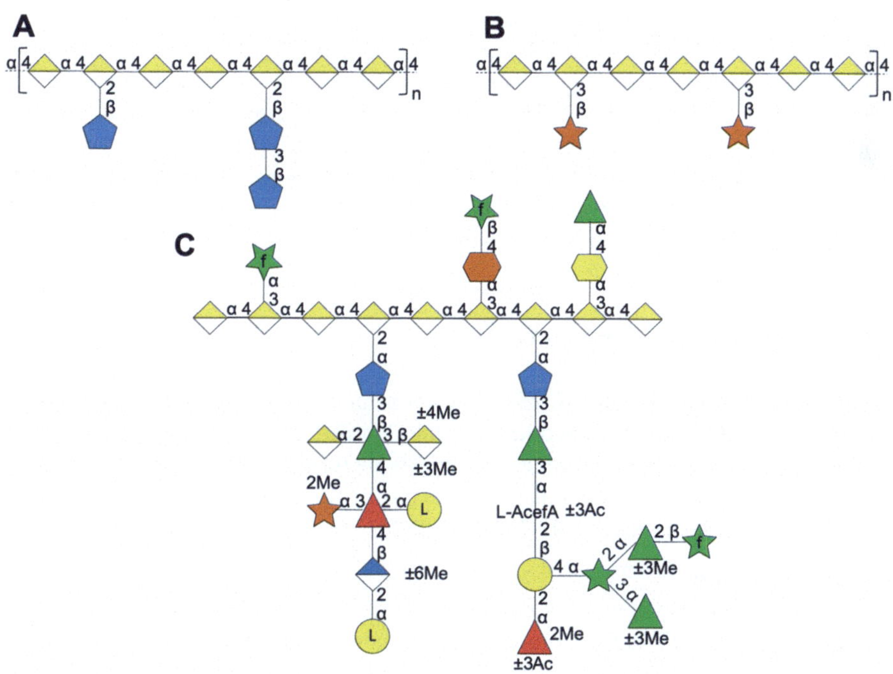

Fig. 2.2 Structure of the side-chains of homogalacturonans (SNFG). (**a**) apiogalacturonan, (**b**) xylogalacturonan, (**c**) rhamnogalacturonan-II. More or less complex side-chains can be attached to homogalacturonan. Rhamnogalacturonan-II is a key substituted galacturonan made of a short backbone branched by up to 5 side-chains varying in complexity from single arabinose unit to highly heterogeneous nonasaccharides

2.2.2.1 Apiogalacturonan

Apiogalacturonan (AGA, Fig. 2.2a) is restricted to some aquatic plants, the duckweeds (*Lemnoideae*) and the marine seagrasses (*Zosteraceae*) (Hart and Kindel 1970; Ovodov et al. 1971; Longland et al. 1989; Gloaguen et al. 2010; Avci et al. 2018). In AGA, β-D-apiofuranosyl residues are linked to *O*-2 and/or *O*-3 of GalA residues as monomers or as the dimer [β-D-Api*f*-(1,3′)-β-D-Api*f*-(1)]. The degree of substitution of HG by apiose varies from 25 to 80% (Hart and Kindel 1970; Ovodov et al. 1971). AGA can represent up to 20% of non-cellulosic cell wall polysaccharides in the green fronds of giant duckweeds (Longland et al. 1989).

2.2.2.2 Xylogalacturonan

Xylogalacturonan (XGA, Fig. 2.2b), an HG substituted solely at *O*-3 by xylose monomers or dimers, has been identified in several plant samples among which duckweeds (Hart and Kindel 1970), apple and watermelon fruits (Schols et al. 1995a; Zandleven et al. 2006; Mort et al. 2008), potato tuber (Zandleven et al. 2006), pea hulls (Le Goff et al. 2001) and *Arabidopsis thaliana* leaves and stalks (Zandleven et al. 2007). Dimeric side chains of xylose have been shown to contain 1,4-linked xylose residues (Zandleven et al. 2006) but 1,2-linked and 1,3-linked xylose residues have also been identified (Le Goff et al. 2001; Nakamura et al. 2002). Methylesterification of XGA has been reported in apple (Schols et al. 1995a). Both the ratio of Xyl monomers and dimers and the degree of substitution of GalA by Xyl vary according to plant samples. A recent study highlighted that the diversification of the *Lemnoideae* was accompanied by a reduction in the abundance of cell wall AGA and an increase in XGA (Avci et al. 2018).

2.2.2.3 Rhamnogalacturonan II

Rhamnogalacturonan II (RG-II, Fig. 2.2c) is a far more complex substituted-HG than AGA or XGA since it encompasses thirteen different sugars and twenty one distinct glycosidic linkages arranged as a backbone formed by nine partially methylesterified GalA residues (from none to three) (Ishii and Kaneko 1998) substituted by different side chains termed A–F (Ndeh et al. 2017).

Large Side-Chains
Side-chains A and B encompass six to nine sugars and are both bound to the HG backbone through a (2,1) β-D-apiofuranose. Although the structure of RG-II is considered as highly conserved throughout the plant kingdom, some heterogeneity may occur within these side-chains. Side-chain A was first described by Stevenson et al. (1988) in *Ficus sycomorus* as an oligosaccharide composed of eight different monosaccharides (Fig. 2.2c). The heterogeneity in the structure of side-chain A comes firstly from the α-L-Gal*p* residue that can be replaced by an α-L-Fuc*p*, not only in

the *Arabidopsis thaliana* mutant *mur1* (Reuhs et al. 2004), but also in the wild-type plants (Pabst et al. 2013). Additionally, side-chain A contains three uronic acids, one α-D-GalA*p*, one β-D-GalA*p* and one β-D-GlcA*p*, that can be methyl-esterified and/or methyl-etherified (Pabst et al. 2013; Buffetto et al. 2014). Side-chain A is involved in RG-II dimerisation through a boron covalent linkage *via* a di-ester bond between two β-D-apiofuranose units (*O*-2 and *O*-3). The three uronic acids residues are able to chelate cations and methyl-esterification that can occur on the β-D-GlcA*p* unit may modulate this chelation property. Side-chain B is particularly heterogeneous (Table 2.1).

Table 2.1 Structural variability of side-chain B in rhamnogalacturonan-II

		Glycosyl substitutions			Chemical modifications			
					Acetyl		Methylether	
Reference	Plant origin	(2-1) α-L-Rha*p*	(2-1) β-L-Ara*f*	(3-1) α-L-Rha*p*	3-α-L-Ace*f*	2-*O*-Me-α-D-Fuc*p*	(2-1) α-L-Rha*p*	(3-1) α-L-Rha*p*
Spellman et al. (1983)	*Acer pseudoplatunus*	X	–	–	–	–	–	–
Thomas et al. (1989)	*Oryza sativa*	X	–	–	–	–	–	–
Whitcombe et al. (1995)	*Acer pseudoplatunus*	X	X/–	–	X/–	X/–	–	–
Kaneko et al. (1997)	*Phyllostachys edulis*	X	–	–	–	–	–	–
Shin et al. (1998)	*Panax ginseng*	X	X/–	X/–	X/–	X/–	–	–
Ishii and Kaneko (1998)	*Beta vulgaris*	–	–	–	X/–	X/–	–	–
Vidal et al. (2000)	Red wine	X	X/–	[a]	X/–	X/–	–	–
Glushka et al. (2003)	*Arabidopsis thaliana*	X	X	X	X/–	X/–	–	–
Matsunaga et al. (2004)	Lycophyte PteridophyteBryophytes	X	X/–	X/–	X/–	X/–	X/–	X/–
Reuhs et al. (2004)	*Arabidopsis thaliana*	X	–	–	–	–	–	–
Seveno et al. (2009)	*Arabidopsis thaliana*	X/–	–	–	–	–	–	–
Pabst et al. (2013)	*Arabidopsis thaliana*	X	X/–	X/–	–	–	–	–
Buffetto et al. (2014)	*Vitis vinifera* Merlot	X	X/–	X/–	X/–	X/–	–	–
Sun et al. (2019)	*Panax ginseng*	X	X/–	X/–	X/–	X/–	–	–

X detected, – not detected

[a]Additional rhamnose was detected but not localised

It was first described in *Ficus sycomorus* as a heptasaccharide, the main chain being made of a β-D-Api*f*-(3,1)-β-L-Rha*p*-(3,1)-α-L-Aceric acid-(2,1)-β-D-Gal*p*-(2,1)-2-*O*-Me-α-L-Fuc*p*. The Gal*p* residue was substituted by a (4,1)-α-L-Ara*p* unit itself substituted by a (2,1)-α-l-Rha*p* residue (Spellman et al. 1983). Depending on the botanical origin, the stage of development and the organ considered, the number of constitutive sugar residues may vary. Hexa- to nonasaccharides were observed based on the substitution of the laterally branched α-L-Ara*p* residue (Fig. 2.2). Whitcombe et al. have described an octasaccharide with a (2,1)-β-L-Ara*f* linked to the laterally branched additional α-L-Rha*p* unit. These authors also described for the first time the occurrence of two acetylation sites on the α-L-Aceric acid and on the 2-*O*-Me α-D-Fuc*p*. Finally, in 1998, Shin et al. described in *Panax ginseng* a nonasaccharide resulting from a second α-l-Rha*p* unit branched at *C*-3 of the β-L-Ara*p* residue. It is difficult to conclude about the existence of the low DP species in vivo, as α-L-Rha*p* residues being very labile according to the extraction conditions and the analytical methods used. As for side-chain A, 2-*O*-Me-α-D-Fuc*p* can be replaced by 2-*O*-Me-L-Gal*p* in the *Arabidopsis thaliana mur1* mutant (Reuhs et al. 2004). In lycophytes, pteridophytes and bryophytes, α-L-Rha*p* units can be methyl-etherified at *O*-3 (Matsunaga et al. 2004).

Short Side-Chains

Side chain E is composed of a unique residue of α-L-Ara*f* linked to a D-GalA*p* at *O*-3 (Buffetto et al. 2014). This modification is termed side chain F when it occurs on a D-GalA unit that is also substituted by side-chain A. Side-chains C and D are dimers that are branched at *O*-3 of a D-GalA units. They are composed of a α-3-deoxy-D-manno-2-octulosonic acid (4,1) α-L-Rha*p* and an α-3-deoxy-D-*lyxo*-2-heptulosonic acid (5,1) β-L-Ara*f*, respectively. In the fractionation scheme used by Buffetto et al. (2014), these side chains are co-eluted with longer ones (co-elution of A and D and of B and C) (Ropartz 2015).

2.3 Rhamnogalacturonan I

RG-I was pictured as a long sequence of alternating L-Rha and D-GalA residues (Fig. 2.3a), Rha residues being substituted with a variety of L-arabinosyl- and D-galactosyl-containing side-chains (O'Neill et al. 1990, Fig. 2.3b–d). Fucose, glucuronic acid and 4-*O*-methyl glucuronic acid residues may also be present in small amounts. RG-I usually accounts for 20–35% of the total pectin amount (Mohnen 2008) but can, for certain plant sources such as soybean, make up to 75% of pectic polysaccharides (Voragen et al. 2009).

2.3.1 Rhamnogalacturonan I Backbone

The backbone of RG-I is made of the alternating diglycosyl repeating unit [2-α-L-Rhap-(1,4)-α-D-GalAp-(1] (McNeil et al. 1980, 1984; Lau et al. 1985, Fig. 2.3a).

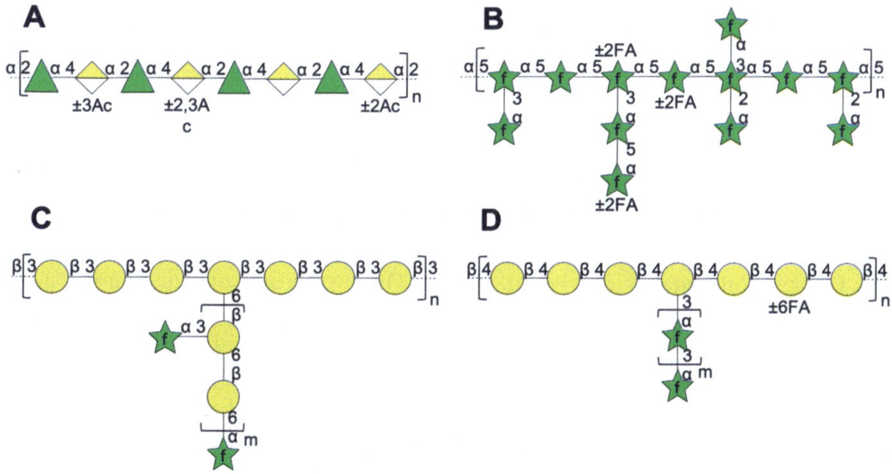

Fig. 2.3 Structure of Rhamnogalacturonan I (SNFG). (**a**) Backbone, (**b**) Arabinans, (**c**) Type I arabinogalactans, (**d**) Type II arabinogalactans. Rhamnogalacturonan I is the second major element of pectin. Its backbone is made of strictly alternating galacturonic acid and rhamnose residues, the latter being branched by arabinose- and galactose-containing side-chains of diverse complexity

Whole RG-I samples isolated from various plant sources usually exhibit Rha/GalA ratios < 1 (Schols and Voragen 1994; Cornuault et al. 2015; Buffetto et al. 2015). Values very close to 1 have however been reported (Konno et al. 1986; Yapo et al. 2007; Ralet and Thibault 2009) and the series of homologous oligomers recovered after controlled acid hydrolysis of pectin samples from apple, sugar beet and citrus all presented strictly alternating sequences (Renard et al. 1995). The presence of RG-I domains in which single Rha residues alternate with two or three consecutive GalA residues is therefore rather implausible. Unlike HG domains, RG-I backbone seems to exhibit a DP that varies depending on the plant source.

An average DP of up to 300 repeats has been estimated for RG-I isolated from suspension-cultured sycamore cell walls (McNeil et al. 1984). Ralet and Thibault (2009) reported values of 60 Rha-GalA repeats for RG-I isolated from citrus albedo and 120 for RG-I isolated from sugar-beet. Acetylation is much more frequent in RG-I than in HG, and RG-I domains isolated from numerous plant species appeared highly acetylated at *O*-2 or *O*-3 of the GalA units (Komalavilas and Mort 1989; Lerouge et al. 1993; Schols and Voragen 1994; Normand et al. 2010; Remoroza et al. 2012). Ishii showed that some GalA residues were 2,3-di-*O*-acetylated in RG-I oligosaccharides isolated from potato tuber and bamboo shoot cell walls (Ishii 1997). In okra, unusual acetylation at *O*-3 of Rha residues has been also shown (Sengkhamparn et al. 2009). Finally, to our knowledge, no evidence for methylesterification of the RG-I domain has ever been reported.

2.3.2 *Rhamnogalacturonan I Side-Chains*

RG-I is also known as the pectin "hairy region" since GalA but mostly Rha residues are substituted by different side-chains varying in their length and composition. GalA residues typically are not substituted and only one study showed that approximately 2% of the GalA residues in the RG-I backbone were substituted at *O*-3 by single β-D-glucuronic acid residues (Renard et al. 1999). In contrast, 20–80% of the Rha residues are substituted at *O*-4 with side-chains in which Ara and Gal predominate (Ridley et al. 2001). The proportion of side-chains, their composition, length and degree of branching vary enormously not only depending on plant sources, organs and tissues (Lerouge et al. 1993) but also on developmental stage (Willats et al. 2001) and on the isolation method used (Schols et al. 1995b). For instance, arabinan-rich pectins are particularly abundant in apple, sugar-beet and carrot while type I arabinogalactan-rich pectins are commonly found in citrus, lupin and potato and type II arabinogalactan-rich pectins have been found in ginseng (Sun et al. 2019). Further, galactan and arabinan epitopes can occur in different zones of organs and even at distinct locations within a single developing organ (Willats et al. 2001). Finally, for a given cell wall material, pectins that were more difficult to extract were more substituted with side-chains than the more easily extractable fractions (Schols et al. 1995b). Side-chains consisting in single Gal units have been reported for RG-I oligomers that have been isolated without using any side-chain degrading enzymes (Schols et al. 1995b; Oosterveld et al. 2000b; Sun et al. 2019). In okra, specifically, α-Gal substitutions were identified (Sengkhamparn et al. 2009). Polymeric side-chains commonly consist in arabinans and type I and II (arabino)-galactans.

2.3.2.1 Arabinans

Arabinans are made of a backbone of α-(1,5)-linked L-Ara*f* residues that are substituted at *O*-2 and/or *O*-3 with single α-L-Ara*f* units or with short chains of α-(1,3)-linked L-Ara*f* units (Yapo 2011, Fig. 2.3b). Sugar-beet arabinans have been particularly studied over the years. They consist of a long backbone of up to sixty or seventy Ara residues (Oosterveld et al. 2002) one third of which being branched at *O*-3, mainly with single Ara*f* units (Guillon and Thibault 1989). The presence of minor amounts of double branching at *O*-2 and *O*-3 has been further evidenced (Westphal et al. 2010). Occasionally, arabinans substituted with single Gal*p* units or with up to four or five β-(1,4)-linked Gal*p* residues have been isolated in soybean, sugar-beet or potato (Sakamoto and Sakai 1995; Nakamura et al. 2002; Øbro et al. 2004). Arabinans branched to a much lower extent than in sugar-beet have been also observed in ginseng root and in potato tuber (Sun et al. 2019; Øbro et al. 2004).

2.3.2.2 Type I (Arabino)-Galactans

Type I (arabino)-galactans encompass a backbone of β-(1,4)-linked D-Gal*p* residues that may be lowly substituted at *O*-3 by β-D-Gal*p* or α-L-Ara*f* single units or by short α-(1,5)-linked L-Ara*f* chains (Aspinall et al. 1967; Morita 1965a, b, Fig. 2.3c). Variants of this basic structure have been reported such as (1) termination of the main chain at *O*-6 with a Gal*p* residue (Lau et al. 1987), (2) termination of the main chain with an Ara*p* residue at *O*-4 or a Fuc residue at *O*-2 (Huisman et al. 2001; O'Neill et al. 1990), (3) insertion of a single Ara*f* or a short α-(1,5)-linked L-Ara*f* chain within the β-(1,4)-linked D-Gal*p* main chain (Huisman et al. 2001; Buffetto et al. 2015), and (4) β-(1,3) galactosyl interruption of the main chain (Hinz et al. 2005). An average chain length in the range of 45–50 was estimated in soybean (Huisman et al. 2001).

2.3.2.3 Type II Arabinogalactans

Type II arabinogalactans are highly branched side-chains characterised by a backbone of β-(1,3)-linked D-Gal*p* residues substituted at *O*-6 with single Gal*p* residues or with short chains of β-(1,6)-linked D-Gal*p* residues, which can in turn be substituted at *O*-3, *O*-4 and/or *O*-6, by Ara-containing chains (Yapo 2011, Fig. 2.3d). Type II arabinogalactans have been recently identified in ginseng (Sun et al. 2019) and it has been shown that the (1,3)-linked galactan backbone was substituted, mainly at *O*-6 but also at *O*-4, (1) with short (1,6)-linked galactan chains terminated by single Ara*f* units, and (2) directly with single Ara*f* residues.

2.3.2.4 Ferulic Acid

In species from the *Amaranthaceae* family such as sugar-beet and spinach RG-I side-chains are esterified by ferulic acid residues (Ishii and Tobita 1993; Colquhoun et al. 1994; Levigne et al. 2004; Quéméner and Ralet 2004). Ferulic acid moieties are mainly ester-linked to *O*-2 of Ara residues from the α-(1,5)-linked L-Ara*f* arabinan backbone and to *O*-6 of Gal residues from the β-(1,4)-linked D-Gal*p* type I galactan backbone but a more peripheral location at *O*-5 of Ara residues from the α-(1,5)-linked L-Ara*f* arabinan backbone has been also identified. Ferulic acid residues can undergo *in vivo* oxidative coupling reactions to form dehydrodimers, thereby covalently cross-linking the polysaccharides they esterify (Fry 1986). Such cross-links have been identified in sugar-beet pectin (Ralet et al. 2005).

2.3.3 Side-Chain Intra- and Inter-Molecular Distribution

Treatment of "hairy regions" with rhamnogalacturonan hydrolase allowed to evidence that the different types of side-chains were not randomly distributed over the RG-I backbone. Indeed, incubating de-esterified apple or ginseng pectin with rhamnogalacturonan hydrolase released RG-I oligosaccharides based on a $GalA_2Rha_2$ and a $GalA_3Rha_3$ backbone (Schols and Voragen 1994; Sun et al. 2019). Most of the RG-I oligomers recovered contained no side-chains and single unit Gal side-chains even though single units or short Ara chains were also detected in ginseng (Sun et al. 2019). Since Rha-GalA oligosaccharides substituted solely by Ara residues were identified (O'Neill et al. 1990; Sun et al. 2019), a direct linkage between Ara and Rha is plausible although a single Gal unit or short type I galactan anchor seem to be usually necessary for Ara and arabinan branching onto RG-I backbone (Buffetto et al. 2015). Next to these lowly branched rhamnogalacturonan hydrolase-degradable regions, highly branched arabinan- and type II arabinogalactan-rich regions that are not rhamnogalacturonan hydrolase-degradable have been identified (Schols and Voragen 1994; Sun et al. 2019). It has been hypothesised that these side-chains are flexible and long enough to wrap around the RG-I backbone, hindering enzymatic breakdown of the RG-I backbone by rhamnogalacturonan hydrolase (Willats et al. 2001; Voragen et al. 2009). The intra- and inter-molecular distribution of the different domains (lowly and highly branched) is extremely difficult to characterise and virtually nothing is presently known about the specific location of individual side-chains along the backbone. The co-existence of different "hairstyles" within a single pectin molecule remains speculative (Voragen et al. 2009).

2.4 Connection Between Pectin Domains

It is now fully admitted that pectin is a highly complex macromolecule in which several domains are covalently linked to each other. It is however particularly challenging to accommodate all available structural information into a universal model structure. Currently, two models are under debate (1) the *smooth* and *hairy* regions model established by de Vries et al. (1981) and updated by Schols and Voragen (1996) (Fig. 2.4) and (2) the RG-I backbone model proposed by Vincken et al. (2003) (Fig. 2.5).

The smooth and hairy regions model proposes that HG domains alternate with "hairy" regions, the latter including three subunits: subunit I: XGA; subunit II: stubs of RG-I backbone substituted by arabinan side-chains (or AG-II, Sun et al. 2019); and subunit III: rather short RG-I stretches (four or six residues) that are unsubstituted or substituted by Gal single units (Schols and Voragen 1996). In contrast, the RG-I backbone model positions HG and XGA as side-chains of RG-I (Vincken et al. 2003). Coenen et al. (2007) are, according to our knowledge, the

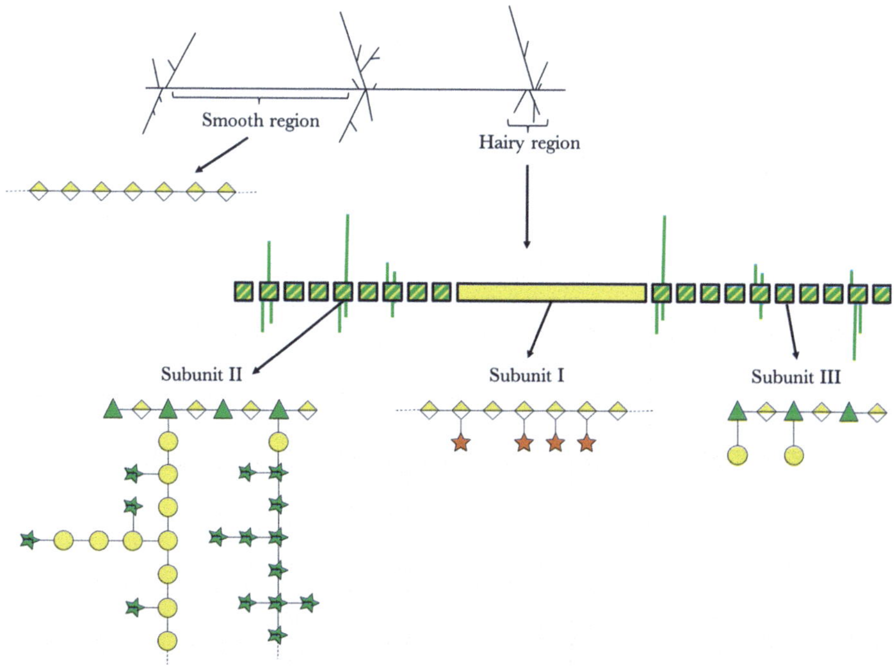

Fig. 2.4 Pectin smooth and hairy region model. The smooth region consists of homogalacturonan. The hairy region consists of three types of subunits: subunit I made of xylogalacturonan, subunit II made of arabinan- and type II arabinogalactan-rich RG-I and subunit III made of lowly substituted RG-I short stretches. Adapted from Schols and Voragen (1996)

only group succeeding in isolating and characterising chimeric oligosaccharides with a HG/XGA segment linked to a RG-I backbone one. A GalA trimer and a Xyl-substituted GalA trimer covalently linked to a short RG-I moiety were identified and H^1-NMR unambiguously showed that the GalA trimer was α-(1,2)-linked to the RG-I moiety.

MS^n experiments further showed that the RG-I moiety was always located at the reducing end of the chimeric oligosaccharides. These results are undoubtedly in favour of the "smooth and hairy regions" model even though this model does not fulfil all the experimental results obtained so far. In particular, it has been calculated that there are in average seventeen HG domains for one RG-I domain in lemon acid-extracted pectin and eight HG domains for one RG-I domain in sugar beet acid-extracted pectin taking into account (1) the partitioning of GalA between HG and RG-I domains, and (2) the length of these domains (Ralet and Thibault 2009). A similar "excess of GalA" was noticed by Yapo et al. (2007) and by Coenen et al. (2007). It is clear that there is a huge surplus of HG domains to build up a pectin macromolecule following the "smooth and hairy regions" model; the "RG-I backbone" model shall therefore not be excluded irrevocably (Coenen et al. 2007).

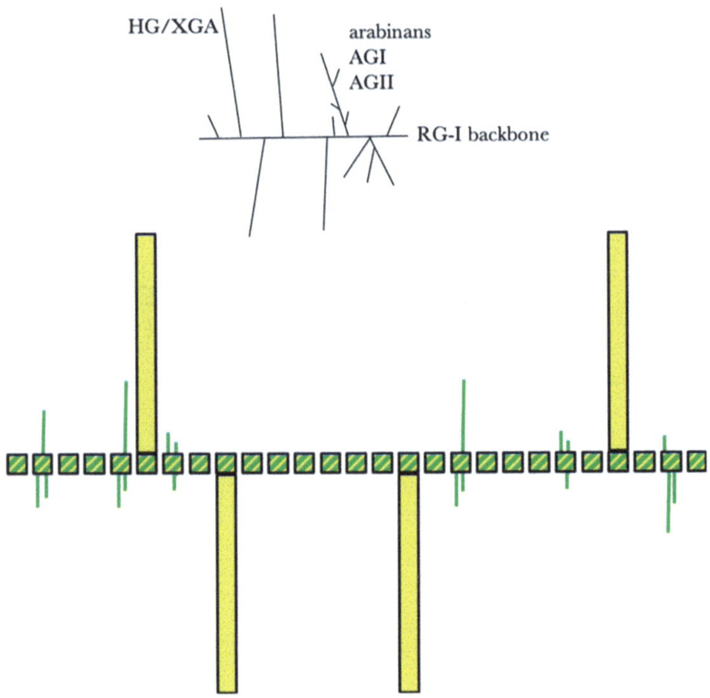

Fig. 2.5 Pectin RG-I backbone model. Homogalacturonan and xylogalacturonan (yellow) together with arabinans, and arabinogalactans (green) are side-chains of the RG-I backbone

2.5 Conclusions

Pectin is a highly complex macromolecule exhibiting a high degree of intra- and inter-molecular heterogeneity. When extracting pectin from a given plant sample, different populations that were located in specific tissues or even in specific spots within a single cell wall are co-extracted and mixed. Studying pectin fine structure implies to be able to master extraction conditions and implement adequate chemical and enzymatic tools to generate oligomers that can be separated and fully analysed. These can further be used to generate probes such as monoclonal antibodies to locate specific motifs *in muro*. To date, complete deconstruction of pectin into analysable oligomers has not been achieved and non-degradable fractions whose structure is unclear always remain. These fractions, encompassing in particular GalA surplus, should be targeted for getting a better understanding of the architecture of pectin macromolecules.

References

Aspinall GO, Begbie R, Hamilton A et al (1967) Polysaccharides of soy-beans. Part III. Extraction and fractionation of polysaccharides from cotyledon meal. J Chem Soc C 170:1065–1070
Atmodjo MA, Hao Z, Mohnen D (2013) Evolving views of pectin biosynthesis. Annu Rev Plant Biol 64:747–779
Avci U, Pena MJ, O'Neill MA (2018) Changes in the abundance of cell wall apiogalacturonan and xylogalacturonan and conservation of rhamnogalacturonan II structure during the diversification of the Lemnoideae. Planta 247:953–971
Bonnin E, Dolo E, Le Goff A et al (2002) Characterisation of pectin subunits released by an optimised combination of enzymes. Carbohydr Res 337:1687–1696
Braconnot H (1825a) Recherches sur un nouvel acide universellement répandu dans tous les végétaux. Ann Chim Phys 28:173–178
Braconnot H (1825b) Nouvelles observations sur l'acide pectique. Ann Chim Phys 30:96–102
Broxterman SE (2018) The architecture of the primary cell walls: the role of pectin reconsidered. PhD thesis, Wageningen University, The Netherland
Buffetto F, Ropartz D, Zhang XJ et al (2014) Recovery and fine structure variability of RGII subdomains in wine (Vitis vinifera Merlot). Ann Bot 114:1327–1337
Buffetto F, Cornuault V, Ridahl MG et al (2015) The deconstruction of pectic rhamnogalacturonan I unmasks the occurrence of a novel arabinogalactan oligosaccharide epitope. Plant Cell Physiol 56:2181–2196
Carpita NC, Gibeaut DM (1993) Structural models of primary cell walls in flowering plants: consistency of molecular structure with the physical properties of the walls during growth. Plant J 3:1–30
Coenen GJ, Bakx EJ, Verhoef RP et al (2007) Identification of the connecting linkage between homo- or xylogalacturonan and rhamnogalacturonan type I. Carbohydr Polym 70:224–235
Colquhoun IJ, de Ruiter GA, Schols HA et al (1990) Identification by n.m.r. spectroscopy of oligosaccharides obtained by treatment of the hairy regions of apple pectin with rhamnogalacturonase. Carbohydr Res 206:131–144
Colquhoun I, Ralet MC, Thibault JF (1994) Structure identification of feruloylated oligosaccharides from sugar-beet pulp by NMR spectroscopy. Carbohydr Res 263:243–256
Cornuault V, Buffetto F, Rydahl MG et al (2015) Monoclonal antibodies indicate low-abundance links between heteroxylan and other glycans of plant cell walls. Planta 242:1321–1334
Daas PJH, Voragen AGJ, Schols HA (2000) Investigation of the galacturonic acid distribution of pectin with enzymes part 2 - characterization of non-esterified galacturonic acid sequences in pectin with endopolygalacturonase. Carbohydr Res 326:120–129
de Vries JA, Voragen AGJ, Rombouts FM et al (1981) Extraction and purification of pectins from alcohol insoluble solids from ripe and unripe apples. Carbohydr Polym 1:117–127
de Vries JA, Rombouts FM, Voragen AGJ et al (1982) Enzymic degradation of apple pectins. Carbohydr Polym 2:25–33
de Vries JA, den Uijl CH, Voragen AGJ et al (1983) Structural features of the neutral sugar side chains of apple pectic substances. Carbohydr Polym 3:193–205
Dong X, Huang YF, Cho BG et al (2018) Advances in mass spectrometry-based glycomics. Electrophoresis 39:3063–3081
Fry SC (1986) Cross linking of matrix polymers in the growing cell walls of angiosperms. Annu Rev Plant Physiol 37:165–186
Gloaguen V, Brudieux V, Closs B et al (2010) Structural characterization and cytotoxic properties of an apiose-rich pectic polysaccharide obtained from the cell wall of the marine phanero- gam *Zostera marina*. J Nat Prod 73:1087–1092
Glushka JN, Terrell M, York WS et al (2003) Primary structure of the 2-O-methyl-alpha-L-fucose-containing side chain of the pectic polysaccharide, rhamnogalacturonan II. Carbohydr Res 338:341–352

Guillon F, Thibault JF (1989) Methylation analysis and mild acid hydrolysis of the "hairy" fragments of sugar-beet pectins. Carbohydr Res 190:85–96
Guillotin SE, Bakx EJ, Boulenguer P et al (2005) Populations having different GalA blocks characteristics are present in commercial pectins which are chemically similar but have different functionalities. Carbohydr Polym 60:391–398
Hart DA, Kindel PK (1970) Isolation and partial characterization of apiogalacturonans from the cell wall of *Lemna minor*. Biochem J 116:569–579
Hellin P, Ralet MC, Bonnin E et al (2005) Homogalacturonans from lime pectins exhibit homogeneous charge density and molar mass distributions. Carbohydr Polym 60:307–317
Hinz SWA, Verhoef R, Schols HA et al (2005) Type I arabinogalactan contains β-D-Gal*p*-(1,3)-β-D-Gal*p* structural elements. Carbohydr Res 340:2135–2143
Huisman MMH, Brüll LP, Thomas-Oates JE et al (2001) The occurrence of internal (1,5)-linked arabinofuranose and arabinopyranose residues in arabinogalactan side chains from soybean pectic substances. Carbohydr Res 330:103–114
Ishii T (1997) O-Acetylated oligosaccharides from pectins of potato tuber cell walls. Plant Physiol 113:1265–1272
Ishii T, Kaneko S (1998) Oligosaccharides generated by partial hydrolysis of the borate-rhamnogalacturonan II complex from sugar beet. Phytochemistry 49:1195–1202
Ishii T, Tobita T (1993) Structural characterization of feruloyl oligo- saccharides from spinach-leafs cell walls. Carbohydr Res 248:179–190
Kaneko S, Ishii T, Matsunaga T (1997) A boron-rhamnogalacturonan-II complex from bamboo shoot cell walls. Phytochemistry 44:243–248
Keenan MHJ, Belton PS, Matthew JA et al (1985) A 13C-n.m.r. study of sugar-beet pectin. Carbohydr Res 138:168–170
Kertesz ZI (1951) The pectic substances. Interscience Publishers Inc., New York
Kohn R, Furda I (1968) Binding of calcium ions to acetyl derivatives of pectin. Collect Czechoslov Chem Commun 33:2217–2225
Kohn R, Malovikova A (1978) Dissociation of acetyl derivatives of pectic acid and intramolecular binding of calcium ions to those substances. Collect Czechoslov Chem Commun 43:1709–1719
Komalavilas P, Mort AJ (1989) The acetylation at O-3 of galacturonic acid in the rhamnose-rich portion of pectins. Carbohydr Res 189:261–272
Konno H, Yamasaki Y, Katoh K (1986) Enzymatic degradation of pectic substances and cell walls purified from carrot cell cultures. Phytochemistry 25:623–627
Körner R, Limberg G, Christensen TMIE et al (1999) Sequencing of partially methyl-esterified oligogalacturonates by tandem mass spectrometry and its use to determine pectinase specificities. Anal Chem 71:1421–1427
Lau JM, Mc Neill M, Darvill AG et al (1985) Structure of the backbone of rhamnogalacturonan I, a pectic polysaccharide in the primary cell walls of plants. Carbohydr Res 137:111–125
Lau JM, McNeil M, Darvill AG et al (1987) Treatment of rhamnogalacturonan I with lithium in ethylenediamine. Carbohydr Res 168:245–274
Le Goff A, Renard CMGC, Bonnin E et al (2001) Extraction, purification and chemical characterisation of xylogalacturonans from pea hulls. Carbohydr Polym 45:325–334
Lerouge P, O'Neill MA, Darvill AG et al (1993) Structural characterization of endo-glucanase-generated oligoglycosyl side chains of rhamnogalacturonan I. Carbohydr Res 243:359–371
Levigne SV, Ralet MCJ, Quéméner B et al (2004) Isolation from sugar beet cell walls of arabinan oligosaccharides esterified by two ferulic acid monomers. Plant Physiol 134:1173–1180
Limberg G, Körner R, Buchholt HC et al (2000) Analysis of different de-esterification mechanisms for pectin by enzymatic fingerprinting using endopectin lyase and endopolygalacturonase II from A. niger. Carbohydr Res 327:293–307
Longland JM, Fry SC, Trewavas AJ (1989) Developmental control of apiogalacturonan biosynthesis and UDP-apiose production in a duckweed. Plant Physiol 90:972–978

Matsunaga T, Ishii T, Matsumoto S et al (2004) Occurrence of the primary cell wall polysaccharide rhamnogalacturonan II in pteridophytes, lycophytes, and bryophytes. Implications for the evolution of vascular plants. Plant Physiol 134:339–351

McNeil M, Darvill AG, Albersheim P (1980) Structure of plant cell walls. X. Rhamnogalacturonan I, a structurally complex pectic polysaccharide in the walls of suspension-cultured sycamore cells. Plant Physiol 66:1128–1134

McNeil M, Darvill AG, Fry SC et al (1984) Structure and function of the primary cell walls of plants. Annu Rev Biochem 53:625–663

Mohnen D (2008) Pectin structure and biosynthesis. Curr Opin Plant Biol 11:266–277

Morita M (1965a) Polysaccharides of soybean seeds. Part 1. Polysaccharide constituents of "hot-water-extract" fraction of soybean seeds and an arabinogalactan as its major component. Agric Biol Chem 29:564–573

Morita M (1965b) Polysaccharides of soybean seeds. Part 2. A methylated arabinogalactan isolated from methylated product of "hot-water-extract" fraction of soybean seed polysaccharides. Agric Biol Chem 29:626–630

Mort A, Zheng Y, Qiu F et al (2008) Structure of xylogalacturonan fragments from watermelon cell-wall pectin. Endopolygalacturonase can accommodate a xylosyl residue on the galacturonic acid just following the hydrolysis site. Carbohydr Res 343:1212–1221

Mucha E, Marianski M, Xu FF et al (2018) Unravelling the structure of glycosyl cations via cold-ion infrared spectroscopy. Nat Commun 9:4174

Nakamura A, Furuta H, Maeda H et al (2002) Structural studies by stepwise enzymatic degradation of the main backbone of soybean soluble polysaccharides consisting of galacturonan and rhamnogalacturonan. Biosci Biotechnol Biochem 66:1301–1313

Ndeh D, Rogowski A, Cartmell A et al (2017) Complex pectin metabolism by gut bacteria reveals novel catalytic functions. Nature 544:65–70

Needs PW, Rigby NM, Colqhoun IJ et al (1998) Conflicting evidence for non-methyl galacturonoyl esters in Daucus carota. Phytochemistry 48:71–77

Ngouemazong DE, Tengweh FF, Duvetter T et al (2011) Quantifying structural characteristics of partially de-esterified pectins. Food Hydrocoll 25:434–443

Normand J, Ralet MC, Thibault JF (2010) Purification, characterization, and mode of action of a rhamnogalacturonan hydrolase from Irpex lacteus, tolerant to an acetylated substrate. Appl Microbiol Biotechnol 86:577–588

O'Neill M, Albersheim P, Darvill A (1990) The pectic polysaccharides of primary cell walls. In: Dey PM, Harborne JB (eds) Methods in plant biochemistry, Carbohydrates, vol 2. Academic Press, London, pp 415–441

Øbro J, Harholt J, Scheller HV et al (2004) Rhamnogalacturonan I in Solanum tuberosum tubers contains complex arabinogalactans structures. Phytochemistry 65:1429–1438

Oosterveld A, Beldman G, Searle-van Leeuwen MJF et al (2000a) Effect of enzymatic deacetylation on gelation of sugar beet pectin in the presence of calcium. Carbohydr Polym 43:249–256

Oosterveld A, Beldman G, Schols HA et al (2000b) Characterization of arabinose and ferulic acid rich pectic polysaccharides and hemicelluloses from sugar beet pulp. Carbohydr Res 328:185–197

Oosterveld A, Beldman G, Voragen AGJ (2002) Enzymatic modification of pectic polysaccharides obtaines from sugar beet pulp. Carbohydr Polym 48:73–81

Ovodov YS, Ovodova RG, Bondarenko OD et al (1971) The pectic substances of *Zosteraceae*: Part IV. Pectinase digestion of zosterine. Carbohydr Res 18:311–318

Pabst M, Fischl RM, Brecker L et al (2013) Rhamnogalacturonan II structure shows variation in the side chains monosaccharide composition and methylation status within and across different plant species. Plant J 76:61–72

Perrone P, Hewage CM, Thomson AR et al (2002) Patterns of methyl and O-acetyl esterification in spinach pectins: new complexity. Phytochemistry 60:67–77

Pippen E, McCready RM, Owens HS (1950) Gelation properties of partly acetylated pectins. J Am Chem Soc 72:813–816

Quéméner B, Ralet MC (2004) Evidence for linkage position determination in known feruloylated mono- and disaccharides using electrospray ion trap mass spectrometry. J Mass Spectrom 39:1153–1160
Quéméner B, Désiré C, Lahaye M et al (2003) Structural characterisation by both positive- and negative-ion electrospray mass spectrometry of partially methyl-esterified oligogalacturonides purified by semi-preparative high-performance anion-exchange chromatography. Eur J Mass Spectrom 9:45–60
Ralet MC, Thibault JF (2009) Hydrodynamic properties of isolated pectic domains: a way to figure out pectin macromolecular structure? In: Schols HA, Visser RGF, Voragen AGJ (eds) Pectins and pectinases. Wageningen Academic Publishers, Wageningen, pp 35–48
Ralet MC, Crépeau MJ, Buchholt HC et al (2003) Polyelectrolyte behaviour and calcium binding properties of sugar beet pectins differing in their degrees of methylation and acetylation. Biochem Eng J 16:191–201
Ralet MC, Cabrera JC, Bonnin E et al (2005) Mapping sugar beet pectin acetylation pattern. Phytochemistry 66:1832–1843
Ralet MC, Crépeau MJ, Bonnin E (2008) Evidence for a blockwise distribution of acetyl groups onto homogalacturonans from a commercial sugar beet (Beta vulgaris) pectin. Phytochemistry 69:1903–1909
Ralet MC, Lerouge P, Quéméner B (2009) Mass spectrometry for pectin structure analysis. Carbohydr Res 344:1798–1807
Ralet MC, Martins W, Tanhatan Nasseri A et al (2012) An innovative enzymatic approach to apprehend the methylesterification pattern of homogalacturonans. Biomacromolecules 13:1615–1624
Remoroza C, Cord-Landwehr S, Leijdekkers AGM et al (2012) Combined HILIC-ELSD/ESI-MSn enables the separation, identification and quantification of sugar beet pectin derived oligomers. Carbohydr Polym 90:41–48
Remoroza C, Buchholt HC, Gruppen H et al (2014) Descriptive parameters for revealing substitution patterns of sugar beet pectins using pectolytic enzymes. Carbohydr Polym 101:1205–1215
Renard C, Jarvis M (1999) Acetylation and methylation of homogalacturonans 2. Effect on ion-binding properties and conformations. Carbohydr Polym 39:209–216
Renard C, Crépeau MJ, Thibault JF (1995) Structure of the repeating units in the rhamnogalacturonic backbone of apple, beet and citrus pectins. Carbohydr Res 275:155–165
Renard C, Crépeau MJ, Thibault JF (1999) Glucuronic acid directly linked to galacturonic acid in the rhamnogalacturonan backbone of beet pectins. Eur J Biochem 266:566–574
Reuhs BL, Glenn J, Stephens SB et al (2004) L-Galactose replaces L-Fucose in the pectic polysaccharide rhamnogalacturonan II synthesized by the L-fucose-deficient mur1 Arabidopsis mutant. Planta 219:147–157
Ridley BL, O'Neill MA, Mohnen D (2001) Pectins: structure, biosynthesis, and oligogalacturonide-related signalling. Phytochemistry 57:929–967
Rombouts FM, Thibault JF (1986) Enzymatic and chemical degradation and the fine structure of pectins from sugar-beet pulp. Carbohydr Res 154:189–203
Ropartz D (2015) Apport des dernières évolutions en spectrométrie de masse pour l'étude structurale des polysaccharides. Université de Nantes, Nantes
Ropartz D, Lemoine J, Giuliani A et al (2014) Deciphering the structure of isomeric oligosaccharides in a complex mixture by tandem mass spectrometry: Photon activation with vacuum ultra-violet brings unique information and enables definitive structure assignment. Anal Chim Acta 807:84–95
Ropartz D, Li P, Fanuel M et al (2016) Charge transfer dissociation of complex oligosaccharides: comparison with collision-induced dissociation and extreme ultraviolet dissociative photoionization. J Am Soc Mass Spectrom 27:1614–1619
Ropartz D, Li P, Jackson GP et al (2017) Negative polarity helium charge transfer dissociation tandem mass spectrometry: Radical-initiated fragmentation of complex polysulfated anions. Anal Chem 89:3824–3828

Ropartz D, Fanuel M, Ujma J et al (2019) Structure determination of large isomeric oligosaccharides of natural origin through multi-pass and multi-stage cyclic traveling wave ion mobility mass spectrometry. Anal Chem 91:12030–12037
Round AN, Rigby NM, MacDougall AJ et al (2010) A new view of pectin structure revealed by acid hydrolysis and atomic force microscopy. Carbohydr Res 345:487–497
Sakamoto T, Sakai T (1995) Analysis of structure of sugar-beet pectin by enzymatic methods. Phytochemistry 39:821–823
Schindler B, Barnes L, Renois G et al (2017) Anomeric memory of the glycosidic bond upon fragmentation and its consequences for carbohydrate sequencing. Nat Commun 8:7
Schols HA, Voragen AGJ (1994) Occurrence of pectic hairy regions in various plant cell wall materials and their degradability by rhamnogalacturonase. Carbohydr Res 256:83–95
Schols HA, Voragen AGJ (1996) Complex pectins: Structure elucidation using enzymes. In: Visser J, Voragen AGJ (eds) Pectins and pectinases. Elsevier, Amsterdam, pp 3–19
Schols HA, Voragen AGJ (2002) The chemical structure of pectins. In: Seymour GB, Knox JP (eds) Pectins and their manipulation. Blackwell Publishing-CRC Press, Oxford, pp 1–29
Schols HA, Bakx EJ, Schipper D et al (1995a) A xylogalacturonan subunit present in the modified hairy regions of apple pectin. Carbohydr Res 279:265–279
Schols HA, Vierhuis E, Bakx EJ et al (1995b) Different populations of pectic hairy regions occur in apple cell walls. Carbohydr Res 275:343–360
Sengkhamparn N, Verhoef R, Bakx EJ et al (2009) Okra pectin contains an unusual substitution of its rhamnosyl residues with actyl and alpha-linked galactosyl groups. Carbohydr Res 334:1842–1851
Seveno M, Voxeur A, Rihouey C et al (2009) Structural characterisation of the pectic polysaccharide rhamnogalacturonan II using an acidic fingerprinting methodology. Planta 230:947–957
Shin KS, Kiyohara H, Matsumoto T et al (1998) Rhamnogalacturonan II dimers cross-linked by borate diesters from the leaves of Panax ginseng C.A. Meyer are responsible for expression of their IL-6 production enhancing activities. Carbohydr Res 307:97–106
Smolenski K (1923) Pectins. Rocznigi Chemki 3:86–152
Spellman MW, McNeil M, Darvill AG et al (1983) Structure of plant cell-walls.14. Characterization of a structurally complex heptasaccharide isolated from the pectic polysaccharide rhamnogalacturonan-II. Carbohydr Res 122:131–153
Stevenson T, Darvill AG, Albersheim P (1988) Structure of plant-cell walls. 23. Structural features of the plant cell-wall polysaccharide rhamnogalacturonan-II. Carbohydr Res 182:207–226
Ström A, Ribelles P, Lundin L et al (2007) Influence of pectin fine structure on the mechanical properties of calcium-pectin and acid-pectin gels. Biomacromolecules 8:2668–2674
Sun L, Ropartz D, Cui LN et al (2019) Structural characterization of rhamnogalacturonan domains from Panax ginseng C. A. Meyer. Carbohydr Polym 203:119–127
Thibault JF, Renard C, Axelos M et al (1993) Studies of the length of homogalacturonic regions in pectins by acid-hydrolysis. Carbohydr Res 238:271–286
Thomas JR, Darvill AG, Albersheim P (1989) Isolation and structural characterization of the pectic polysaccharide rhamnogalacturonan II from walls of suspension-cultured rice cells. Carbohydr Res 185:261–277
Ujma J, Ropartz D, Giles K et al (2019) Cyclic ion mobility mass spectrometry distinguishes anomers and open-ring forms of pentasaccharides. J Am Soc Mass Spectrom 30:1028–1037
van Alebeek GJWM, Zabotina O, Beldman G et al (2000) Structural analysis of (methyl-esterified) oligogalacturonides using post-source decay matrix-assisted laser desorption/ionization time-of-flight mass spectrometry. J Mass Spectrom 35:831–840
Varki A, Cummings RD, Aebi M et al (2015) Symbol nomenclature for graphical representations of glycans. Glycobiology 25:1323–1324
Vauquelin LN (1790) Analyse du tamarin. Ann Chim 5:92–106
Vidal S, Doco T, Williams P et al (2000) Structural characterization of the pectic polysaccharide rhamnogalacturonan II: evidence for the backbone location of the aceric acid-containing oligoglycosyl side chain. Carbohydr Res 326:277–294

Vincken JP, Schols HA, Oomen RJFJ et al (2003) If homogalacturonan were a side-chain of rhamnogalacturonan I. Implication for cell wall architecture. Plant Physiol 132:1781–1789
Voragen AGJ, Pilnik W, Thibault JF et al (1995) Pectins. In: Stephen AM (ed) Food polysaccharides and their applications. Marcel Dekker Inc, New York, pp 287–339
Voragen AGJ, Coenen GJ, Verhoef RP et al (2009) Pectin, a versatile polysaccharide present in plant cell walls. Struct Chem 20:263–275
Westphal Y, Kühnel S, de Waard P et al (2010) Branched arabino-oligosaccharides isolated from sugar beet arabinan. Carbohydr Res 345:1180–1189
Whitcombe AJ, Oneill MA, Steffan W, Albersheim P, Darvill AG (1995) Structural characterization of the pectic polysaccharide, rhamnogalacturonan-II. Carbohydr Res 271:15–29
Willats WGT, McCartney L, Mackie W et al (2001) Pectin: cell biology and prospects for functional analysis. Plant Mol Biol 47:9–27
Willats WGT, Knox JP, Mikkelsen JD (2006) Pectin: new insights into an old polymer are starting to gel. Trends Food Sci Technol 17:97–104
Yapo BM (2011) Rhamnogalacturonan-I: A structurally puzzling and functionally versatile polysaccharide from plant cell walls and mucilages. Polym Rev 51:391–413
Yapo BM, Lerouge P, Thibault JF et al (2007) Pectins from citrus plant cell walls contain homogalacturonan homogeneous with respect to molar mass, rhamnogalacturonan I and rhamnogalacturonan II. Carbohydr Polym 69:426–435
Zandleven JS, Beldman G, Bosveld M et al (2006) Enzymatic degradation studies of xylogalacturonans from apple and potato, using xylogalacturonan hydrolase. Carbohydr Polym 65:495–503
Zandleven J, Sorensen SO, Harholt J et al (2007) Xylogalacturonan exists in cell walls from various tissues of Arabidopsis thaliana. Phytochemistry 68:1219–1226

Chapter 3
Pectin Degrading Enzymes

Estelle Bonnin and Jérôme Pelloux

3.1 Introduction

Pectins are complex polysaccharides present in the primary cell wall, their main characteristic being their high content of partly methyl-esterified galacturonic acid (GalA), and the presence of neutral sugars on the side chains. The seventeen different monosaccharides and more than twenty different linkages make it a very complex macromolecule, whose enzymatic degradation requires many different enzymes related to the narrow specificity of each. Enzymes involved in pectin degradation catalyse different types of bonds. Hydrolases split a covalent bond by adding a molecule of water while the lyases catalyse β-elimination reactions without the addition of a water molecule. The former comprise particularly the glycosidases and the esterases, and belong to the group 3 of the Enzyme Commission (EC 3.). The latter belong to the group 4 (EC 4.) and produce a double bond on the newly formed non-reducing end. Polysaccharide-degrading enzymes can be distinguished as endo- or exo-enzymes. The endo-enzymes cut linkages randomly inside the chain and most often are less active when the degree of polymerisation decreases. In contrast, exo-enzymes have the highest affinity for the terminal bond on the reducing- or non-reducing end of the chain. Enzymes related to synthesis, degradation, or modification of oligo- and polysaccharides are classified in the CAZy (for Carbohydrate-Active enZymes) database (www.cazy.org). Glycoside-hydrolases (GH), polysaccharide-lyases (PL), carbohydrate-esterases (CE), glycoside-transferases and carbohydrate-binding modules are grouped into structurally- and catalytically-related families.

E. Bonnin (✉)
INRAE, UR BIA Biopolymères Interactions Assemblages, Nantes, France
e-mail: estelle.bonnin@inrae.fr

J. Pelloux
BIOPI Biologie des Plantes et Innovation SFR Condorcet, Université de Picardie, Amiens, France
e-mail: jerome.pelloux@u-picardie.fr

V. Kontogiorgos (ed.), *Pectin: Technological and Physiological Properties*,
https://doi.org/10.1007/978-3-030-53421-9_3

This chapter focuses on highlighting the diversity of pectin-related enzymes by describing their structural diversity, the regulation of their expression, their substrate specificity and modes of action. Considering the diversity of pectin structure, enzymes will target a large diversity of linkages and/or polysaccharides, acting either on the backbone or on the side chains of pectin. These changes in pectin chemistry have dramatic consequences on polymer rheology, and ultimately the cell wall, affecting plant development, plant susceptibility to pathogens as well as the quality of end-products. Despite having being studied for a number of years, understanding the diversity of pectin-related enzymes is still a major challenge.

3.2 Homogalacturonan-Degrading Enzymes

Homogalacturonan (HG) backbone is depolymerised by different enzymes according to the recognition site on the GalA chain (Table 3.1; Fig. 3.1a). The linkage between two non-esterified GalA is recognised by polygalacturonases (PG) and pectate lyases (PL), whereas the linkage harboring an esterified GalA at the right side is recognised by pectin lyases (PNL). PL and PNL are often referred as to PLL (Pectin Lyase-Like).

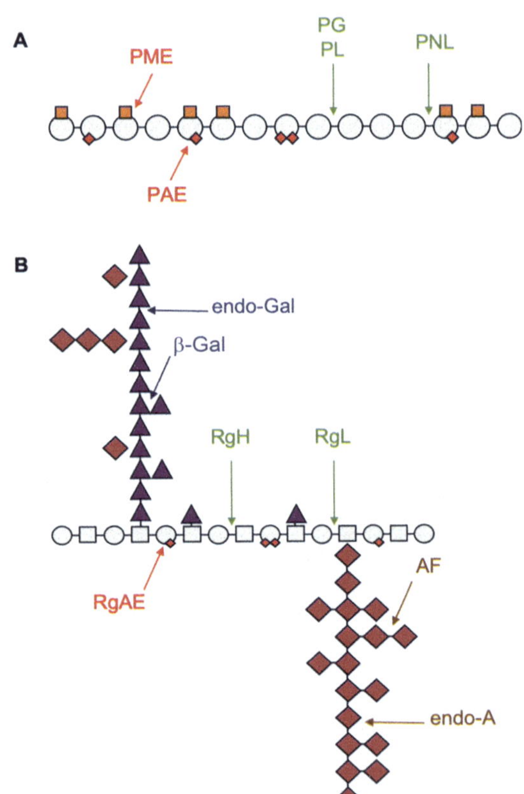

Fig. 3.1 Site of action of pectin degrading enzymes. (**a**) Homogalacturonan degrading enzymes, (**b**) Rhamnogalacturonan degrading enzymes. Grey circle: galacturonic acid, grey square: rhamnose, purple triangle: galactose, brown diamond: arabinose, orange square: methyl group, and red diamond: acetyl group. *PG* polygalacturonase (endo/exo), *PL* pectate lyase (endo/exo), *PNL* pectin lyase, *PME* pectin methylesterase, *PAE* pectin acetylesterase, *endo-Gal* endogalactanase, *β-Gal* β-galactosidase, *endo-A* endo-arabinanase, *AF* arabinofuranosidase, *RgH* RG-hydrolase, *RgL* RG-lyase, *RgAE* RG-acetylesterase

Table 3.1 Enzymes involved in pectin degradation

Enzymes	EC number	CAZy[a] family	Substrate[a]	Product
Hydrolases				
Endopolygalacturonase	3.2.1.15	GH28	HG–LM Pectin	OligoGalA
Exopolygalacturonase	3.2.1.67	GH28	HG–LM Pectin	GalA
Exo-poly-α-D-galacturonosidase	3.2.1.82	GH28	HG–LM Pectin	$GalA_2$
Rhamnogalacturonan hydrolase	3.2.1.171	GH28	RG-I–Pectin	OligoRG
Rhamnogalacturonan rhamnohydrolase	3.2.1.174	GH28	RG-I–Pectin	Rhamnose
Rhamnogalacturonan galacturonohydrolase	3.2.1.173	GH28	RG-I–Pectin	GalA
Xylogalacturonanhydrolase	3.2.1.–	GH28	Xylogalacturonan	Xylosylated OligoGalA
Endo (1,4) galactanase	3.2.1.89	GH53	1,4 Galactan	1,4 Oligogalactosides
Endo (1,3) galactanase	3.2.1.181	GH16	1,3-6 Galactan	1,3-6 Oligogalactosides
Endo (1,6) galactanase	3.2.1.164	GH5-30	1,3-6 Galactan	1,3-6 Oligogalactosides
β-galactosidase	3.2.1.23	GH1-2-35-42	Oligogalactosides	Galactose
Endoarabinanase	3.2.1.99	GH43	Arabinan	Oligoarabinosides
α-L-arabinofuranosidase	3.2.1.55	GH43-51-54-62	Oligoarabinosides	Arabinose
Lyases				
Endopectate lyase	4.2.2.2	PL1-2-9	HG	Unsaturated OligoGalA
Exopectate lyase	4.2.2.9	PL1-2-9	HG	Unsaturated OligoGalA
Endopectin lyase	4.2.2.10	PL1	HM Pectin	Unsaturated MeOligoGalA
Rhamnogalacturonan endolyase	4.2.2.23	PL4	RG-I	Unsaturated OligoRG
Rhamnogalacturonan exolyase	4.2.2.24	PL11-26	RG-I	Unsaturated OligoGalA
Esterases				
Pectin methyl esterase	3.1.1.11	CE8	Pectin	Pectic acid + methanol
Pectin acetyl esterase	3.1.1.–	CE12	Pectin	Pectic acid + acetic acid
Rhamnogalacturonan acetyl esterase	3.1.1.–	CE12	Pectin	RG-I + acetic acid
Feruloyl esterase	3.2.1.73	CE1	Feruloylated oligosides	Oligosides + ferulic acid

[a]*GH* glycosyl hydrolase, *PL* polysaccharide lyase, *CE* carbohydrate esterase, *HG* homogalacturonan, *RG-I* rhamnogalacturonan I, *LM* low methoxy, *HM* high methoxy, *GalA* galacturonic acid

3.2.1 *Polygalacturonases (PG)*

Endo-polygalacturonases (Endo-PG, EC 3. 2.1.15) and exo-polygalacturonases (Exo-PG, EC 3.2.1.67) generally catalyse the hydrolysis of the α-(1–4) glycosidic bond between two adjacent D-GalA units in non-methylated stretches of HG. In plants, they are synthesised in the endoplasmic reticulum before being modified in the Golgi apparatus and secreted in the cell wall where they contribute to the changes in the pectin chemistry and mechanics during plant growth. In tomato fruit, they notably contribute to ripening and to the loss of firmness during storage (Van Dijk et al. 2006). In contrast, apple pectins do not appear to be major targets for the ripening-involved enzymes. Nevertheless, PG1 expression levels have been associated with softening in a range of cultivars (Wakasa et al. 2006) and suppression of PG1 gene expression in transgenic apple plants results in a firmer fruit (Atkinson et al. 2012). PG play a key role in the regulation of plant development such as root/hypocotyl growth, stomata functioning, and pollen formation, as has recently been highlighted in a number of studies in *Arabidopsis* (Rhee et al. 2003; Ogawa et al. 2009; Xiao et al. 2014, 2017; Rui et al. 2017).

Phytopathogenic organisms, including microorganisms and parasitic plants, also produce PG. They contribute to host colonisation by degrading the physical barrier of cell wall. In herbivorous insects, although it has long been assumed that pectin degradation is ensured by enzymes from the digestive microflora, advances in genome sequencing and bioinformatics analysis have recently revealed the presence of endogenous pectin-related enzymes (Kirsh et al. 2016). In the CAZy database, the GH family 28 includes all the PG. As reviewed by Yang et al. (2018), they all share the four amino-acids segments NTD, DD, GHG and RIK in their active site. The aspartic acid residues in the two first segments are the catalytic acids. From the reported structures of microbial PG, they display a β-helix 3D structure. The active site of a typical PG is a tunnel-like binding cleft with which the enzyme binds the polysaccharide to produce oligosaccharides with variable degrees of polymerisation (Shimizu et al. 2002). In contrast, in a typical exo-PG, inserted stretches of amino acids transform this open-ended channel in a closed pocket only able to attack the non-reducing end of the pectic oligomers (Abbott and Boraston 2007).

In the genomes of the mammalian intestine bacteria, enzyme-coding sequences are grouped together in polysaccharide utilisation loci (PUL) (Flint et al. 2008). In the gut bacteria *Bacteroides xylanisolvens*, six PUL are devoted to pectin degradation, including one particularly involved in HG breakdown (Despres et al. 2016). Whatever the source considered, PG are most often the products of large multigenic families. For instance, even in monocotyledonous plants, such as *Brachypodium dystachion* where pectin content is low, more than forty sequences have been annotated as putative PG (The International Brachypodium Initiative 2010). All these isozymes exhibit different specificities, which are likely to play a key role in the adaptation to different environments. As an example, all PG isoforms in the model plant *Arabidopsis thaliana* have a predicted pI between 4.55 and 10.26 (data extracted from www.arabidopsis.org). This would regulate their activity according to the pH and ionic microenvironment of the cell wall. To a lesser extent, the filamentous fungus *Aspergillus niger* produces seven endo-PG with different optimal conditions for pH and temperature. In most cases, optimal pH for PG is in the acidic range, from 3 to 6. This optimum can be modulated by the presence of ions. Monovalent cations such as Na^+ affect the optimal pH by screening the charges of the carboxylic groups. Divalent cations such as calcium interact with the HG chains to form dimers and thus modify the availability of the substrate to the enzyme. Even if the main feature of PG is to break down linkages between non-substituted GalA, the isoforms of *A. niger* vary in their tolerance towards methyl substituents on the GalA chain. Exo- and endo-PGs can also vary in their tolerance towards acetyl substituents (Sakamoto et al. 2002; Bonnin et al. 2003). Xylose also occurs as substituent in xylogalacturonan. Some exo-PG are able to release the dimer Xyl-GalA (Sakamoto et al. 2002) while endo-enzymes named xylogalacturonases work within the chain (Zandleven et al. 2005, 2006).

When attacked by pathogens the host plant secretes in the cell wall polygalacturonase-inhibiting proteins (PGIP) dedicated to the inhibition of polygalacturonases. From the current knowledge, PGIP are not involved in the regulation of plant PG activity but target the microbial PG with various specificities. PGIP belong to a large family of leucine-rich proteins and the residues involved in the β-strand/β-turn motif were shown to be crucial for their affinity and specificity to PGs. The 3D structure analysis of the complex formed by *Fusarium moniliforme* PG and *Phaseolus vulgaris* PGIP highlighted the role of some active site-residues in the recognition of the enzyme by the PGIP (Federici et al. 2001). However, it was further shown that the complex formation varies according to PG and that PGIP inhibits the enzyme activity although not covering the active site (Bonivento et al. 2008). The presence of PGIP was demonstrated in many plants including those containing low pectin level such as wheat (Kalunke et al. 2015).

3.2.2 Pectin and Pectate Lyases

Pectin Lyases-Like (PLLs), include Pectate Lyases (PL: endo-PLs, EC 4.2.2.2 and exo-PLs, EC 4.2.2.9) and PectiN Lyases (PNL: endo-PNLs, EC 4.2.2.10), and catalyse the degradation of HG through a β-elimination process, thus producing unsaturated oligogalacturonides (OGs) that differ from that generated by PG. In the CAZy database, they belong to the PL1, 2 and 3 families. PLLs were identified in plants as well as in bacteria and fungi. A number of studies, in particular on tomato and strawberry, have shown that PLLs play a key role in the regulation of fruit softening (Wang et al. 2018). This was notably demonstrated through the use of transgenic lines silencing or overexpressing fruit-specific PL isoforms, showing that changes in pectin chemistry can affect fruit firmness as well as shelf life and resistance to pathogens (Santiago-Doménech et al. 2008; Posé et al. 2013; Yang et al. 2017). Consequently, the engineering of PLL levels in fruit appears as a strategy of interest to control the quality of the end product. Until now, the determination of the roles of PLL in controlling plant development has remained scarce although a relatively large number of genes encoding PLL are expressed in several plant species, including *Arabidopsis* (Sun and Van Nocker 2010; Cao 2012; McCarthy et al. 2014). Considering the diversity of the expression patterns of PLL, the enzymes are likely to be involved in a number of developmental processes including, for instance, cell separation. A number of PLL from various sources (mainly bacterial and fungal ones) were biochemically characterised, which showed that PL and PNL differ with regards to their substrate specificity, pH-dependence and affinity for cations (Mayans et al. 1997; Herron et al. 2000). Pectate lyases have indeed more affinity for low methyl esterified pectins and an alkaline pH optimum while pectin lyases require high DM pectins and low pH. In contrast, the biochemical characterisation of plant PLLs has proven to be more difficult. However, PLLs from *Hevea brasiliensis*, *Zinnia elegans* and mango fruit expressed in heterologous system show optimal activities at rather neutral to alkaline pH (Domingo et al. 1998; Chourasia et al. 2006; Chotigeat et al. 2009). The structure of pectin lyase from *Aspergillus niger* was determined, showing that the enzyme folds into a parallel β-helix and that it shares structural similarities with pectate lyases albeit low sequence identities (Pickersgill et al. 1994; Mayans et al. 1997). The contribution of pathogen PLLs to virulence and plant infection was recently highlighted (Expert et al. 2018; Yang et al. 2018; Voxeur et al. 2019). Furthermore, some studies demonstrated the role of plant PLLs in generating resistance to pathogens, likely by the modification of cell wall pectins of the host (Vogel et al. 2002).

3.2.3 Pectin Methylesterases (PMEs)

Pectin Methyl Esterases (PMEs, EC 3.1.1.11) belong to the CE8 family of CAZy, and catalyse the specific demethylesterification of methyl ester residues at the C6 of GalA, releasing non-methyl esterified HG, protons and methanol. Non-methyl esterified stretches of HG can be the target of pectin-degrading enzymes such as PG, or cross-linking divalent ions, in particular calcium, to form the so-called egg-box structures. In either case, the PME-mediated changes in pectin structure have dramatic consequences on mechanical properties of plant cell walls (Hocq et al. 2017a). In plants, PMEs are divided into two groups depending on their protein sequence. Group 1 PMEs contain a mature active part while group 2 PMEs have, in addition to the mature part, an N-terminal extension (PRO-region), that can be cleaved by specific subtilisin proteases at dedicated conserved dibasic motifs (RRLL, RKLL…). PMEs are synthesised in the endoplasmic reticulum and the maturation of the protein, including glycosylation and processing of the PRO part, can occur in the Golgi. The processing of group 2 PMEs was reported to play a key role in the export of mature active protein at the cell wall (Wolf et al. 2009). PMEs were shown to control fruit development and fruit ripening, in particular in tomato and in apple (Frenkel et al. 1998; Brummell et al. 2004; De Freitas et al. 2012) but in other species as well (Ross et al. 2011; Wen et al. 2013; Zhang et al. 2019). PMEs play a central role in regulating developmental processes as diverse as primordia emergence at the shoot apical meristem (Peaucelle et al. 2008, 2011), adventitious root formation (Guénin et al. 2011), plant-pathogen interactions (Hewezi et al. 2008; Raiola et al. 2011), pollen tube elongation (Jiang et al. 2005; Leroux et al. 2015), seed development (Müller et al. 2013), mucilage production (Turbant et al. 2016), wood and stem mechanical properties (Siedlecka et al. 2008; Hongo et al. 2012), and dark-grown hypocotyl elongation (Derbyshire et al. 2007; Pelletier et al. 2010; Wolf et al. 2012). Using transgenic lines impaired in the activity of PMEs and multidisciplinary approaches including atomic force microscopy (AFM), it was shown *in planta* that increased PME activity surprisingly induced a decrease in cell wall stiffness, thus questioning the above-mentioned egg-box model (Peaucelle et al. 2011; Andres-Robin et al. 2018). This highlights the complexity of the regulation of pectin chemistry and could be related to distinct modes of action of plant PMEs and/or the role of pectin degrading enzymes acting synergistically with PMEs (Hocq et al. 2017b). PMEs belong to rather large multigenic families in all plant species sequenced to date (both in monocot and dicot species), which questions the specificity of isoforms and their roles in the context of plant development. PMEs have distinct pH optimum; plant PMEs were reported to have neutral to alkaline optimum while fungal enzymes are more active in the acidic pH range (Jolie et al. 2010; Dixit et al. 2013; Sénéchal et al. 2014). PME activity can be regulated by a number of cations and the pH might influence the mode of action of PMEs, as determined for apple PME (Bordenave 1996; Denès et al. 2000). The mode of action of fungal and plant PMEs differs. Indeed plant PMEs have been largely described as having a processive mode of action, where the enzyme removes all contiguous methyl esters from a single chain of HG before dissociating from the substrate. In contrast, fungal isoforms are often described as having a random mode of action (Jolie et al. 2010). These distinct modes of action may have different consequences on pectin structure, which might partially explain the links between pectin chemistry and cell wall mechanical properties *in planta*. The differences in the mode of action of the PMEs are likely to be related to specific structural features. The 3D-structure was resolved for plant, bacterial, fungal and insect PMEs (Jenkins et al. 2001; Johansson et al. 2002; Di Matteo et al. 2005; Kent et al. 2016). All PMEs share the same distinctive parallel β-helix structure with highly conserved amino acids GxYxE, QAVAL, QDTL, DFIFG, LGRPW, which have been shown to be functionally important (Pelloux et al. 2007). In particular, the active site contains two aspartic acid, two glutamine and an arginine residues and the mechanism of action has been determined: one of the aspartic acid residues, stabilised by the hydrogen bond with arginine, performs the nucleophilic attack on the carboxymethyl group of the HG (Johansson et al. 2002). Recently, some specific features have been shown to be of importance in determining the processive versus non-processive mode of action, shedding new light on the diversity of these enzymes (Kent et al. 2016). One of the specific features of plant PMEs is their ability to be regulated by proteinaceous inhibitors named Pectin Methylesterase Inhibitors (PMEIs) (Di Matteo et al. 2005). PMEIs are encoded by a large multigenic family, which suggest some specificities in the interaction between isoforms. PME and PMEIs can form a stoichiometric 1:1 complex and the inhibition has been reported to be pH-dependent (Sénéchal et al. 2015). However, some PMEI can inhibit PME activity in a pH-independent manner (Hocq et al. 2017b) and molecular dynamics (MD) simulation allowed determining the amino acids, which are of importance for the pH-dependence (Sénéchal et al. 2017). Therefore, cell wall pH, modulated by PME activity itself, appears as a key element of enzyme activity regulation, presumably through a feedback-loop mechanism (Hocq et al. 2017b). In contrast, pathogen PMEs are not inhibited by any of the PMEI tested so far, and the structural determinants of the lack of inhibition have been also determined (Jolie et al. 2010). A range of chemical compounds has been shown to inhibit either plant or phytopathogen PMEs (Lewis et al. 2008; L'Enfant et al. 2015, 2019). Compared to what has been observed for PG, fungal PME do not appear to be a primary element of the pathogenicity but the plant-PME mediated changes in pectin structure can affect the colonisation of the host plant by fungal or nematode pathogens as well as by viruses (Chen and Citovsky 2003; Lionetti et al. 2007; Hewezi et al. 2008). The PME-mediated production of methanol appears to be of primary importance in the resistance to herbivorous insects acting as a long-distance signal during plant herbivore interactions (Dorokhov et al. 2012).

3.2.4 Pectin Acetylesterases (PAE)

In addition to being methyl esterified, pectins can be acetylated, the degree of acetylation varying from 5 to 30% depending on the species and organs considered. Pectin acetyl esterases (PAEs, EC 3.1.1.6) cleave the acetyl ester bond at *O*-2 or *O*-3 positions of GalA in both HG and RG-I (Williamson 1991). The control of the degree of acetylation of pectin by PAE affects the physico-chemical properties of pectins, and is of importance for the activity of other pectin-remodelling enzymes. For instance, the activity of PAE is higher if the pectins have been previously demethyl esterified. PAE have been identified in a wide range of organisms, including plants, fungi and bacteria, but the full biochemical characterisation of plant PAEs have, until now, been impaired by the lack of purified enzymes. Nevertheless, purified plant PAE from mung bean and orange show optimal activity for pH in the range of 5.5 to 6.5. A PAE from *Dickeya dadantii*, a phytopathogenic bacterium was active on both synthetic substrates and highly acetylated sugar beet pectin (Shevchik and Hugouvieux-Cotte-Pattat 2003). However, most of the biochemical characterization of PAEs used synthetic substrates such as triacetin and *p*-nitrophenyl acetate, which questions the specificity of the enzyme towards pectic substrates *in planta*. PAEs belong to the CE13 family of the CAZy database, which is characterised by the presence of the conserved residues serine (S), glycine (G), asparagine (N) and histidine (H), where S and H, together with an aspartic acid (D) form the catalytic triad (Alalouf et al. 2011; Oh et al. 2016). Considering this, it is refereed as belonging to the SGNH protein family. Although until now no 3D crystallographic structure was resolved for PAEs, homology modelling showed that PAEs share structural similarities with human notum that has a palmitoleoyl-protein carboxylesterase activity (Philippe et al. 2017). However, the structural specificities of pathogen and plant PAEs have yet to be determined, notably to understand the biochemical specificity of PAEs as compared to PME, but also compared to rhamnogalacturonan acetylesterases. As other pectin remodelling enzymes, PAEs are encoded by a multigenic family in plants albeit to a smaller extent. Studies related to the roles of

PAEs in the regulation of pectin structure and plant development are relatively scarce. The contribution of two *Arabidopsis* PAEs to inflorescence growth (de Souza and Pauly 2015) and the role of pectin deacetylation in pollen tube elongation and reproductive process in tobacco (Gou et al. 2012) have been revealed using mutants. A link has been further established between PAE-mediated deacetylation and the mechanical properties of potato tuber (Orfila et al. 2012). Although not formally demonstrated using functional genomics approaches, PAEs are likely to be involved in the control of cell wall structure during pathogen infection as a number of genes encoding PAEs has shown an increased expression following pathogen attack (Philippe et al. 2017).

3.3 Rhamnogalacturonan-Degrading Enzymes

3.3.1 Rhamnogalacturonan-Hydrolases

The backbone of rhamnogalacturonan I (RG-I) is made of repeating disaccharide [→2)-α-L-Rhap-(1→4)-α-D-GalpA-(1→] units (see Chap. 2). Splitting these two types of linkages involves hydrolases/lyases and endo/exo enzymes (Table 3.1, Fig. 3.1b). The first reported RG-hydrolase (EC 3.2.1.171) was purified from *Aspergillus aculeatus* and hydrolysed the α-D-Gal*p*A-(1→2)-α-L-Rha*p* glycosidic bond with an endo mode of action, to release oligosaccharides with a reducing β-D-Gal*p*A (Schols et al. 1990). Several similar enzymes were subsequently reported in various strains (Kofod et al. 1994; Sakamoto et al. 1994; Suykerbuyk et al. 1997). To date, *Aspergillus* genus is recognised as the main producer of rhamnogalacturonan-degrading enzymes. Only three enzymes are reported from other sources: *Botrytis cinerea* (Chen et al. 1997), *Irpex lacteus* (Normand et al. 2010) and *Penicillium chrysogenum* (Matsumoto et al. 2017). The mode of action of rhamnogalacturonan-hydrolase has been further studied. It requires five repeating disaccharides to bind to its substrate (Mutter et al. 1998a) and was hindered by methyl and acetyl esters on the backbone (Schols et al. 1990; Kofod et al. 1994). Conversely, the enzyme from *I. lacteus* was highly tolerant to acetyl groups (Normand et al. 2010). *P. chrysogenum* RGH activity was hindered by the presence of galactose on the backbone (Matsumoto et al. 2017). The first solved 3D crystallographic structure of a GH28 enzyme was the RG-hydrolase from *Aspergillus aculeatus* (Petersen et al. 1997). It forms a right helix of 13 turns of beta strands. Its active site is located in a groove almost perpendicular to the helix axis. It is *N*- and *O*-glycosylated at 2 and 18 sites, respectively, for a total of 6 kDa.

The exo-hydrolases active on RG-I are of two types: (a) the rhamnogalacturonan galacturonohydrolases (EC 3.2.1.173, GH28) allow the exo-hydrolysis of the α-D-GalA-(1→2)-α-L-Rha bond releasing single D-galacturonic acid from the non-reducing end of rhamnogalacturonan oligosaccharides (Mutter et al. 1998b), and (b) the rhamnogalacturonan rhamnohydrolases (EC 3.2.1.174, GH78 and 106) allow the exo-hydrolysis of the α-L-Rha-(1→4)-α-D-GalA bond releasing single β-L-rhamnose from the non-reducing end of rhamnogalacturonan oligosaccharides

(Mutter et al. 1994). The former is usually more active on small RG-I substrates whereas the latter is generally hindered by a galactose substitution on the terminal rhamnose (Mutter et al. 1998b). All the enzymes gathered in GH78 and GH106 release rhamnose from different substrates. They differ in their 3D structures, which have been resolved for α-L-rhamnosidases of each family. The first crystal structure of a GH78 member has been determined for *Bacillus* sp. GL1 α-L-rhamnosidase B (Cui et al. 2007). It is a five-module structure in which the catalytic one is an $(\alpha/\alpha)_6$-barrel. The rhamnosidase from *Bacteroides thetaiotaomicron* is the only 3D structure resolved in GH106. It follows a $(\beta/\alpha)_8$-barrel domain with two additional β-strands and it needs a calcium ion to bind the substrate (Ndeh et al. 2017). The most complicated structure of rhamnogalacturonan II (RG-II) is recognised and degraded by the bacteria of the digestive tract (Ndeh et al. 2017). In particular, in *Bacteroides thetaiotaomicron,* different PUL encode numerous enzymes with narrow specificity for each glycosidic linkages present in this polysaccharide, except 2-*O*-methyl-D-xylose- α1,3-L-fucose.

3.3.2 Rhamnogalacturonan-Lyases

Rhamnogalacturonan-lyases allow the eliminative cleavage of L-α-Rha*p*-(1–4)-α-D-Gal*p*A bonds of RGI domains leaving L-Rha*p* at the reducing end and 4, 5-unsaturated D-Gal*p*A at the non-reducing end according to an endo (EC 4.2.2.23) or an exo (EC 4.2.2.24) mechanism. The tetrasaccharide Δ4,5GalA-α-1,2-Rha-α-1,4-GalA-α-1,2-Rha where Δ means that the non-reducing end is unsaturated, is the major limit product of the RG endo-lyases. The disaccharide Δ4,5GalA-α-1,2-Rha can be further split by unsaturated rhamnogalacturonyl hydrolase (EC 3.2.1.172, GH105). Very few RG-lyases are reported in microorganisms (Azadi et al. 1995; Mutter et al. 1996; Pages et al. 2003), as well as in plants (carrot, Stratilova et al. 1998; cotton seed, Naran et al. 2007). Occurrence and roles of RG-lyases in plants have been recently reviewed (Ochoa-Jiménez et al. 2018). It shows that plant RG-lyases have only around 25% identity with the enzyme from *Aspergillus aculeatus* and that many of them contain a carbohydrate binding module at the C-terminal side. The endo RG-lyases are grouped in three different polysaccharide-lyase (PL) families, PL4, PL9 and PL11. PL4 is a monospecific family containing only RG-endolyases. It is divided in five sub-classes, PL4-1 and PL4-4 contain mainly bacterial enzymes, PL4-2 contains exclusively enzymes from plant origin, and PL4-3 and PL4-5 contain exclusively enzymes from Eukariota. The enzymes in PL9 are almost exclusively from bacterial origin. This family is divided into four sub-classes and contains only two RG-lyases, both from *Bacteroides thetaiotaomicron* in sub-class PL9-1. The 3D structure of one of these two has been shown to have the same lysine residue and calcium ion in the active site, as in the pectate lyase from the bacterium *Dickeya dadantii* (Luis et al. 2018). PL11contains endo- and exo-lyases in the two sub-classes, only from bacterial origin except two fungal enzymes in sub-class PL11-2 (*Aspergillus nidulans* and *Neocallimastix frontalis*), showing again the versatility of the genus *Aspergillus* for the pectin degrading enzymes. Apart from the PL9 family, exo-RG-lyases are also found in the PL26 family. It is a monospecific family containing only exo-RG-lyases, essentially from bacterial origin but the only one for which the activity has been experimentally demonstrated is from *Penicillium chrysogenum* (Iwai et al. 2015). The enzyme mainly released unsaturated galacturonosyl rhamnose disaccharide. It bypassed the Gal side chains of RG backbone but did not require Gal decoration for degradation. The enzyme is active in the dimeric form and on the reducing end of RG (Kunishige et al. 2018). Each unit is composed of three domains. Domain III displays a $(\alpha/\alpha)_6$-barrel fold and contains the substrate-binding site.

3.3.3 *Rhamnogalacturonan-Acetylesterase*

RG acetylesterases (EC 3.1.1.86, CE 12), required for the deacetylation of the RG-I backbone have been firstly described in *Aspergillus* species (Searle-van Leeuwen et al. 1992, 1996; Kauppinen et al. 1995). As pectins carry acetyl-esters on both homogalacturonan and rhamnogalacturonan (Ralet et al. 2005), it is sometimes difficult to distinguish between HG- and RG-acetylesterase specificity. It needs to have the appropriate substrates corresponding to the two different structural domains (Bonnin et al. 2008). Only the 3D structure of the *Aspergillus aculeatus* RG-acetylesterase has been resolved (Mølgaard et al. 2000). The enzyme folds into an α/β/α structure with an open cleft active site.

3.4 Side-Chain Degrading Enzymes

Pectin side chains are formed by the neutral sugars arabinose and galactose linked by different linkages depending on whether they belong to type I or type II arabinogalactans (See Chap. 2). The enzymatic degradation of neutral side chains in pectin involves endo- and exo-arabinanases, arabinofuranosidases, endo- and exo-galactanases and β-galactosidases (Table 3.1; Fig. 3.1b). As the neutral sugars are sometimes substituted by ferulic acid, the enzymatic degradation also requires feruloyl esterases.

3.4.1 *Arabinanases and α-Arabinofuranosidases*

α-L-arabinofuranose is present in plant cell wall polysaccharides, primarily in pectin arabinan but also in some hemicelluloses (xyloglucan, arabinoxylan). Its release from polysaccharides requires endo- and exo-arabinanases and α-arabinofuranosidase activities. Endo-arabinanases (EC 3.2.1.99, GH43) randomly cleave the α-1,5-linkages in the internal region of the arabinan backbone, preferably on polysaccharides and they are mostly found in microorganisms. In phytopathogenic fungi they have long been shown to play a role in pathogenesis (Stranne and Sakuragi 2016). As an example, in *Botrytis cinerea* a GH43 endo-arabinanase is a host-specific virulence factor (Nafisi et al. 2014). On the bacteria side, the genus *Bacteroides* exhibits an arabinan utilization locus with two GH43 arabinanases and two GH51 arabinofuranosidases (Arnal et al. 2015). So the corresponding encoding sequences are grouped in the genome in a locus dedicated to the arabinan breakdown, together with other sequences dedicated to arabinose binding, transport and utilisation. Plant arabinanases are poorly documented except those from seeds where they would allow the use of arabinan as storage reserve (Gomez et al. 2009). Exo-arabinanases (no EC number) hydrolyse arabinan from the non-reducing end to release arabinotriose (Kaji and Shimokawa 1984) or arabinobiose (Sakamoto and Thibault 2001). To date they are only described in bacteria and fungi, and belong to the mono-specific CAZy family GH93. Arabinofuranosidases (EC 3.2.1.55) release α-L-arabinofuranose linked to other sugars with (1,2), (1,3) and (1,5) linkages, in arabinan, arabinoxylan, and arabinogalactan. They have been classified according to their substrate specificity and mode of action (Beldman et al. 1997): class A contains arabinofuranosidases not active on polysaccharides, class B contains those active on polysaccharides, and the last class contains those specific for the hemicellulosic arabinoxylan. They are found in several GH families: 43, 51, 54, and 62. This reflects not only differences in their primary sequences and protein structures but also differences in their substrate specificity. Those from GH43 cleave α-(1,5) linkages in linear arabinan and arabino-oligosaccharides while those from GH62 are specific for arabinoxylan. The arabinofuranosidases from GH51 and GH54 have generally broader specificities and can be involved in the hydrolysis of pectin as well as hemicelluloses (Beylot et al. 2001; Yang et al. 2015). Arabinan-degrading systems are extensively described in few fungi and bacteria, notably *Penicillium purpurogenum* (Mardones et al. 2015; Ravanal and Eyzaguirre 2015; Vilches et al. 2018), *Penicillium chrysogenum* (Sakamoto and Thibault 2001; Sakamoto et al. 2011a, 2013a; Shinozaki et al. 2015), and the thermophilic bacterium *Caldanaerobius polysaccharolyticus* (Wefers et al. 2017). As they are produced by each strain, the three types of enzymes involved in the degradation of arabinan may act synergistically and thus be more efficient, including on branched arabinan. Due to their action towards the different arabinosyl linkages, arabinofuranosidases are key elements in this synergistic process, since they cleave the linkages in the arabinose side chains, allowing arabinanases to cleave the linkages in the backbone.

3.4.2 Galactanases and β-Galactosidases

Galactan-degrading enzymes have been reviewed by Sakamoto and Ishimaru (2013). Endo-galactanases include endo-β-1,4-galactanases (EC 3.2.1.89, GH53), endo-β-1,3-galactanases (EC 3.2.1.181, GH16) and endo-β-1,6-galactanases (EC 3.2.1.164, GH5 and GH30). They all randomly cleave the galactan chain in shorter oligosaccharides, and endo-β-1,3-galactanases and endo-β-1,6-galactanases are also involved in the hydrolysis of galactose from arabinogalactan-proteins. Endo-β-1,4-galactanases and endo-β-1,6-galactanases have been described in both fungi and bacteria. In *Aspergillus niger*, galacturonic acid and arabinose, but not galactose, induce the expression of endo-β-1,4-galactanase (de Vries et al. 2002), suggesting that the gene is co-expressed with genes encoding other pectinolytic enzymes. Another mechanism of expression control seems to predominate in bacteria since galactooligosaccharides induce endo-β-1,4-galactanase expression in *Geobacillus stearothermophilus* (Tabachnikov and Shoham 2013). To date, two 3D-structures of bacterial endo-β-1,4-galactanases and four 3D structures of fungal endo-β-1,4-galactanases are available (cazy.org/GH53). As expected for enzymes belonging to GH53, their structure reveals a (β/α)(8)-barrel architecture. The enzyme from *Bacteriodes thetaiotaomicron* presents a relatively open active site cleft compare to that in the enzyme from *Bacillus subtilis* (Böger et al. 2019). The catalytic residues of the two enzymes align although the enzyme from *B. thetaiotaomicron* is more active on short oligosaccharides than on polymeric galactan. Endo-β-1,4-galactanases generally accumulate monomers and dimers of galactose. A β-1,3-galactanase acting with an endo mode of action was first evidenced in the winter mushroom *Flammulina velutipes* (Kotake et al. 2011). More recently two other endo-β-1,3-galactanases were shown in *Aspergillus flavus* and in *Neurospora crassa* (Yoshimi et al. 2017). All three enzymes belong to GH16 family and β-1,3-galactotriose has been identified as the smallest possible substrate. The hydrolysis of the galactan structures is further realised by the action of exo-enzymes. Plant, fungi and bacteria produce exo-β-1,4-galactanases (EC 3.2.1.) belonging to GH35 family. *Penicillium chrysogenum* was the first microorganism reported to produce such an enzyme (Sakamoto et al. 2013b). Its activity was the highest on galactan, decreased with the degree of polymerisation of the substrate and released only galactose. Its synergistic action with the endo-β-1,4-galactanase from the same strain fully degraded lupin galactan into galactose. Due to their specific mode of action, the exo-β-1,4-galactanases from *Bacillus subtilis* and *Aspergillus niger* are able to keep the galactose moiety in their active site to add it on an acceptor, thus catalysing a transfer reaction (Nakano et al. 1991; Bonnin and Thibault 1996). Similarly, the hydrolysis of type II arabinogalactan is realised by the action of exo-β-1,3-galactanases (EC 3.2.1.145, GH43). They break the last linkage on the non-reducing end of the polysaccharide even if the last galactose residue is substituted by a side chain (Sakamoto et al. 2011b). Most of the exo-β-1,3-galactanases described so far from microorganisms contain a CBM at their C-terminus: CBM35 in fungal enzymes from *F. oxysporum, I. lacteus*, and *P. chrysosporium*, and CBM13 in bacterial enzymes from *C. thermocellum, S. avermitilis*, and *Streptomyces* sp. (Sakamoto and Ishimaru 2013). The difference between exo-galactanases and galactosidases is that the former are highly active on the polymeric galactan while the latter are not. The β-galactosidases (EC 3.2.1.23) belong to several GH. They are mainly grouped in GH1, 2, 35 and 42, but some individuals are also present in GH59, 147 and 165. As already mentioned for arabinofuranosidases, it is sometimes difficult to know precisely the function of the galactosidases in the hydrolysis of cell wall polysaccharides as galactose is present not only in pectin but also in hemicelluloses (xyloglucan, galactomannan). The β-galactosidase from *Bacteroides ovatus* is more likely dedicated to pectin degradation as it has been recently shown to belong to a galactan polysaccharide utilizing locus (PUL) (Luis et al. 2018). However, this enzyme was particularly active on galactan and would therefore be better referenced by the term exo-β-1,4-galactanase. The family GH165 was recently created to classify a new β-galactosidase from the bacterial strain *Sinorhizobium meliloti* exhibiting low sequence identity to other known GHs (Cheng et al. 2017).

3.4.3 Feruloylesterases

Ferulic acid is a hydroxycinnamic acid frequently found in certain pectins. Its release is catalysed by feruloyl esterases, also termed ferulic acid esterases, FAE (EC 3.1.1.73, CE1), which hydrolyse the ester bond between ferulic acid and the carrier sugar. They are produced by many bacteria and fungi. In the CAZy database, they are all members of the carbohydrate esterase family 1 (CE1), and the few crystal structures available to date show the classical eight-strand α/β fold of lipase/esterase structure. As the carrier sugars and the linkage types are diverse, different classification systems have been proposed. Crepin et al. (2004) proposed to classify FAE into 4 groups (A to D) according to their substrate specificity. The ability to release 5,5′-diferulic acid was also considered (Table 3.2).

Interestingly, this classification was supported by amino acid sequence identity. However, this system has been built on a limited number of enzymes and sequences known at that time, and a new sub-class E has been further created to classify the FAEs that do not belong to any of the four previously defined groups. The classification of fungal FAEs has been further refined according to gene sequence (Benoit et al. 2008). The resulting seven sub-families contain FAEs from the three types A, B and C according to Crepin's classification. The most recent classification of fungal FAEs proposed thirteen sub-families based on both phylogeny and substrate specificity (Dilokpimol et al. 2016). Only one classification attempted to group FAEs from fungal, bacterial and plant origins (Udatha et al. 2011). This lead to twelve sub-families based on amino acid composition and secondary structure, but without any biochemical data. However, the high sequence identity between FAEs and other enzymes such as acetyl-esterases or lipases makes this classification without functional data quite inappropriate. The biochemical features of the microbial FAEs reported so far were recently reviewed (Oliveira et al. 2019). Most of them are active at 40–50 °C and pH 6–7. They are most often characterised on synthetic substrate (methylferulate) and this does not allow anticipating easily they action on natural substrates.

Table 3.2 Classification of feruloyl esterases (adapted from Crepin et al. 2004)

	Type A	Type B	Type C	Type D
Key enzyme	*A. niger* FaeA	*P. funiculosum* FaeB *N. crassa* Fae-1	*A. niger* FaeB *T. stipitatus* FaeC	*Piromyces equi* EstA
Synthetic substrates	MFA, MSA, MpCa	MFA, MpCA, MCA	MFA, MSA, MpCA, MCA	MFA, MSA, MpCA, MCA
Targeted linkages in natural substrates	FA-O-5-Ara (xylan) FA-O-6-Gal (pectin)		FA-O-5-Ara FA-O-2-Ara (xylan, pectin)	
Release of 5-5′ diferulic acid	Yes	No	No	Yes
Sequence similarity	Lipase	CE1 acetyl xylan esterase	Chlorogenate esterase, tannase	Xylanase

3.5 Synergistic Action of Pectin Degrading Enzymes

The enzymatic synergy is defined as the cooperation of two or more enzymes to produce a combined reaction greater than the sum of their individual action. Synergistic activities of the numerous hydrolases, lyases and esterases presented above are needed for full degradation of pectin. For this reason, the microorganisms coordinate the production of the different necessary enzymes. As an example, the genome of *Aspergillus niger* contains twenty one genes encoding GH28 enzymes endo- and exo-PG as well as endo- and exo-rhamnogalacturonases (Martens-Uzunova et al. 2006). In *Bacteroides* species, 30 GHs and PLs are devoted to the degradation of the major pectin domains (Luis et al. 2018). In the PUL unravelled in the *Bacteroidetes* genome, the physically linked genes are co-expressed and target a unique substrate to improve the efficiency of the degradation (Arnal et al. 2015; Despres et al. 2016). Similarly in the thermophilic bacteria *Caldanaerobius polysaccharolyticus*, a gene cluster of six GH has been identified (Wefers et al. 2017). The enzymes belonged to GH 27, 43, 51 and 127 and were all involved in arabinan degradation. As already mentioned in the previous sections, each activity is produced by a multiple-gene family that results in different proteins with different conditions of action (pH, temperature, recognition site on the substrate etc.). Behind the possible synergy for each strain, this large enzyme set also allows cross-feeding between different species, as demonstrated in the complex microbial community of human gut microbiota (Luis et al. 2018). It is likely that *in planta* pectin remodelling enzymes can act in a synergistic manner to fine-tune pectin structure and control cell wall mechanical properties that allow growth or differentiation. Although difficult to address *in planta*, it has been shown that both PME and PG are needed to degrade cell wall in *Arabidopsis* pollen tetrad, thus allowing proper pollen grain development (Francis et al. 2006; Ogawa et al. 2009).

3.6 Conclusion

To date, the literature about pectin-degrading and modifying enzymes is already very large. However, this is a complex area and there are still enzymes to discover, as some pectin structural motifs are still resistant to the hydrolysis by enzymes discovered so far. This is particularly the case of the RG-II structures. Moreover the growing knowledge about these enzymes allows thinking of a number of applications for the industrial processing of plants, either by using microbial enzymes as technological aids or by controlling the activity of endogenous plant enzymes. Thus, the knowledge of these enzymes paves the way to applications as common as the processing of fruit juices, the extraction of oil or the retting of plant fibres.

References

Abbott DW, Boraston AB (2007) The structural basis for exopolygalacturonase activity in a family 28 glycoside hydrolase. J Mol Biol 368:1215–1222

Alalouf O, Balazs Y, Volkinshtein M, Grimpel Y, Shoham G, Shoham Y (2011) A new family of Carbohydrate Esterases is represented by a GDSL hydrolase/acetylxylan esterase from *Geobacillus stearothermophilus*. J Biol Chem 286:41993–42001

Andres-Robin A, Reymond MC, Dupire A, Battu V et al (2018) Evidence for the regulation of *Gynoecium morphogenesis* by ETTIN via cell wall dynamics. Plant Physiol 178:1222–1232

Arnal G, Bastien G, Monties N et al (2015) Investigating the function of an arabinan utilization locus isolated from a termite gut community. Appl Environ Microbiol 81:31–39

Atkinson RG, Sutherland PW, Johnston SL et al (2012) Down-regulation of POLYGALACTURONASE 1 alters firmness, tensile strength and water loss in apple (*Malus x domestica*) fruit. BMC Plant Biol 12:129

Azadi P, O'Neill MA, Bergmann C, Darvill AG, Albersheim P (1995) The backbone of the pectic polysaccharide rhamnogalacturonan I is cleaved by an endohydrolase and an endolyase. Glycobiology 5:783–789

Beldman G, Schols HA, Pitson SM, Searle-van Leeuwen MJF, Voragen AGJ (1997) Arabinans and arabinan degrading enzymes. Adv Macromol Carbohydr Res 1:1–64

Benoit I, Danchin EGJ, Bleichrodt R-J, de Vries RP (2008) Biotechnological applications and potential of fungal feruloyl esterases based on prevalence, classification and biochemical diversity. Biotechnol Lett 30:387–396

Beylot M-H, McKie V, Voragen AGJ, Doeswijk-Voragen CHL, Gilbert HJ (2001) The *Pseudomonas cellulosa* glycoside hydrolase family 51 arabinofuranosidase exhibits wide substrate specificity. Biochem J 358:607–614

Böger M, Hekelaar J, van Leeuwen SS, Dijkhuizen L, Lammerts van Bueren A (2019) Structural and functional characterization of a family GH53 β-1,4-galactanase from *Bacteroides thetaiotaomicron* that facilitates degradation of prebiotic galactooligosaccharides. J Struct Biol 205:1–10

Bonivento D, Pontiggia D, Di Matteo A et al (2008) Crystal structure of the endopolygalacturonase from the phytopathogenic fungus *Colletotrichum lupini* and its interaction with polygalacturonase-inhibiting proteins. Proteins 70:294–299

Bonnin E, Thibault J-F (1996) Galactooligosaccharide production by transfer reaction of an exogalactanase. Enzym Microb Technol 19:99–106

Bonnin E, Le Goff A, van Alebeek GJWM, Voragen AGJ, Thibault J-F (2003) Mode of action of *Fusarium moniliforme* endopolygalacturonase towards acetylated pectin. Carbohydr Polym 52:381–388

Bonnin E, Clavurier K, Daniel S, Kauppinen S, Mikkelsen JD, Thibault J-F (2008) Pectin acetylesterases from *Aspergillus* are able to deacetylate homogalacturonan as well as rhamnogalacturonan. Carbohydr Polym 74:411–418

Bordenave M (1996) Analysis of pectin methyl esterases. Plant Cell Wall Anal 17:165–180

Brummell DA, Cin VD, Crisosto C, Labavitch J (2004) Cell wall metabolism during maturation, ripening and senescence of peach fruit. J Exp Bot 55:2029–2039

Cao J (2012) The pectin lyases in *Arabidopsis thaliana*: evolution, selection and expression profiles. PLoS ONE 7:e46944

Chen M, Citovsky V (2003) Systemic movement of a tobamovirus requires host cell pectin methylesterase. Plant J 35:386–392

Chen C, Chen JL, Lin TY (1997) Purification and characterization of a xylanase from *Trichoderma longibrachiatum* for xylooligosaccharide production. Enzym Microb Technol 21:91–96

Cheng J, Romantsov T, Engel K, Doxey AC, Rose DR, Neufeld JD, Charles TC (2017) Functional metagenomics reveals novel β-galactosidases not predictable from gene sequences. PLoS ONE 12:e0172545

Chotigeat W, Duangchu S, Wititsuwannakun R, Phongdara A (2009) Cloning and characterization of pectate lyase from *Hevea brasiliensis*. Plant Physiol Biochem 47:243–247

Chourasia A, Sane VA, Nath P (2006) Differential expression of pectate lyase during ethylene-induced postharvest softening of mango (*Mangifera indica* var. *Dashehari*). Physiol Plant 128:546–555

Crepin VF, Faulds CB, Connerton I (2004) Functional classification of the microbial feruloylesterases. Appl Microbiol Biotechnol 63:647–652

Cui Z, Maruyama Y, Mikami B, Hashimoto W, Murata K (2007) Crystal structure of glycoside hydrolase family 78 α-L-rhamnosidase from *Bacillus* sp GL1. J Mol Biol 374:384–398

De Freitas ST, Handa AK, Wu Q, Park S, Mitcham EJ (2012) Role of pectin methylesterases in cellular calcium distribution and blossom-end rot development in tomato fruit. Plant J 71:824–835

De Souza AJ, Pauly M (2015) Comparative genomics of pectin acetylesterases: insight on function and biology. Plant Signal Behav 10:e1055434

De Vries RP, Parenicova L, Hinz SWA, Kester H, Beldman G, Benen JAE, Visser J (2002) The β-1,4-endogalactanase A gene from *Aspergillus niger* is specifically induced on arabinose and galacturonic acid and plays an important role in the degradation of pectic hairy regions. Eur J Biochem 269:4985–4993

Denès J-M, Baron A, Renard CMGC, Péan C, Drilleau J-F (2000) Different action patterns for apple pectin methylesterase at pH 7.0 and 4.5. Carbohydr Res 327:385–393

Derbyshire P, McCann MC, Roberts K (2007) Restricted cell elongation in *Arabidopsis* hypocotyls is associated with a reduced average pectin esterification level. BMC Plant Biol 7:1–12

Despres J, Forano E, Lepercq P et al (2016) Unraveling the pectinolytic function of *Bacteroides xylanisolvens* using a RNA-seq approach and mutagenesis. BMC Genomics 17:147

Di Matteo A, Giovane A, Raiola A et al (2005) Structural basis for the interaction between pectin methylesterase and a specific inhibitor protein. Plant Cell 17:849–858

Dilokpimol A, Mäkelä MR, Aguilar-Pontes MV et al (2016) Diversity of fungal feruloyl esterases: updated phylogenetic classification, properties, and industrial applications. Biotechnol Biofuels 9:231

Dixit S, Upadhyay S, Singh H, Pandey B, Chandrashekar K, Verma P (2013) Pectin Methylesterase of Datura species, purification, and characterization from *Datura stramonium* and its application. Plant Signal Behav 8:e25681

Domingo C, Roberts K, Stacey NJ, Connerton I, Ruíz-Teran F, McCann MC (1998) A pectate lyase from *Zinnia elegans* is auxin inducible. Plant J 13:17–28

Dorokhov YL, Komarova TV, Petrunia I (2012) Airborne signals from a wounded leaf facilitate viral spreading and induce antibacterial resistance in neighboring plants. PLoS Pathol 8:e1002640

Expert D, Patrit O, Shevchik VE, Perino C, Boucher V, Creze C, Wenes E, Fagard M (2018) *Dickeya dadantii* pectic enzymes necessary for virulence are also responsible for activation of the *Arabidopsis thaliana* innate immune system. Mol Plant Pathol 19:313–327

Federici L, Caprari C, Mattei B et al (2001) Structural requirements of endopolygalacturonase for the interaction with PGIP (polygalacturonase-inhibiting protein). Proc Natl Acad Sci U S A 98:13425–13430

Flint HJ, Bayer EA, Rincon MT, Lamed R, White BA (2008) Polysaccharide utilization by gut bacteria: potential for new insights from genomic analysis. Nat Rev Microbiol 6:121–131

Francis KE, Lam SY, Copenhaver GP (2006) Separation of *Arabidopsis* pollen tetrads is regulated by QUARTET1, a pectin methylesterase gene. Plant Physiol 142:1004–1013

Frenkel C, Peters JS, Tieman DM, Tiznado ME, Handa AK (1998) Pectin methylesterase regulates methanol and ethanol accumulation in ripening tomato (*Lycopersicon esculentum*) fruit. J Biol Chem 273:4293–4295

Gomez LD, Steele-King CG, Jones L, Foster JM, Vuttipongchaikij S, McQueen-Mason SJ (2009) Arabinan metabolism during seed development and germination in *Arabidopsis*. Mol Plant 2:966–976

Gou J, Miller LM, Hou G, Yu X, Chen X, Liu C (2012) Acetylesterase-mediated deacetylation of pectin impairs cell elongation, pollen germination, and plant reproduction. Plant Cell 24:50–65

Guénin S, Mareck A, Rayon C et al (2011) Identification of pectin methylesterase 3 as a basic pectin methylesterase isoform involved in adventitious rooting in *Arabidopsis thaliana*. New Phytol 192:114–126

Herron SR, Benen JA, Scavetta RD, Visser J, Jurnak F (2000) Structure and function of pectic enzymes: virulence factors of plant pathogens. Proc Natl Acad Sci U S A 97:8762–8769

Hewezi T, Howe P, Maier TR, Hussey RS et al (2008) Cellulose binding protein from the parasitic nematode *Heterodera schachtii* interacts with *Arabidopsis* pectin methylesterase: cooperative cell wall modification during parasitism. Plant Cell 20:3080–3093

Hocq L, Pelloux J, Lefebvre V (2017a) Connecting homogalacturonan-type pectin remodeling to acid growth. Trends Plant Sci 22:20–29

Hocq L, Sénéchal F, Lefebvre V, Lehner A, Domon JM et al (2017b) Combined experimental and computational approaches reveal distinct pH dependence of pectin methylesterase inhibitors. Plant Physiol 173:1075–1109

Hongo S, Sato K, Yokoyama R, Nishitani K (2012) Demethylesterification of the primary wall by pectin methylesterase 35 provides mechanical support to the *Arabidopsis* stem. Plant Cell 24:2624–2634

Iwai M, Kawakami T, Ikemoto T et al (2015) Molecular characterization of a *Penicillium chrysogenum* exo-rhamnogalacturonan lyase that is structurally distinct from other polysaccharide lyase family proteins. Appl Microbiol Biotechnol 99:8515–8525

Jenkins JA, Mayans O, Smith D, Worboys K, Pickersgill R (2001) Three-dimensional structure of *Erwinia chrysanthemi* pectin methylesterase reveals a novel esterase active site. J Mol Biol 305:951–960

Jiang S, Yang SL, Xie LF, San Puah CH, Zhang XQ et al (2005) VANGUARD1 encodes a pectin methylesterase that enhances pollen tube growth in the *Arabidopsis* style and transmitting tract. Plant Cell 17:584–596

Johansson K, El-Ahmad M, Friemann R, Jörnvall H, Markovic O, Eklund H (2002) Crystal structure of plant pectin methylesterase. FEBS Lett 514:243–249

Jolie RP, Duvetter T, Van Loey AM, Hendrickx ME (2010) Pectin methylesterase and its proteinaceous inhibitor: a review. Carbohydr Res 345:2583–2595

Kaji A, Shimokawa K (1984) New exo-type arabinase from *Erwinia carotovora* IAM 1024. Agric Biol Chem 48:67–72

Kalunke RM, Tundo S, Benedetti M, Cervone F, De Lorenzo G, D'Ovidio R (2015) An update on polygalacturonase-inhibiting protein (PGIP), a leucine-rich repeat protein that protects crop plants against pathogens. Front Plant Sci 6:146

Kauppinen S, Christgau S, Kofod LV, Halkier T, Dörreich K, Dalbøge H (1995) Molecular cloning and characterization of a rhamnogalacturonan acetylesterase from *Aspergillus aculeatus*. J Biol Chem 270:27172–27178

Kent LM, Loo TS, Melton LD, Mercadante D, Williams MA et al (2016) Structure and properties of a non-processive, salt-requiring, and acidophilic pectin methylesterase from *Aspergillus niger* provide insights into the key determinants of processivity control. J Biol Chem 291:1289–1306

Kirsh R, Heckel DG, Pauchet Y (2016) How the rice we evil breaks down the pectin network: enzymatic synergism and sub-functionalization. Insect Biochem Mol Biol 71:72–82

Kofod LV, Kauppinen S, Christgau S, Andersen LN, Heldt-Hansen H-P, Dörreich K, Dalbøge H (1994) Cloning and characterization of two structurally and functionally divergent rhamnogalacturonases from *Aspergillus aculeatus*. J Biol Chem 269:29182–29189

Kotake T, Hirata N, Degi Y et al (2011) Endo-β-1,3-galactanase from winter mushroom *Flammulina velutipes*. J Biol Chem 286:27848–27854

Kunishige Y, Iwai M, Nakazawa M, Ueda M, Tada T, Nishimura S, Sakamoto T (2018) Crystal structure of exo-rhamnogalacturonan lyase from *Penicillium chrysogenum* as a member of polysaccharide lyase family 26. FEBS Lett 592:1378–1388

L'Enfant M, Domon JM, Rayon C, Desnos T, Ralet MC, Bonnin E, Pelloux J, Pau-Roblot C (2015) Substrate specificity of plant and fungi pectin methylesterases: identification of novel inhibitors of PMEs. Int J Biol Macromol 81:681–691

L'Enfant M, Kutudila P, Rayon C, Domon JM, Shin WH, Kihara D, Wadouachi A, Pelloux J, Pourceau G, Pau-Roblot C (2019) Lactose derivatives as potential inhibitors of pectin methylesterases. Int J Biol Macromol 132:1140–1146

Leroux C, Bouton S, Kiefer-Meyer MC, Fabrice TN et al (2015) Pectin methylesterase48 is involved in *Arabidopsis* pollen grain germination. Plant Physiol 167:367–380

Lewis KC, Selzer T, Shahar C, Udi Y, Tworowski D, Sagi I (2008) Inhibition of pectin methyl esterase activity by green tea catechins. Phytochemistry 69:2586–2592

Lionetti V, Raiola A, Camardella L, Giovane A, Obel N, Pauly M et al (2007) Overexpression of pectin methylesterase inhibitors in Arabidopsis restricts fungal infection by *Botrytis cinerea*. Plant Physiol 143:1871–1880

Luis AS, Briggs J, Zhang X et al (2018) Dietary pectic glycans are degraded by coordinated enzyme pathways in human colonic *Bacteroides*. Nat Microbiol 3:210–219

Mardones W, Callegari E, Eyzaguirre J (2015) Heterologous expression of a *Penicillium purpurogenum* exo-arabinanase in *Pichia pastoris* and its biochemical characterization. Fungal Biol 119:1267–1278

Martens-Uzunova ES, Zandleven JS, Benen JAE et al (2006) A new group of exo-acting family 28 glycoside hydrolases of *Aspergillus niger* that are involved in pectin degradation. Biochem J 400:43–52

Matsumoto S, Yamada H, Kunishige Y, Takenaka S, Nakazawa M, Ueda M, Sakamoto T (2017) Identification of a novel Penicillium chrysogenum rhamnogalacturonan rhamnohydrolase and the first report of a rhamnogalacturonan rhamnohydrolase gene. Enzym Microb Technol 98:76–85

Mayans O, Scott M, Connerton I, Gravesen T, Benen J et al (1997) Two crystal structures of pectin lyase A from *Aspergillus* reveal a pH driven conformational change and striking divergence in the substrate-binding clefts of pectin and pectate lyases. Structure 5:677–689

McCarthy TW, Der JP, Honaas LA, de Pamphilis CW, Anderson CT (2014) Phylogenetic analysis of pectin-related gene families in *Physcomitrella patens* and nine other plant species yields evolutionary insights into cell walls. BMC Plant Biol 14:79

Mølgaard A, Kauppinen S, Larsen S (2000) Rhamnogalacturonan acetylesterase elucidates the structure and function of a new family of hydrolases. Structure 8:373–383

Müller K, Levesque-Tremblay G, Bartels S, Weitbrecht K et al (2013) Demethylesterification of cell wall pectins in *Arabidopsis* plays a role in seed germination. Plant Physiol 161:305–316

Mutter M, Beldman G, Schols HA, Voragen AGJ (1994) Rhamnogalacturonan a-L-rhamnopyranohydrolase: a novel enzyme specific for the terminal non reducing rhamnosyl unit in rhamnogalacturonan region of pectins. Plant Physiol 106:241–250

Mutter M, Colquhoun IJ, Schols HA, Beldman G, Voragen AGJ (1996) Rhamnogalacturonase B from *Aspergillus aculeatus* is a rhamnogalacturonan α -L-rhamnosyl-(1,4)-α-D-galactopyranosyluronide lyase. Plant Physiol 110:73–77

Mutter M, Renard CMGC, Beldman G, Schols HA, Voragen AGJ (1998a) Mode of action of RG-hydrolase and RG-lyase toward rhamnogalacturonan oligomers. Characterization of degradation products using RG-rhamnohydrolase and RG-galacturonohydrolase. Carbohydr Res 311:155–164

Mutter M, Beldman G, Pitson SM, Schols HA, Voragen AGJ (1998b) Rhamnogalacturonan α-D-galactopyranosyluronohydrolase, an enzyme that specifically removes the terminal nonreducing galacturonosyl residue in rhamnogalacturonan regions of pectin. Plant Physiol 117:153–163

Nafisi M, Stranne M, Zhang L, van Kan JAL, Sakuragi Y (2014) The endo-arabinanase BcAra1 is a novel host-specific virulence factor of the necrotic fungal phytopathogen *Botrytis cinerea*. Mol Plant Microbe Interact 27:781–792

Nakano H, Kitahata H, Watanabe Y, Fujimoto H, Ajisaka K, Takenishi S (1991) Transfer reaction catalysed by exo-β-1,4-galactanase from *Bacillus subtilis*. Agric Biol Chem 55:2075–2082

Naran R, Pierce ML, Mort AJ (2007) Detection and identification of rhamnogalacturonan lyase activity in intercellular spaces of expanding cotton cotyledons. Plant J 50:95–107

Ndeh D, Rogowski A, Cartmell A, Luis AS, Baslé A, Gray J, Venditto I, Briggs J, Zhang X, Labourel A, Terrapon N, Buffetto F, Nepogodiev S, Xiao Y, Field RA, Zhu Y, O'Neill MA, Urbanowicz B, York WS, Davies GJ, Abbott DW, Ralet M-C, Martens EC, Henrissat B, Gilbert HJ (2017) Complex pectin metabolism by gut bacteria reveals novel catalytic functions. Nature 544:65–73

Normand J, Ralet M-C, Thibault J-F, Rogniaux H, Delavault P, Bonnin E (2010) Purification, characterization and mode of action of a rhamnogalacturonan-hydrolase from *Irpex lacteus*, tolerant to an acetylated substrate. Appl Microbiol Biotechnol 86:577–588

Ochoa-Jiménez V-A, Berumen-Valera G, Fernandez-Valle R, Tiznado M-E (2018) Rhamnogalacturonan lyase: a pectin modification enzyme of higher plants. Emirate J Food Agric 30:910–917

Ogawa M, Kay P, Wilson S, Swain SM (2009) Arabidopsis dehiscence zone polygalacturonase1 (ADPG1), ADPG2, and QUARTET2 are polygalacturonases required for cell separation during reproductive development in *Arabidopsis*. Plant Cell 21:216–233

Oh C, Ryu BH, An DR, Nguyen DD, Yoo W, Kim T et al (2016) Structural and biochemical characterization of an octameric carbohydrate acetylesterase from *Sinorhizobium meliloti*. FEBS Lett 2016(590):1242–1252

Oliveira DM, Mota TR, Oliva B, Segato F, Marchiosi R, Ferrarese-Filho O, Faulds CB, dos Santos WD (2019) Feruloyl esterases: biocatalysts to overcome biomass recalcitrance and for the production of bioactive compounds. Bioresour Technol 278:408–423

Orfila C, Degan FD, Jørgensen B, Scheller HV, Ray PM, Ulvskov P (2012) Expression of mung bean pectin acetyl esterase in potato tubers: effect on acetylation of cell wall polymers and tuber mechanical properties. Planta 236:185–196

Pages S, Valette O, Abdou L, Belaich A, Belaich J-P (2003) A rhamnogalacturonan lyase on the *Clostridium cellulolyticum* cellulosome. J Bacteriol 185:4727–4733

Peaucelle A, Louvet R, Johansen JN, Höfte H, Laufs P, Pelloux J, Mouille G (2008) *Arabidopsis* phyllotaxis is controlled by the methyl-esterification status of cell-wall pectins. Curr Biol 18:1943–1948

Peaucelle A, Braybrook SA, Le Guillou L, Bron E et al (2011) Pectin-induced changes in cell wall mechanics underlie organ initiation in *Arabidopsis*. Curr Biol 21:1720–1726

Pelletier S, Van Orden J, Wolf S et al (2010) A role for pectin de-methylesterification in a developmentally regulated growth acceleration in dark-grown *Arabidopsis* hypocotyls. New Phytol 188:726–739

Pelloux J, Rusterucci C, Mellerowicz EJ (2007) New insights into pectin methylesterase structure and function. Tr Plant Sci 12:267–277

Petersen TN, Christgau S, Kofod LV, Kauppinen MS, Dalbøge H, Johnson AH (1997) Crystallization and preliminary X-ray studies of rhamnogalacturonase A from *Aspergillus aculeatus*. Acta Crystallogr D 53:105–107

Philippe F, Pelloux J, Rayon C (2017) Plant pectin acetylesterase structure and function: new insights from bioinformatic analysis. BMC Genomics 18:456

Pickersgill R, Jenkins J, Harris G, Nasser W, Robert-Baudouy J (1994) The structure of Bacillus subtilis pectate lyase in complex with calcium. Nat Struct Biol 1:717–723

Posé S, Kirby AR, Paniagua C, Waldron KW, Morris VJ, Quesada MA, Mercado JA (2013) The nanostructural characterization of strawberry pectins in pectate lyase or polygalacturonase silenced fruits elucidates their role in softening. Carbohydr Polym 132:134–145

Raiola A, Lionetti V, Elmaghraby I, Immerzeel P, Mellerowicz E et al (2011) Pectin methylesterase is induced in *Arabidopsis* upon infection and is necessary for a successful colonization by necrotrophic pathogens. Mol Plant Microbe Interact 24:432–440

Ralet M-C, Cabrera J-C, Bonnin E, Quemener B, Hellin P, Thibault J-F (2005) Mapping sugar beet pectin acetylation pattern. Phytochemistry 66:1832–1843

Ravanal MC, Eyzaguirre J (2015) Heterologous expression and characterization of α-L-arabinofuranosidase 4 from *Penicillium purpurogenum* and comparison with the other isoenzymes produced by the fungus. Fungal Biol 119:641–647

Rhee SY, Osborne E, Poindexter P, Somerville C (2003) Microspore separation in the quartet 3 mutants of *Arabidopsis* is impaired by a defect in a developmentally regulated polygalacturonase required for pollen mother cell wall degradation. Plant Physiol 133:1170–1180

Ross HA, Wright KM, McDougall GJ, Roberts AG et al (2011) Potato tuber pectin structure is influenced by pectin methyl esterase activity and impacts on cooked potato texture. J Exp Bot 62:371–381

Rui Y, Xiao C, Yi H, Kandemir B, Wang JZ, Puri VM, Anderson CT (2017) Polygalacturonase involved in expansion3 functions in seedling development, rosette growth, and stomatal dynamics in *Arabidopsis thaliana*. Plant Cell 29:2413–2432

Sakamoto T, Ishimaru M (2013) Peculiarities and applications of galactanolytic enzymes that act on type I and II arabinogalactans. Appl Microbiol Biotechnol 97:5201–5213

Sakamoto T, Thibault J-F (2001) Exo-arabinanase of *Penicillium chrysogenum* able to release arabinobiose from α-1,5-L-arabinan. Appl Environ Microbiol 67:3319–3321

Sakamoto M, Shirane Y, Naribayashi I, Kimura K, Morishita N, Sakamoto T, Sakai T (1994) Purification and characterization of a rhamnogalacturonase with protopectinase activity from *Trametes sanguinea*. Eur J Biochem 226:285–291

Sakamoto T, Bonnin E, Thibault J-F (2002) Purification and characterisation of two exo-polygalacturonases from *Aspergillus niger* able to degrade xylogalacturonan and acetylated homogalacturonan. Biochim Biophys Acta 1572:10–18

Sakamoto T, Ogura A, Inui M, Tokuda S, Hosokawa S, Ihara H, Kasai N (2011a) Identification of a GH62 α-L-arabinofuranosidase specific for arabinoxylan produced by *Penicillium chrysogenum*. Appl Microbiol Biotechnol 90:137–146

Sakamoto T, Tanaka H, Nishimura Y, Ishimaru M, Kasai N (2011b) Characterization of an exo-β-1,3-D-galactanase from *Sphingomonas* sp. 24T and its application to structural analysis of larch wood arabinogalactan. Appl Microbiol Biotechnol 90:1701–1710

Sakamoto T, Inui M, Yasui K, Hosokawa S, Idhara H (2013a) Substrate specificity and gene expression of two *Penicillium chrysogenum* α-L-arabinofuranosidases (AFQ1 and AFS1) belonging to glycoside hydrolase families 51 and 54. Appl Microbiol Biotechnol 97:1121–1130

Sakamoto T, Nishimura Y, Makino Y, Sunagawa Y, Harada N (2013b) Biochemical characterization of a GH53 endo-β-1,4-galactanase and a GH35 exo-β-1,4-galactanase from *Penicillium chrysogenum*. Appl Microbiol Biotechnol 97:2895–2906

Santiago-Doménech N, Jiménez-Bemúdez S, Matas AJ, Rose JK, Muñoz-Blanco J, Mercado JA, Quesada MA (2008) Antisense inhibition of a pectate lyase gene supports a role for pectin depolymerization in strawberry fruit softening. J Exp Bot 59:2769–2779

Schols HA, Geraeds CCJM, Searle-van Leeuwen MJF, Kormelink FJM, Voragen AGJ (1990) Rhamnogalacturonase : a novel enzyme that degrades the hairy regions of pectins. Carbohydr Res 206:105–115

Searle-van Leeuwen MJF, van der Broek LAM, Schols HA, Beldman G, Voragen AGJ (1992) Rhamnogalacturonan acetylesterase: a novel enzyme from *Aspergillus aculeatus*, specific for the deacetylation of hairy regions of pectins. Appl Microbiol Biotechnol 38:347–349

Searle-van Leeuwen MJF, Vincken J-P, Schipper D, Voragen AGJ, Beldman G (1996) Acetyl esterases of *Aspergillus niger*: purification and mode of action on pectins. In: Visser J, Voragen AGJ (eds) Pectins and pectinases. Elsevier, Amsterdam, pp 793–798

Sénéchal F, Wattier C, Rustérucci C, Pelloux J (2014) Homogalacturonan-modifying enzymes: structure, expression, and roles in plants. J Exp Bot 65:5125–5560

Sénéchal F, L'Enfant M, Domon JM, Rosiau E, Crépeau MJ et al (2015) Tuning of pectin methylesterification: Pectin methylesterase inhibitor 7 modulates the processive activity of co-expressed pectin methylesterase 3 in a pH-dependent manner. J Biol Chem 290:23320–22335

Sénéchal F, Habrylo O, Hocq L, Domon JM, Marcelo P, Lefebvre V, Pelloux J, Mercadante D (2017) Structural and dynamical characterization of the pH-dependence of the pectin methylesterase-pectin methylesterase inhibitor complex. J Biol Chem 292:21538–21547

Shevchik VE, Hugouvieux-Cotte-Pattat N (2003) PaeX, a second pectin acetylesterase of *Erwinia chrysanthemi* 3937. J Bacteriol 185:3091–3100

Shimizu T, Nakatsu T, Miyairi K, Okuno T, Kato H (2002) Active-site architecture of endopolygalacturonase I from *Stereum purpureum* revealed by crystal structures in native and ligand-bound forms at atomic resolution. Biochemist 41:6651–6659

Shinozaki A, Hosokawa S, Nakazawa M, Ueda M, Sakamoto T (2015) Identification and characterization of three *Penicillium chrysogenum* α-L-arabinofuranosidases (PcABF43B, PcABF51C, and AFQ1) with different specificities toward arabino-oligosaccharides. Enzym Microb Technol 73-74:65–71

Siedlecka A, Wiklund S, Péronne MA, Micheli F, Lesniewska J et al (2008) Pectin methyl esterase inhibits intrusive and symplastic cell growth in developing wood cells of *Populus*. Plant Physiol 146:554–565

Stranne M, Sakuragi Y (2016) Arabinan: biosynthesis and a role in host-pathogen interactions. In: Collinge DB (ed) Plant pathogen resistance biotechnology. Wiley, Hoboken, pp 91–103

Stratilova E, Markovic O, Dzurova M, Malovikava A, Capek P, Omelkova J (1998) The pectolytic enzymes of carrots. Biologia 53:731–738

Sun L, Van Nocker S (2010) Analysis of promoter activity of members of the pectate lyase-like (PLL) gene family in cell separation in *Arabidopsis*. BMC Plant Biol 10:1–13

Suykerbuyk MEG, Kester HCM, Schaap PJ, Stam H, Musters W, Visser J (1997) Cloning and characterization of two rhamnogalacturonan hydrolase genes from *Aspergillus niger*. Appl Environ Microbiol 63:2507–2515

Tabachnikov O, Shoham Y (2013) Functional characterization of the galactan utilization system of Geobacillus stearothermophilus. FEBS J 280:950–964

The International Brachypodium Initiative (2010) Genome sequencing and analysis of the model grass *Brachypodium distachyon*. Nature 463(7282):763–768

Turbant A, Fournet F, Lequart M, Zabijak L, Pageau K, Bouton S, Van Wuytswinkel O (2016) PME58 plays a role in pectin distribution during seed coat mucilage extrusion through homogalacturonan modification. J Exp Bot 67:2177–2190

Udatha DBRKG, Kouskoumvekaki I, Olsson L, Panagiotou G (2011) The interplay of descriptor-based computational analysis with pharmacophore modeling builds the basis for a novel classification scheme for feruloyl esterases. Biotechnol Adv 29:94–110

Van Dijk C, Boeriu C, Stolle-Smits T, Tijskens LMM (2006) The firmness of stored tomatoes (cv. Tradiro). 2. Kinetic and near infrared models to describe pectin degrading enzymes and firmness loss. J Food Eng 77:585–593

Vilches F, Ravanal MC, Bravo-Moraga F, Gonzalez-Nilo D, Eyzaguirre J (2018) *Penicillium purpurogenum* produces a novel endo-1,5-arabinanase, active on debranched arabinan, short arabinooligosaccharides and on the artificial substrate *p*-nitrophenyl arabinofuranoside. Carbohydr Res 455:106–113

Vogel J, Raab T, Schiff C, Somerville S (2002) PMR6, a pectate lyase–like gene required for powdery mildew susceptibility in *Arabidopsis*. Plant Cell 14:2095–2106

Voxeur A, Habrylo O, Guénin S, Miart F, Soulié M-C, Rihouey C, Pau-Roblot C, Domon J-M, Gutierrez L, Pelloux J, Mouille G, Fagard M, Höfte H, Vernhettes S (2019) Oligogalacturonide production upon *Arabidopsis thaliana–Botrytis cinerea* interaction. Proc Natl Acad Sci U S A 116(39):19743–19752

Wakasa Y, Kudo H, Ishikawa R, Akada S, Senda M, Niizeki M, Harada T (2006) Low expression of an *ENDOPOLYGALACTURONASE* gene in apple fruit with long-term storage potential. Postharvest Biol Technol 39:193–198

Wang D, Yeats TH, Uluisik S, Rose JKC, Seymour GB (2018) Fruit softening: revisiting the role of pectin. Trends Plant Sci 23:302–310

Wefers D, Dong J, Abdel-Hamid A, Paul HM, Pereira GV, Han Y, Dodd D, Baskaran R, Mayer B, Mackie RI, Cann IKO (2017) Enzymatic mechanism for arabinan degradation and transport in the thermophilic bacterium *Caldanaerobius polysaccharolyticus*. Appl Environ Microbiol 83:e00794–e00717

Wen B, Ström A, Tasker A, West G, Tucker GA (2013) Effect of silencing the two major tomato fruit pectin methylesterase isoforms on cell wall pectin metabolism. Plant Biol 15:1025–1032

Williamson G (1991) Purification and characterization of pectin acetylesterase from orange peel. Phytochemistry 30:445–449

Wolf S, Rausch T, Greiner S (2009) The N-terminal pro region mediates retention of unprocessed type-I PME in the Golgi apparatus. Plant J 58:361–375

Wolf S, Mravec J, Greiner S, Mouille G, Höfte H (2012) Plant cell wall homeostasis is mediated by brassinosteroid feedback signaling. Curr Biol 22:1732–1737

Xiao C, Somerville C, Anderson CT (2014) Polygalacturonase involved in expansion1 functions in cell elongation and flower development in *Arabidopsis*. Plant Cell 26:1018–1035

Xiao C, Barnes WJ, Zamil MS, Yi H, Puri VM, Anderson CT (2017) Activation tagging of *Arabidopsis* polygalacturonase involved in expansion2 promotes hypocotyl elongation, leaf expansion, stem lignification, mechanical stiffening, and lodging. Plant J 89:1159–1173

Yang Y, Zhang L, Guo M, Sun J, Matsukawa S, Xie J, Wei D (2015) Novel α-L-Arabinofuranosidase from *Cellulomonas fimi* ATCC 484 and its substrate-specificity analysis with the aid of computer. J Agric Food Chem 63:3725–3733

Yang L, Huang W, Xiong F, Xian Z, Su D, Ren M, Li Z (2017) Silencing of SlPL, which encodes a pectate lyase in tomato, confers enhanced fruit firmness, prolonged shelf-life and reduced susceptibility to grey mould. Plant Biotechnol J 12:1544–1555

Yang Y, Yu Y, Liang Y, Anderson CT, Cao J (2018) A profusion of molecular scissors for pectins: classification, expression, and functions of plant polygalacturonases. Front Plant Sci 9:1208

Yoshimi Y, Yaguchi K, Kaneko S, Tsumuraya Y, Kotake T (2017) Properties of two fungal endo-β-1,3-galactanases and their synergistic action with an exo-β-1,3-galactanase in degrading arabinogalactan-proteins. Carbohydr Res 453-454:26–35

Zandleven JS, Beldman G, Bosveld M, Benen JAE, Voragen AGJ (2005) Mode of action of xylogalacturonan hydrolase towards xylogalacturonan and xylogalacturonan oligosaccharides. Biochem J 387:719–725

Zandleven JS, Beldman G, Bosveld M, Schols HA, Voragen AGJ (2006) Enzymatic degradation studies of xylogalacturonans from apple and potato, using xylogalacturonan hydrolase. Carbohydr Polym 65:495–503

Zhang L, Wang P, Chen F, Lai S, Yu H, Yang H (2019) Effects of calcium and pectin methylesterase on quality attributes and pectin morphology of jujube fruit under vacuum impregnation during storage. Food Chem 289:40–48

Chapter 4
Isolation and Characterisation of Pectin

Gordon A. Morris and Hana A. S. Binhamad

4.1 Introduction

There are a large number of fruits and vegetables which have been reported to contain pectins (Table 4.1) including for example, blackcurrants (Alba et al. 2018), broccoli florets (Houben et al. 2011), carrots (Houben et al. 2011), melon (Denman and Morris 2015), okra pods (Alba et al. 2015; Kpodo et al. 2017), pumpkin (Jun et al. 2006; Košťálová et al. 2010, Košťálová et al. 2013a, b, Košťálová et al. 2014, Košťálová et al. 2016; Simpson and Morris 2014; Sukhenko et al. 2020), sunflower heads (Anger and Berth 1985; Muthusamy et al. 2019), tomatoes (Round et al. 1997; Zhang et al. 2019) and several other sources (Ahmadi Gavlighi et al. 2018; Burana-osot et al. 2010; Cardoso et al. 2002; Deng et al. 2019; Grønhaug et al. 2010; Happi Emaga et al. 2008; Huang et al. 2011. 2016; Inngjerdingen et al. 2007, 2008; Jacobo-Valenzuela et al. 2011; Kazemi et al. 2019a, b; Khodaei and Karboune 2013; Kliemann et al. 2009; Koubala et al. 2008; Kratchanova et al. 1991; Levigne et al. 2002a, b; Liang et al. 2012; Manrique and Lajolo 2002; Meersman et al. 2017; Mikshina et al. 2015; Minjares-Fuentes et al. 2014; Muthukumaran et al. 2017; Ognyanov et al. 2016; Oliveira et al. 2016; Pagán and Ibarz 1999; Pagán et al. 1999; Pereira et al. 2016; Robert et al. 2008; Rojas et al. 2015; Sabater et al. 2020; Sari et al. 2018; Seixas et al. 2014; Sun et al. 2019; Urias-Orona et al. 2010; Wang et al. 2016; Yapo 2009a, b; Zarei et al. 2017; Zhang et al. 2020).

However, many of these "pectins" from novel sources contain large amounts of neutral sugars (e.g., arabinose, galactose and rhamnose) and therefore often contain less than the legal minimum requirement of 65% galacturonic acid and cannot be defined as "pectin" for food and pharmaceutical applications (May 1990). This therefore limits the potential commercial sources of food and pharmaceutical grade pectins (May 1990) and at the present time 85% of all commercial pectins are

G. A. Morris (✉) · H. A. S. Binhamad
Department of Chemical Sciences, University of Huddersfield, Huddersfield, UK
e-mail: g.morris@hud.ac.uk

V. Kontogiorgos (ed.), *Pectin: Technological and Physiological Properties*,
https://doi.org/10.1007/978-3-030-53421-9_4

Table 4.1 Some examples of pectin sources extraction techniques and characterisation methods discussed in this chapter

Characteristic sources of pectin	Extraction methods	Characterisation methods
Apple pomace	Solvent extraction	Phenol sulfuric acid assay (Dubois)
Banana peel	Microwave assisted extraction	*m*-hydroxydiphenyl assay
Carrot waste	Ultrasound assisted extraction	Equivalent weight
Citrus fruits (orange, lemon, grapefruit etc.)	Enzyme assisted extraction	Titrations
Mango peel	Supercritical fluid extraction	Infrared (IR) spectroscopy
Melon	Electromagnetic induction heating extraction	High performance anion exchange chromatography coupled to pulsed amperometric detector (HPAEC-PAD)
Okra pods	Subcritical water extraction	Nuclear magnetic resonance (NMR)
Passion fruit peel		Gas chromatography coupled to mass spectrometry (GC-MS)
Peach pomace		Molar mass and molar mass distribution
Pumpkin waste		Intrinsic viscosity
Soy beans		ζ-potential
Sugar beet pulp		Rheology
Tomato waste		Capillary zone electrophoresis (CZE)
Watermelon rinds		Atomic force microscopy (AFM)

extracted from citrus peels, 14% from apple pomace and 1% from sugar beet pulp (Ciriminna et al. 2016) (see Fig. 4.1).

It is also important to note that yield, chemical composition and molecular weight for example will depend on a number of variables including genetic variety (Kpodo et al. 2017), ripening stage (Deng et al. 2019; Prasanna et al. 2007; Seymour et al. 1987) and extraction techniques.

4.2 Extraction of Pectin

There are several methods which have been used for the extraction of pectins including, acid extraction, isolation with enzymes (e.g., cellulases, hemicellulases, polygalacturonases), electromagnetic induction heating, microwave and ultrasound associated extraction (Adetunji et al. 2017; Bagherian et al. 2011; Kashyap et al. 2001; Mishra et al. 2012; Wang et al. 2007; Vasco-Correa and Zapata Zapata 2017; Marić et al. 2018; Rodsamran and Sothornvit 2019).

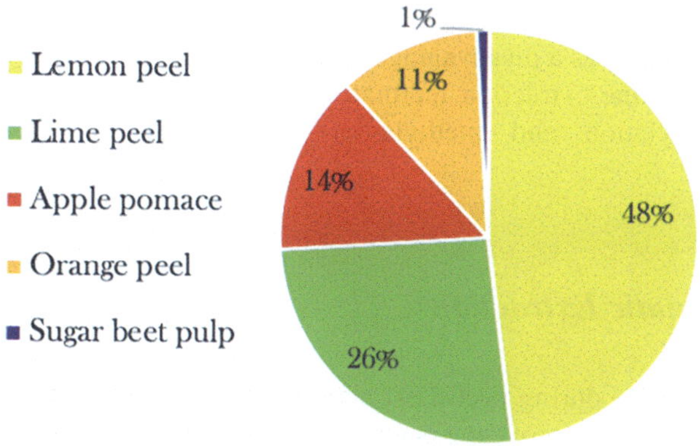

Fig. 4.1 Major commercial sources of pectin (adapted from (Ciriminna et al. 2016). Lemon peel (48%), lime peel (26%), apple pomace (14%), orange peel (11%) and sugar beet pulp (1%)

4.2.1 Conventional Extraction Methods

Extraction of pectin is a process which involves the physical-chemical hydrolysis and extraction of pectin macromolecules from plant tissues (Seixas et al. 2014) under the influence of different factors, including temperature, pH and extraction time (Adetunji et al. 2017; Pagán et al. 2001). The first step being solubilisation of the protopectin followed by a second hydrolysis reaction where the pectin is degraded, lowering the yield at longer extraction times, and once all the "extractable" pectin has been exhausted the hydrolysis predominates (Pagán and Ibarz 1999; Pagán et al. 1999). Therefore, the temperature, pH, duration of extraction, type of acid, number of extraction cycles, the ratio of water to raw material and volumes of organic solvent can modify the quantity as well as the quality of the extracted pectins (Denman and Morris 2015; Kliemann et al. 2009; Kumar et al. 2010; Levigne et al. 2002a; Pagán and Ibarz 1999; Pagán et al. 1999; Samavati 2013; Sudhakar and Maini 2000). In general, the highest yields of pectins are obtained by the most suitable method for industrial extraction of pectin which is hot acid (pH 1.5–2.5) extractions at temperatures around 70–90 °C (May 1990). Acid extraction has another advantage which is that the pectin obtained is ordinarily enriched with galacturonic acid residues. This however can result in pectin degradation (Pagán and Ibarz 1999; Pagán et al. 1999) and environmental issues surrounding the disposal of hazardous chemicals (Minjares-Fuentes et al. 2014) which has driven a large amount of research in alternative non-conventional extraction techniques (Adetunji et al. 2017; Bagherian et al. 2011; Fishman et al. 2006; Košťálová et al. 2016; Minjares-Fuentes et al. 2014; Qiu et al. 2019; Rodsamran and Sothornvit 2019; Rojas et al. 2015; Sabater et al. 2020; Sari et al. 2018; Seixas et al. 2014; Vasco-Correa and Zapata Zapata 2017; Wang et al. 2007, 2016; Zarei et al. 2017; Marić et al. 2018). The isolation of pectins has also be carried out on a laboratory scale using of chelating agents (e.g., potassium oxalate), dilute sodium hydroxide solutions, water (hot or cold) and buffered solutions (Alba 2015). The extraction with chelating agents has a disadvantage in that it is difficult to remove the residual chelates, also alkaline extraction facilitates reduction of chain length and in the degree of acetylation and methylation by β-elimination (Rombouts and Thibault 1986).

4.2.2 Enzymatic Extraction

Enzymes can be used during pectin extraction to hydrolyse the matrix of the plant cell wall and therefore improve the permeability of the plant cell wall to solvent and improve mass transfer. The most commonly enzymes are cellulases, proteases, hemicellulases and xylanases or commercial formulations which contain one or more of these enzymes for example Celluclast®, Econase®, Neutrase® or Viscoferm® (Adetunji et al. 2017; Marić et al. 2018; Wikiera et al. 2015; Panouillé et al. 2006). The yield of pectin will depend on extraction time, enzyme activity, temperature and pH and therefore extraction conditions need to be optimised for specific enzymes (Adetunji et al. 2017; Marić et al. 2018) and some "typical" extraction conditions have been outlined previously (Adetunji et al. 2017; Marić et al. 2018; Wikiera et al. 2015). Enzyme assisted extraction usually requires lower temperatures and less solvents than conventional acid extraction, however there are disadvantages in terms of the high cost of enzymes and issues around scale-up (Adetunji et al. 2017; Marić et al. 2018).

4.2.3 Microwave-Assisted Extraction

The exposure of a sample matrix and/or solvent to microwave energy causes a disorganised rotation of any polar molecules which results in heat (Adetunji et al. 2017; Bagherian et al. 2011; Marić et al. 2018), as with other non-conventional pectin extraction technologies this can reduce extraction times and solvent usage (Adetunji et al. 2017; Marić et al. 2018). There have been a number of studies on the extraction of pectins from a number of sources (Adetunji et al. 2017; Bagherian et al. 2011; Fishman et al. 2000, 2006; Kazemi et al. 2019a; Košťálová et al. 2016; Marić et al. 2018; Rodsamran and Sothornvit 2019; Rojas et al. 2015; Samavati 2013; Sari et al. 2018; Seixas et al. 2014; Wang et al. 2007; Zarei et al. 2017). Experimental conditions usually require the optimisation of microwave assisted extraction conditions (power/temperature, time and solid: liquid ratio). However as with acid extraction prolonged extraction under "harsh" conditions may degrade the pectin chain (Fishman et al. 2006). For more information see for example (Adetunji et al. 2017; Marić et al. 2018).

4.2.4 Ultrasonic Extraction

Ultrasonic waves travelling through a liquid medium, for example acidified (pH 1.5) water (Bagherian et al. 2011), causes bubbles/cavities to form, expand and collapse in a process called cavitation (Adetunji et al. 2017; Marić et al. 2018). This process occurs over a very short time period (~400 μs) and can lead to very high temperatures and pressures which will enhance solubility in the solvent and increase the rate of mass transfer (Adetunji et al. 2017; Marić et al. 2018). Rather than a conventional ultrasonic bath pectin extractions are usually investigated using an ultrasonic probe, as this results in more efficient cavitation (Adetunji et al. 2017) and there have recently been a number of ultrasound assisted extractions of pectins form different sources (Bagherian et al. 2011; Košťálová et al. 2010; Minjares-Fuentes et al. 2014; Sabater et al. 2020; Wang et al. 2016). Like acid and microwave assisted extraction, experimental conditions (power/temperature, time and solid: liquid ratio) are important in terms of optimisation (Adetunji et al. 2017). As with other extraction/solubilisation methodologies longer extraction times under more "harsh" conditions have been reported to increase yield whilst decreasing polymer chain lengths (Bagherian et al. 2011; Kök et al. 2009; Mohod and Gogate 2011).

4.2.5 Sub-Critical Water Extraction/Accelerated Solvent Extraction

Sub-critical water extraction uses liquid water at temperatures above its usual boiling point without a change in phase, this method is also called pressurised hot water extraction or superheated water extraction or more generally accelerated solvent extraction. The use of conventional solvents and elevated temperatures and pressures which facilitates the penetration of the solvent in to the sample matrix, increases solubility and mass transfer whilst reducing solvent viscosity and surface tension. Again the yield and quality of the pectin is correlated with the extraction conditions (pressure, temperature, time and solid: liquid ratio) and optimisation is required (Chen et al. 2015; Le Normand et al. 2011). In this section some of the methods of pectin extraction have been discussed, but it is important to note that these extraction techniques could be used sequentially for example ultrasonic extraction prior to microwave assisted extraction (Bagherian et al. 2011) or ultrasonic extraction combined with subcritical water extraction (Chen et al. 2015). After extraction, the crude pectin extracts are then separated by a filtration, centrifugation, precipitation with alcohol (methanol, ethanol or isopropanol) or by precipitation with an insoluble salt e.g., aluminium (May 1990). The precipitate obtained is washed with alcohol and pressed to remove soluble contaminants, and finally dried and milled to get powdered pectin (Joye and Luzio 2000).

4.3 Characterisation of Pectin

After extraction and isolation, it is essential to fully characterise pectin polymers. This section describes current methods widely used in the physico-chemical characterisation of pectins.

4.3.1 Phenol-Sulfuric Acid/Total Carbohydrate Assay

Upon the addition of sulphuric acid, pectins when subjected to heat will hydrolyse and form furan derivatives which may further react with a phenol solution to produce a coloured complex (λ_{max} ~ 490 nm), although different monosaccharides react differently and uronic acids, such as galacturonic acid have a relatively low colour response (Dubois et al. 1956). The intensity of the colour produced will depend on the amount of pectin. However, it is important to note that this is a relative technique and the quality of the results will depend the reaction time and the choice of an appropriate standard i.e., galacturonic acid for pectin analysis.

4.3.2 Galacturonic Acid Content and Degree of Methyl Esterification (DM)

4.3.2.1 *m*-Hydroxydiphenyl Assay

Upon the addition of sulphuric acid, pectins when subjected to heat will hydrolyse and form furan derivatives which may further react with a m-hydroxdiphenyl solution to produce a pink complex (λ_{max} ~ 520 nm). The galacturonic acid content can then be determined from a calibration curve of galacturonic acids standards typically over the concentration range 10–100 μg mL^{-1} (Brummer and Cui 2016; Filisetti-Cozzi and Carpita 1991). This method is less sensitive to the presence of other monosaccharides or protein molecules than earlier methods (Brummer and Cui 2016).

4.3.2.2 Equivalent Weight

A pectin sample (0.5% w/v) is dissolved in a 5% v/v ethanol solution and the acidic solution is titrated with sodium hydroxide (0.1N) in the presence of an appropriate indicator e.g., methyl red (Alfa Ezhil Rose and Abilasha 2016) while sodium chloride can be added to sharpen the end point (Owens et al. 1952). This titration can be used to give an estimate of total anhydrouronic acid content and the neutralised solution can then be used to estimate degree of methyl esterification (Owens et al. 1952).

$$\text{Equivalent weight} = \frac{1000 \times \text{Weight of sample}(\text{g})}{\text{Volume of NaOH}(\text{mL}) \times \text{Normality of NaOH}} \quad (4.1)$$

After having neutralised the free galacturonic acid the degree of methyl esterification (methoxyl content) can determined from a further titration. The pectin is saponified by the addition of 25 mL of 0.25N sodium hydroxide solution in left at room temperature for 30 min after which time 25 mL of 0.25N hydrochloric acid is added and the excess acid is back titrated with 0.1N sodium hydroxide as per the equivalent weight titration described previously. From which total anhydrouronic acid content (AUA) and %DM can be calculated (see Eqs. (4.2), (4.3), and (4.4)):

$$\text{Methoxyl content}(\%\text{MeO}) = \frac{\text{Volume of NaOH}(\text{mL}) \times \text{Normality of NaOH} \times 3.1}{\text{Weight of sample}(\text{g})} \quad (4.2)$$

$$\%\text{AUA} = \frac{\text{Volume of NaOH}(\text{mL})\text{from Eq.}(1) \times \text{Normality of NaOH} \times 100 \times 176}{\text{Weight of sample}(\text{g}) \times 1000} + \frac{\text{Volume of NaOH}(\text{mL})\text{from Eq.}(2) \times \text{Normality of NaOH} \times 100 \times 176}{\text{Weight of sample}(\text{g}) \times 1000} \quad (4.3)$$

$$\%\text{DM} = \frac{176 \times \%\text{MeO}}{31 \times \%\text{AUA}} \times 100 \quad (4.4)$$

where 176 and 31 are the formula weights of AUA and MeO, respectively.

4.3.2.3 Infra-Red (IR) Spectroscopy

The degree of methyl esterification can also be determined directly from infra-red spectroscopy (IR) (Chatjigakis et al. 1998; Filippov and Kohn 1975; Gnanasambandam and Proctor 2000; Manrique and Lajolo 2002; Chylińska et al. 2016). This can be done quite simply by measuring the absorbance of the stretching vibrations for a free carboxylic acid group (COO^-) at approximately 1630 cm^{-1} and of the corresponding methylated group (COOMe) at approximately 1740 cm^{-1} (Chatjigakis et al. 1998; Filippov and Kohn 1975; Gnanasambandam and Proctor 2000; Manrique and Lajolo 2002; Szymańska-Chargot et al. 2016). This is usually done by plotting the calibration curve for pectins of known degree of esterification versus the area (Chatjigakis et al. 1998; Manrique and Lajolo 2002) using the following equation:

$$DM\% = \frac{(\text{Area of COOMe peak})}{(\text{Area of COOMe peak}) + (\text{Area of COO}^{-}\text{ peak})} \tag{4.5}$$

4.3.3 *Sugar Composition*

4.3.3.1 High Performance Anion Exchange Chromatography Coupled with Pulsed Amperometric Detection (HPAEC-PAD)

This method allows the direct determination of pectic monosaccharides without any need for prior derivatisation. In brief, pectin samples are hydrolysed to their monosaccharide constituent sugars using a strong acid usually trifluoroacetic acid (TFA) as this is easier to remove by evaporation. The monosaccharides are then separated using a high performance anion exchange column (Corradini et al. 2012) at a pH of ~12 using for example 10 mM sodium hydroxide, for the separation of uronic acids sodium acetate is often added to the mobile phase to increase the ionic strength and decrease their retention time. This high pH will impart either a fully or partially ionised negative charge on the individual monosaccharides and they can be separated depending their respective acid dissociation constants, pKa (Table 4.2). The column effluent is then detected using a pulsed amperometric detector (PAD), however the detector response is not equivalent for the different monosaccharides, therefore calibration curves generated from purified monosaccharide standards are required for quantification (Brummer and Cui 2016).

4.3.3.2 Nuclear Magnetic Resonance (NMR)

NMR is powerful technique in the analysis of pectins and allows detailed structural information (Cheng and Neiss 2012). This technique is based on the net spin of atomic nuclei in an applied magnetic field and nuclei with nuclear spins of 1/2 (e.g., ^{1}H and ^{13}C) are widely used in the analysis of pectins. The magnetic field experienced by the nucleus is not equal to the applied magnetic field due to shielding from electrons and this results in a chemical shift (relative to a reference). This chemical shift

Table 4.2 pKa values for typical pectic monosaccharides at 25 °C adapted from (Bhattacharyya and Rohrer 2012; Kaijanen et al. 2015)

Monosaccharide	pKa
Glucose	12.28–12.35
Galactose	12.39
Xylose	12.15
Mannose	12.08
Rhamnose	11.84
Arabinose	12.34–12.43
Galacturonic acid	3.48

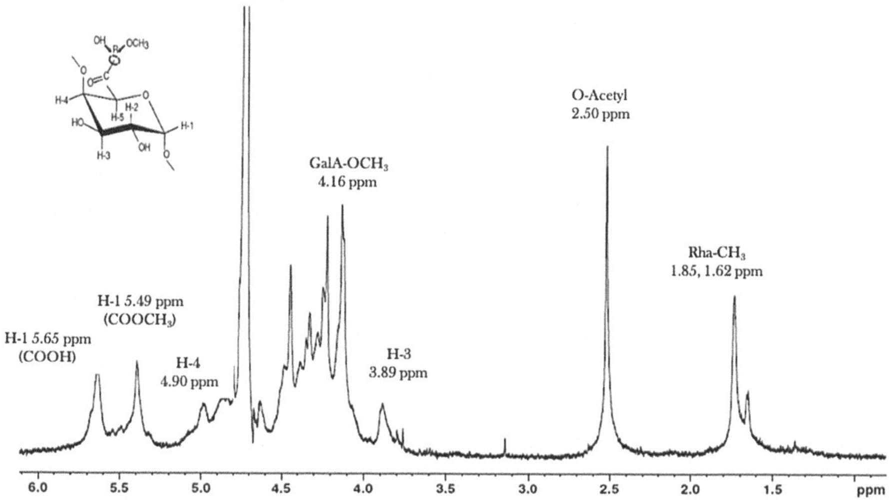

Fig. 4.2 Typical ^{1}H NMR spectrum for okra pectin in D_2O at 70 °C where some of the peaks have assignments have been identified. A methylated galacturonic acid residue is shown for clarification (Binhamad 2018)

is a function of the nucleus and its local environment (Cheng and Neiss 2012) and can be used to identify specific protons on each monosaccharide unit (Fig. 4.2). Pectins are usually dissolved in a deuterated solvent, for example D_2O to allow the –OH groups to be fully deuterium exchanged (Cheng and Neiss 2012). This may require the pectin being dissolved in D_2O then freeze-dried prior to re-dissolution in D_2O with this process being repeated on several occasions. Furthermore, due to high molecular weight/ viscosity of pectin solutions it is often advisable to run pectin samples at higher temperatures and/or higher magnetic field strengths for greater spectral resolution.

Integration of the acetic and methanol peaks at ~1.92 and 3.36 ppm after saponification can also be used to quantify the degree of methylation and acetylation (Müller-Maatsch et al. 2016). Further detailed structural information can be obtained from 2D-NMR methods for example correlated spectroscopy (COSY), correlated spectroscopy (TOCSY), heteronuclear single quantum correlated spectroscopy (HSQC) and heteronuclear multiple quantum correlated spectroscopy (HMBC).

4.3.3.3 Gas Chromatography Coupled to Mass Spectrometry (GC-MS)

Gas chromatography coupled to mass spectrometry is often used in pectin analysis as it is one of the few techniques which has the capability of identifying the linkages which form the polysaccharide chain (Carpita and Shea 1989; Sparkman et al. 2011). GC-MS is often the preferred method of analysis as it requires less material than other techniques e.g., NMR (Lindhorst 2000). In order to prepare the sample for linkage analysis, methylation and acetylation are required to identify where the constituent monosaccharides are linked to one another via glycosidic bonds. This analysis is done by methylating all the free –OH groups in the sample and then using acid hydrolysis to break the glycosidic bonds. Partially methylated alditol acetates (PMAAs) are then produced by reducing and acetylating the methylated sugars. The positions of these groups on the individual monosaccharide can be determined using GC-MS and it is therefore possible to identify the types of links in the chain i.e., it will be possible to tell if a hexose is 1–4 linked or an end terminal etc. (Sims et al. 2018). For molecules containing uronic acids the uronic acid residues can be reduced using sodium borodeuteride ($NaBD_4$) prior to methylated, hydrolysis and acetylation (for more details see (Sims et al. 2018)).

4.3.4 Molar Mass and Molar Mass Distribution

One of the most important parameters when characterising pectins (or any other polymers) is the molar mass or molecular weight, expressed in g mol^{-1} or Daltons (Da). Pectins, like other polysaccharides, are poly(disperse) and it is therefore important to consider the molar mass distribution (Fujita 1962) and the molar mass "averages": M_n (number average), M_η (viscosity average), M_w (weight average) and M_z (z-average), where for heterogeneous systems $M_n < M_\eta \sim M_w < M_z$. The type of average calculated will depend on the type of instrumentation method used. Number average molecular weight is calculated from end group analysis and osmotic pressure (Devine 1974; Harding et al. 1991b) and are limited in their applicability to high molecular weight polyelectrolyte molecules such as pectin and have been to some extent been superseded by other techniques (Harding et al. 1991b). Viscosity average molecular weight is calculated from intrinsic viscosity measurements (see the section on intrinsic viscosity for more details) and usually requires some knowledge or assumptions about the dilute solution conformation of the pectin. In the last 40 years size exclusion chromatography coupled to multi-angle light scattering (SEC-MALS) has become the method of choice in the determination of the molecular weight of pectins (Harding et al. 2016; Morris et al. 2014). In brief, size exclusion chromatography is based on the separation of molecules due to their hydrodynamic radius using a chromatographic column consisting of a matrix of porous polymer beads. Pectin molecules depending on their size will partition in and out of these pores. The partition between pores can be described using the distribution coefficient, K_D ($0 \leq K_D \leq 1$). Therefore, the total volume accessible to the pectin molecules is the retention volume V_R. If $K_D = 0$, then $V_R = V_o$ and the molecule is too large to diffuse into the column matrix, this is known as the void volume, and when $K_D = 1$ the polymer can penetrate in to the entire bead matrix and $V_R = V_T$, which is called total volume. However, the major disadvantage of a standalone SEC system is that one can only assign relative molecular weights through calibration with standards of known molecular weight which relies on both the standards and sample of interest behaving at least similarly in the SEC columns. However absolute estimates of molecular weight can be calculated with the appropriate detection system for example multi-angle light scattering (MALS). Light scattering enables the absolute determination of molecular weight for pectin molecules. Pectin molecules are often of relatively high molecular weight (>100,000 g mol^{-1}) and therefore a single molecular can have several scattering points and this reduces the scattering intensity at all angles other than zero. Therefore, scattering is usually measured at several angles (up to 18) and extrapolated to zero angle (Debye 1944; Wyatt 1993; Zimm 1948). Therefore, with the addition of a suitable on-line concentration detector, usually refractive index for pectins, one can calculate the molecular weight distribution and the radius of gyration distribution without any prior assumptions of pectin conformation (Tanford 1961). The MALS detector can also be coupled to asymmetric flow field-flow fractionation (AF4-MALS) which can also be used in the molecular weight determination of pectins (Thielking and Kulicke 1998). Asymmetric flow field-flow fractionation (AF4) separates pectin molecules based on their differences in the diffusion coefficient, which is related to their size and shape (Nilsson et al. 1996). AF4 is a non-chromatographic separation technique and separation is achieved by flow through a channel to which a perpendicular force is applied (Fig. 4.3). The channel consists of an upper solid plate which is imperme-

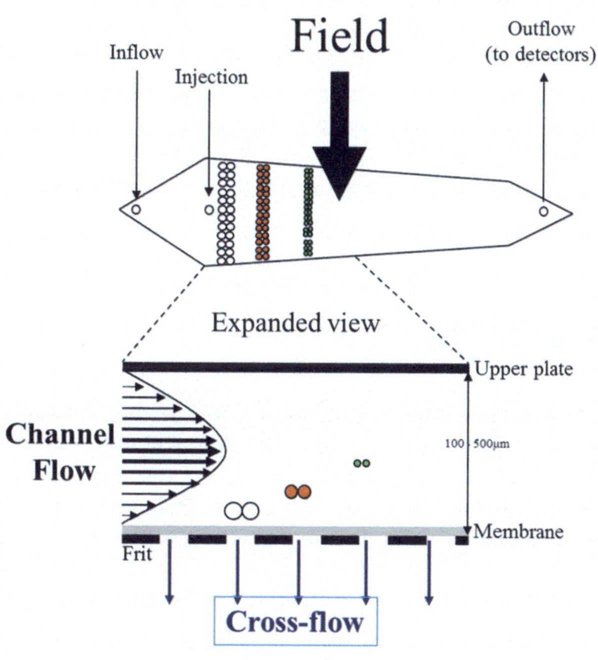

Fig. 4.3 Schematic representation of an asymmetric flow field flow fractionation system where pectin molecules of smaller (filled green circle) hydrodynamic volumes (larger diffusion coefficient) are experiencing a greater flow velocity than those with intermediate diffusion coefficients (filled brown circle) and larger (open circle) hydrodynamic volumes (smaller diffusion coefficient)

able to solvent and a lower (accumulation) plate which is permeable to solvents (Pauck and Cölfen 1998; Wahlund and Giddings 1987). The laminar flow of the mobile phase creates a parabolic flow profile within the channel; therefore, the stream moves more slowly closer to the lower plate wall than it does in the channel centre (Fig. 4.3). Since separation is based on diffusion coefficient, the smaller molecules tends to elute faster than the larger molecules as they are able to diffuse more readily in to the higher flow zones in the channel (Runyon et al. 2014). The use online MALS and refractive index detectors can then be used to estimate distributions of molecular weight and radius of gyration.

Despite the convenience of SEC-MALS and AF4-MALS, there are several limitations. For example, for large molecular weight polysaccharides SEC columns can give poor separation or non-size exclusion interactions with the columns can give anomalous results. Whereas in the case of field-flow fractionation systems problems of non-inertness through anomalous interactions with the membranes can also lead to erroneous results (Harding et al. 2016). The molecular weight of pectins can also be determined using either sedimentation equilibrium in the analytical ultracentrifuge (SE-AUC) and with prior knowledge of pectin conformation using sedimentation velocity in the analytical ultracentrifuge (SV-AUC) (Harding et al. 1991a, 2011a, b; Morris et al. 2000, 2014) which offers a complementary orthogonal approach to MALS with inherent fractionation ability without the need for columns or membranes. Although probably being the "gold standard" in molecular weight determination of polymers these methods have not been extensively thus far in the molecular weight determination of pectins and the theory and appropriate methodologies have been described and reviewed previously (see for example, (Devine 1974; Harding 1995a, b, 2005a, b; Harding et al. 2015; Harding et al. 2010, 2011a; Harding et al. 2016; Harding et al. 1991a; Harding et al. 2011b; Morris et al. 2014; Morris et al. 2000).

4.3.5 *Intrinsic Viscosity*

The intrinsic viscosity is an important parameter which gives information about the size and conformation of a pectin polymer in dilute solution. Intrinsic viscosity is usually measured using an Ostwald capillary viscometer, the flow of pectin molecules in solution through the capillary will follow Poiseuille's law. From this the ratio of viscosities can be calculated—relative viscosity:

$$\eta_{rel} = \left(\frac{t}{t_0}\right)\left(\frac{\rho}{\rho_0}\right) \tag{4.6}$$

where t is the flow time for the pectin solution, to is the flow time for the solvent (e.g., water). Due to the measurements being commonly made at low concentrations, ρ/ρ_o can often be ignored (see e.g. (Harding 1997)), although densities of pectin solutions can be measured routinely and for analytical centrifugation based techniques are essential. The specific (η_{sp}), viscosity is defined as follows:

$$\eta_{sp} = \eta_{rel} - 1 \tag{4.7}$$

and this, divided by concentration, c (g mL^{-1}) is known as the reduced specific viscosity, η_{sp}/c (mL g^{-1}). To minimise non-ideality effects, measurements that are made at several concentrations are extrapolated to infinite dilution using the Huggins or Kraemer approaches, or both (Huggins 1942; Kraemer 1938):

$$\frac{\eta_{sp}}{c} = [\eta](1 + K_H[\eta]c) \tag{4.8}$$

$$\frac{\ln(\eta_{rel})}{c} = [\eta](1 - K_K[\eta]c) \tag{4.9}$$

where the intrinsic viscosity [η] is taken as the is the mean of the intercepts from Eqs. (4.8) and (4.9) and K_H and K_K are the Huggins and Kraemer constants respectively. A useful method for measuring intrinsic viscosities is to calculate the relative and specific viscosities at one concentration and utilise the Solomon-Ciutâ or Solomon-Gottesman approximation (Solomon and Ciută 1962; Solomon and Gottesman 1968). The intrinsic viscosity can then be accurately estimated (error generally ~1%) by a single measurement at low concentration (see for example (Morris 2001)).

$$[\eta] \approx \frac{\left(2\eta_{sp} - 2\ln(\eta_{rel})\right)^{1/2}}{c} \tag{4.10}$$

Differential pressure viscometers (DPV) based on the 4-capilliary bridge design (Haney 1985a, b) can be utilised on-line and integrated with the SEC-MALS system and the specific viscosity can then be calculated as follows:

$$\eta_{sp} = \frac{4\Delta P}{P_i - 2\Delta P} \tag{4.11}$$

where Pi is the inlet pressure and ΔP the differential pressure across the midpoint of the bridge

There the intrinsic viscosity can then be estimated as a function of elution volume using the Solomon-Ciutâ (Gotesman) approximation (Eq. (4.9)) where the weight average viscosity across the entire peak area corresponds to the bulk solution intrinsic viscosity measured using a traditional Ostwald viscometer. The intrinsic viscosity can then be related to the molecular weight via the Mark-Houwink power law relationship.

$$[\eta] = \kappa_\eta M^a \quad (4.12)$$

where κ_η and α are obtained from the intercept and slope of the double log plot of $[\eta]$ vs. M_w The value of a can be used as an estimation of gross macromolecular conformation and hence a values of ~0 correspond to spheres, 0.5–0.8 to random coils, and up to 1.8 to rigid rods (see (Harding 1997)). Typically, pectin has a Mark-Houwink α value in the range 0.62–0.94 (Anger and Berth 1985; Axelos et al. 1987; Axelos and Thibault 1991; Berth et al. 1977; Fishman et al. 2001, 2006; Garnier et al. 1993; Harding et al. 1991a, b; Malovikova et al. 1993; Masuelli 2011; Morris et al. 2000, 2002, 2008; Qiu et al. 2019), although this will depend on the source, genetic variety and extraction method, etc.

4.3.6 Rheology

The physical behaviour (deformation and flow) of substances, is classified in two categories: viscous liquid or elastic solid. Materials, like pectins, which exhibit both elastic and viscous behaviours are referred to viscoelastic materials (Mezger 2006) and these properties are important in terms of their functionality as gelling and thickening agents (May 1990).

4.3.6.1 Viscosity

Depending on viscous behaviour of fluids as a function of shear rate and stress, they are categorised as either Newtonian or non-Newtonian systems (Chhabra and Richardson 1999). This relationship is defined according to Newton's law:

$$\eta = \frac{\tau}{\dot{\gamma}} \quad (4.13)$$

When temperature is constant for a Newtonian fluid the relationship between the applied shear stress and shear rate is linear and there will be no change in viscosity when shear is applied, for example: water, oils, alcohol (Chhabra and Richardson 1999). In contrast, for non-Newtonian fluids the relationship between shear stress and shear rate is non-linear and viscosity will change when shear is applied and three types of flows are considered: shear-thickening (dilatant), shear-thinning (pseudo-plastic) and Bingham plastic (Chhabra and Richardson 1999). Pectin solutions behave as a Newtonian liquid up to a certain concentration, although this concentration will depend on the physical properties of the pectin (M_w, $[\eta]$, radius of gyration), above this concentration pectin solutions are shear-thinning and therefore viscosity decreases with shear rate which is important from a processing as well as an organoleptic point of you. Viscosity is usually measured as a function of shear rate (e.g., 0.1–1000 s^{-1}) using a rheometer with a cone and plate geometry and extrapolated to zero-shear (η_o) using for example the Cross model (Morris 1990). Furthermore, as the pectin concentration increases the physical entanglement between polymer chains will increase and above a critical concentration (c*) pectin chains can no longer distribute themselves randomly throughout the solution and adjacent chains will interpenetrate. This is called the "semi-dilute" regime and a further transition to the "concentrated" regime occurs at c**. Viscosity and the critical coil overlap concentration (c* and c**) will depend on degree of esterification, solvent environment (i.e., salt concentration, sugar concentration and pH) and temperature (Oakenfull 1991).

4.4 Conclusions

Pectins are a family of chemically diverse biopolymers which are used extensively in the food and pharmaceutical industries as for example gelling/thickening agents, emulsifiers or excipients. There are a number of physico-chemical properties which are important in these applications including monosaccharide composition, degree of esterification and molecular weight all which will depend on the sources of pectin and the extraction conditions. In order to extract and characterise a novel pectin there are a number of recommendations which may be of use to a researcher new to the field of pectin chemistry. I would suggest a simple acid based extraction with an experimental design to measure the effects of pH, extraction temperature and time (Denman and Morris 2015; Kumar et al. 2010; Levigne et al. 2002a). There are then a number of important characterisations which would need to be undertaken including galacturonic acid content, degree of methyl esterification and the weight average molecular weight. The determination of galacturonic acid content using the m-hydroxydiphenyl assay is a simple colourimetric method which is easy to perform and gives reliable results even in the presence of interfering protein molecules. Degree of esterification can also be estimated using titrimetric analysis which is simple, robust and available in every chemistry laboratory. Further characterisation with HPAEC, GC-MS and NMR will give much more detailed compositional and structural information. Finally, molecular weight determination is also very important and requires the use of more specialised equipment (SEC-MALS, AF4-MALS or AUC) which may be not be available in a number of laboratories. However, with some knowledge of the pectin conformation, molecular weight can be estimated using a simple capillary viscometer and the Mark-Houwink power law relationship.

References

Adetunji LR, Adekunle A, Orsat V, Raghavan V (2017) Advances in the pectin production process using novel extraction techniques: a review. Food Hydrocoll 62:239–250

Ahmadi Gavlighi H, Tabarsa M, You S, Surayot U, Ghaderi-Ghahfarokhi M (2018) Extraction, characterization and immunomodulatory property of pectic polysaccharide from pomegranate peels: enzymatic vs conventional approach. Int J Biol Macromol 116:698–706

Alba K (2015) Isolation, characterization and functional properties of okra pectin. PhD, University of Huddersfield

Alba K, Laws AP, Kontogiorgos V (2015) Isolation and characterization of acetylated LM-pectins extracted from okra pods. Food Hydrocoll 43:726–735

Alba K, MacNaughtan W, Laws AP, Foster TJ, Campbell GM, Kontogiorgos V (2018) Fractionation and characterisation of dietary fibre from blackcurrant pomace. Food Hydrocoll 81:398–408

Alfa Ezhil Rose P, Abilasha D (2016) Extraction and characterization of pectin from lemon peel. Int J Adv Sci Res 1(12):12–15

Anger H, Berth G (1985) Gel permeation chromatography of sunflower pectin. Carbohydr Polym 5(4):241–250

Axelos MAV, Thibault JF (1991) The chemistry of low-methoxyl pectin gelation. Chem Technol Pectin 1991:109–118

Axelos MAV, Lefebvre J, Thibault JF (1987) Conformation of a low methoxyl citrus pectin in aqueous solution. Top Catal 1(5-6):569–570

Bagherian H, Zokaee Ashtiani F, Fouladitajar A, Mohtashamy M (2011) Comparisons between conventional, microwave- and ultrasound-assisted methods for extraction of pectin from grapefruit. Chem Eng Process 50(11-12):1237–1243

Berth G, Anger H, Linow F (1977) Streulichtphotometrische und viskosimetrische Untersuchungen an Pektinen in wäßrigen Lösungen zur Molmassenbestimmung. FoodReview 21(10):939–950

Bhattacharyya L, Rohrer JS (2012) Applications of ion chromatography for pharmaceutical and biological products. Wiley, Hoboken

Binhamad H (2018) Extraction, characterisation and properties of polysaccharides from novel sources. PhD, University of Huddersfield

Brummer Y, Cui SW (2016) Detection and determination of polysaccharides in foods. In: Food polysaccharides and their applications, 2nd edn. Wiley, New York, pp 675–712

Burana-osot J, Soonthornchareonnon N, Chaidedgumjorn A, Hosoyama S, Toida T (2010) Determination of galacturonic acid from pomelo pectin in term of galactose by HPAEC with fluorescence detection. Carbohydr Polym 81(2):461–465

Cardoso SM, Silva AMS, Coimbra MA (2002) Structural characterisation of the olive pomace pectic polysaccharide arabinan side chains. Carbohydr Res 337(10):917–924

Carpita NC, Shea EM (1989) Linkage structure of carbohydrates by gas chromatography-mass spectrometry (GC-MS) of partially methylated alditol acetates. In: Analysis of carbohydrates by GLC and MS. CRC Press, Boca Raton, pp 157–216

Chatjigakis AK, Pappas C, Proxenia N, Kalantzi O, Rodis P, Polissiou M (1998) FT-IR spectroscopic determination of the degree of esterification of cell wall pectins from stored peaches and correlation to textural changes. Carbohydr Polym 37(4):395–408

Chen H-M, Fu X, Luo Z-G (2015) Properties and extraction of pectin-enriched materials from sugar beet pulp by ultrasonic-assisted treatment combined with subcritical water. Food Chem 168:302–310

Cheng HN, Neiss TG (2012) Solution NMR spectroscopy of food polysaccharides. Polym Rev 52(2):81–114

Chhabra RP, Richardson JF (1999) Non-Newtonian flow in the process industries: fundamentals and engineering applications. Butterworth-Heinemann, Oxford

Chylińska M, Szymańska-Chargot M, Zdunek A (2016) FT-IR and FT-Raman characterization of non-cellulosic polysaccharides fractions isolated from plant cell wall. Carbohydr Polym 154:48–54

Ciriminna R, Fidalgo A, Delisi R, Ilharco LM, Pagliaro M (2016) Pectin production and global market. Agro Food Ind Hi Tech 27(5):17–20
Corradini C, Cavazza A, Bignardi C (2012) High-performance anion-exchange chromatography coupled with pulsed electrochemical detection as a powerful tool to evaluate carbohydrates of food interest: principles and applications. Int J Carbohydr Chem 487564:13
Debye P (1944) Light scattering in solutions. J Appl Phys 15(4):338–342
Deng LZ, Pan Z, Zhang Q, Liu ZL, Zhang Y, Meng JS, Gao ZJ, Xiao HW (2019) Effects of ripening stage on physicochemical properties, drying kinetics, pectin polysaccharides contents and nanostructure of apricots. Carbohydr Polym 222:114980
Denman LJ, Morris GA (2015) An experimental design approach to the chemical characterisation of pectin polysaccharides extracted from *Cucumis melo Inodorus*. Carbohydr Polym 117:364–369
Devine WC (1974) Physico-chemical studies on pectins. PhD, University of Edinburgh
Dubois M, Gilles KA, Hamilton JK, Rebers PA, Smith F (1956) Colorimetric method for determination of sugars and related substances. Anal Chem 28(3):350–356
Filippov MP, Kohn R (1975) Determination of the esterification degree of carboxyl groups of pectin with methanol by means of infrared spectroscopy. Chem Zvesti 29(1):88–91
Filisetti-Cozzi TMCC, Carpita NC (1991) Measurement of uronic acids without interference from neutral sugars. Anal Biochem 197(1):157–162
Fishman ML, Chau HK, Hoagland P, Ayyad K (2000) Characterization of pectin, flash-extracted from orange albedo by microwave heating, under pressure. Carbohydr Res 323:126
Fishman ML, Chau HK, Kolpak F, Brady J (2001) Solvent effects on the molecular properties of pectins. J Agric Food Chem 49(9):4494–4501
Fishman ML, Chau HK, Hoagland PD, Hotchkiss AT (2006) Microwave-assisted extraction of lime pectin. Food Hydrocoll 20(8):1170–1177
Fujita H (1962) Mathematical theory of sedimentation analysis. Academic, New York
Garnier C, Axelos MAV, Thibault JF (1993) Phase diagrams of pectin-calcium systems: influence of pH, ionic strength, and temperature on the gelation of pectins with different degrees of methylation. Carbohydr Res 240:219–232
Gnanasambandam R, Proctor A (2000) Determination of pectin degree of esterification by diffuse reflectance Fourier transform infrared spectroscopy. Food Chem 68(3):327–332
Grønhaug TE, Ghildyal P, Barsett H, Michaelsen TE, Morris G, Diallo D, Inngjerdingen M, Paulsen BS (2010) Bioactive arabinogalactans from the leaves of Opilia celtidifolia Endl. ex Walp. (Opiliaceae). Glycobiology 20(12):1654–1664
Haney MA (1985a) A differential viscometer: part two. Am Lab 17(4):116–126
Haney MA (1985b) A differential viscometer: part one. Am Lab 17(1):41–56
Happi Emaga T, Ronkart SN, Robert C, Wathelet B, Paquot M (2008) Characterisation of pectins extracted from banana peels (Musa AAA) under different conditions using an experimental design. Food Chem 108(2):463–471
Harding SE (1995a) Some recent developments in the analytical ultracentrifugation of food proteins. FoodReview 39(5-6):375–395
Harding SE (1995b) Some recent developments in the size and shape analysis of industrial polysaccharides in solution using sedimentation analysis in the analytical ultracentrifuge. Carbohydr Polym 28(3):227–237
Harding SE (1997) The intrinsic viscosity of biological macromolecules. Progress in measurement, interpretation and application to structure in dilute solution. Prog Biophys Mol Biol 68(2-3):207–262
Harding SE (2005a) Analysis of polysaccharides by ultracentrifugation. Size, conformation and interactions in solution. Adv Polym Sci 186:211–254
Harding SE (2005b) Challenges for the modern analytical ultracentrifuge analysis of polysaccharides. Carbohydr Res 340(5):811–826

Harding SE, Berth G, Ball A, Mitchell JR, de la Torre JG (1991a) The molecular weight distribution and conformation of citrus pectins in solution studied by hydrodynamics. Carbohydr Polym 16(1):1–15

Harding SE, Vårum KM, Stokke BT, Smidsrød O (1991b) Molecular weight determination of polysaccharides. Adv Carbohydr Anal 1:63–144

Harding SE, Abdelhameed AS, Morris GA (2010) Molecular weight distribution evaluation of polysaccharides and glycoconjugates using analytical ultracentrifugation. Macromol Biosci 10(7):714–720

Harding SE, Abdelhameed AS, Morris GA (2011a) On the hydrodynamic analysis of conformation in mixed biopolymer systems. Polym Int 60(1):2–8

Harding SE, Schuck P, Abdelhameed AS, Adams G, Kök MS, Morris GA (2011b) Extended Fujita approach to the molecular weight distribution of polysaccharides and other polymeric systems. Methods 54(1):136–144

Harding SE, Abdelhameed AS, Gillis RB, Morris GA, Adams GG (2015) Characterization of capsular polysaccharides and their glycoconjugates by hydrodynamic methods. Methods Mol Biol 1331:211–227

Harding SE, Adams GG, Gillis RB, Almutairi FM, Morris GA (2016) Aspects of the analytical ultracentrifuge determination of the molar mass distribution of polysaccharides. In: Analytical ultracentrifugation: instrumentation, software, and applications. pp 375–386

Houben K, Jolie RP, Fraeye I, Van Loey AM, Hendrickx ME (2011) Comparative study of the cell wall composition of broccoli, carrot, and tomato: structural characterization of the extractable pectins and hemicelluloses. Carbohydr Res 346(9):1105–1111

Huang PH, Lu HT, Wang YT, Wu MC (2011) Antioxidant activity and emulsion-stabilizing effect of pectic enzyme treated pectin in soy protein isolate-stabilized oil/water emulsion. J Agric Food Chem 59(17):9623–9628

Huang JH, Kortstee A, Dees DCT, Trindade LM, Schols HA, Gruppen H (2016) Modification of potato cell wall pectin by the introduction of rhamnogalacturonan lyase and β-galactosidase transgenes and their side effects. Carbohydr Polym 144:9–16

Huggins ML (1942) The viscosity of dilute solutions of long-chain molecules. IV. Dependence on concentration. J Am Chem Soc 64(11):2716–2718

Inngjerdingen KT, Patel TR, Chen X, Kenne L, Allen S, Morris GA, Harding SE, Matsumoto T, Diallo D, Yamada H, Michaelsen TE, Inngjerdingen M, Paulsen BS (2007) Immunological and structural properties of a pectic polymer from *Glinus oppositifolius*. Glycobiology 17(12):1299–1310

Inngjerdingen M, Inngjerdingen KT, Patel TR, Allen S, Chen X, Rolstad B, Morris GA, Harding SE, Michaelsen TE, Diallo D, Paulsen BS (2008) Pectic polysaccharides from Biophytum petersianum Klotzsch, and their activation of macrophages and dendritic cells. Glycobiology 18(12):1074–1084

Jacobo-Valenzuela N, Zazueta-Morales JJ, Gallegos-Infante JA, Aguilar-Gutierrez F, Camacho-Hernández IL, Rocha-Guzman NE, Gonzalez-Laredo RF (2011) Chemical and physicochemical characterization of winter squash (Cucurbita moschata D.). Not Bot Horti Agrobot Cluj-Napoca 39(1):34–40

Joye DD, Luzio GA (2000) Process for selective extraction of pectins from plant material by differential pH. Carbohydr Polym 43(4):337–342

Jun HI, Lee CH, Song GS, Kim YS (2006) Characterization of the pectic polysaccharides from pumpkin peel. LWT- Food Sci Technol 39(5):554–561

Kaijanen L, Metsämuuronen S, Reinikainen SP, Pietarinen S, Jernström E (2015) Profiling of water-soluble carbohydrates in pine and spruce extracts by capillary zone electrophoresis with direct UV detection. Wood Sci Technol 49(4):795–809

Kashyap DR, Vohra PK, Chopra S, Tewari R (2001) Applications of pectinases in the commercial sector: a review. Bioresour Technol 77(3):215–227

Kazemi M, Khodaiyan F, Hosseini SS (2019a) Utilization of food processing wastes of eggplant as a high potential pectin source and characterization of extracted pectin. Food Chem 294:339–346

Kazemi M, Khodaiyan F, Hosseini SS, Najari Z (2019b) An integrated valorization of industrial waste of eggplant: Simultaneous recovery of pectin, phenolics and sequential production of pullulan. Waste Manag 100:101–111

Khodaei N, Karboune S (2013) Extraction and structural characterisation of rhamnogalacturonan I-type pectic polysaccharides from potato cell wall. Food Chem 139(1-4):617–623

Kliemann E, De Simas KN, Amante ER, Prudêncio ES, Teófilo RF, Ferreira MMC, Amboni RDMC (2009) Optimisation of pectin acid extraction from passion fruit peel (Passiflora edulis flavicarpa) using response surface methodology. Int J Food Sci Technol 44(3):476–483

Kök MS, Abdelhameed AS, Ang S, Morris GA, Harding SE (2009) A novel global hydrodynamic analysis of the molecular flexibility of the dietary fibre polysaccharide konjac glucomannan. Food Hydrocoll 23(7):1910–1917

Košťálová Z, Hromádková Z, Ebringerová A (2010) Isolation and characterization of pectic polysaccharides from the seeded fruit of oil pumpkin (Cucurbita pepo L. var. Styriaca). Ind Crop Prod 31(2):370–377

Košťálová Z, Hromádková Z, Ebringerová A (2013a) Structural diversity of pectins isolated from the Styrian oil-pumpkin (*Cucurbita pepo var. styriaca*) fruit. Carbohydr Polym 93(1):163–171

Košťálova Z, Hromádková Z, Ebringerová A, Polovka M, Michaelsen TE, Paulsen BS (2013b) Polysaccharides from the Styrian oil-pumpkin with antioxidant and complement-fixing activity. Ind Crop Prod 41(1):127–133

Košťálová Z, Hromádková Z, Ebringerová A (2014) Erratum: Structural diversity of pectins isolated from the Styrian oil-pumpkin (*Cucurbita pepo var. styriaca*) fruit (Carbohydr Polym (2013) 93 (163-171)). Carbohydr Polym 99:831

Košťálová Z, Aguedo M, Hromádková Z (2016) Microwave-assisted extraction of pectin from unutilized pumpkin biomass. Chem Eng Process 102:9–15

Koubala BB, Kansci G, Mbome LI, Crépeau MJ, Thibault JF, Ralet MC (2008) Effect of extraction conditions on some physicochemical characteristics of pectins from "Améliorée" and "Mango" mango peels. Food Hydrocoll 22(7):1345–1351

Kpodo FM, Agbenorhevi JK, Alba K, Bingham RJ, Oduro IN, Morris GA, Kontogiorgos V (2017) Pectin isolation and characterization from six okra genotypes. Food Hydrocoll 72:323–330

Kraemer EO (1938) Molecular weights of celluloses. Ind Eng Chem 30(10):1200–1203

Kratchanova M, Bénémou C, Kratchanov C (1991) On the pectic substances of mango fruits. Carbohydr Polym 15(3):271–282

Kumar M, Chauhan AKR, Kumar S, Kumar A, Malik S (2010) Design and evaluation of pectin based metrics for transdermal patches of meloxicam. J Pharm Res Health Care 2(3):244–247

Le Normand M, Edlund U, Ek M (2011) Spruce bark hemicelluloses and pectins: extraction and characterization. pp 103–106

Levigne S, Ralet MC, Thibault JF (2002a) Characterisation of pectins extracted from fresh sugar beet under different conditions using an experimental design. Carbohydr Polym 49(2):145–153

Levigne S, Thomas M, Ralet MC, Quemener B, Thibault JF (2002b) Determination of the degrees of methylation and acetylation of pectins using a C18 column and internal standards. Food Hydrocoll 16(6):547–550

Liang RH, Chen J, Liu W, Liu CM, Yu W, Yuan M, Zhou XQ (2012) Extraction, characterization and spontaneous gel-forming property of pectin from creeping fig (*Ficus pumila L.*) seeds. Carbohydr Polym 87(1):76–83

Lindhorst TK (2000) Essentials of carbohydrate chemistry and biochemistry. Wiley, New York

Malovikova A, Rinaudo M, Milas M (1993) On the characterization of polygalacturonate salts in dilute solution. Carbohydr Polym 22(2):87–92

Manrique GD, Lajolo FM (2002) FT-IR spectroscopy as a tool for measuring degree of methyl esterification in pectins isolated from ripening papaya fruit. Postharvest Biol Technol 25(1):99–107

Marić M, Grassino AN, Zhu Z, Barba FJ, Brnčić M, Rimac Brnčić S (2018) An overview of the traditional and innovative approaches for pectin extraction from plant food wastes and by-products: ultrasound-, microwaves-, and enzyme-assisted extraction. Trends Food Sci Technol 76:28–37

Masuelli MA (2011) Viscometric study of pectin. Effect of temperature on the hydrodynamic properties. Int J Biol Macromol 48(2):286–291

May CD (1990) Industrial pectins: sources, production and applications. Carbohydr Polym 12(1):79–99

Meersman E, Struyf N, Kyomugasho C, Jamsazzadeh Kermani Z, Santiago JS, Baert E, Hemdane S, Vrancken G, Verstrepen KJ, Courtin CM, Hendrickx M, Steensels J (2017) Characterization and degradation of pectic polysaccharides in cocoa pulp. J Agric Food Chem 65(44):9726–9734

Mezger TG (2006) The rheology handbook: for users of rotational and oscillatory rheometers. Vincentz Network GmbH and Co KG, Hannover

Mikshina PV, Idiyatullin BZ, Petrova AA, Shashkov AS, Zuev YF, Gorshkova TA (2015) Physicochemical properties of complex rhamnogalacturonan I from gelatinous cell walls of flax fibers. Carbohydr Polym 117:853–861

Minjares-Fuentes R, Femenia A, Garau MC, Meza-Velázquez JA, Simal S, Rosselló C (2014) Ultrasound-assisted extraction of pectins from grape pomace using citric acid: a response surface methodology approach. Carbohydr Polym 106(1):179–189

Mishra RK, Banthia AK, Majeed ABA (2012) Pectin based formulations for biomedical applications: a review. Asian J Pharm Clin Res 5(4):1–7

Mohod AV, Gogate PR (2011) Ultrasonic degradation of polymers: effect of operating parameters and intensification using additives for carboxymethyl cellulose (CMC) and polyvinyl alcohol (PVA). Ultrason Sonochem 18(3):727–734

Morris ER (1990) Shear-thinning of 'random coil' polysaccharides: characterisation by two parameters from a simple linear plot. Carbohydr Polym 13(1):85–96

Morris GA (2001) Hydrodynamic investigation of polysaccharides and their interactions with casein. PhD, University of Nottingham

Morris GA, Foster TJ, Harding SE (2000) The effect of the degree of esterification on the hydrodynamic properties of citrus pectin. Food Hydrocoll 14(3):227–235

Morris GA, Foster TJ, Harding SE (2002) A hydrodynamic study of the depolymerisation of a high methoxy pectin at elevated temperatures. Carbohydr Polym 48(4):361–367

Morris GA, de Torre JC, Ortega A, Castile J, Smith A, Harding SE (2008) Molecular flexibility of citrus pectins by combined sedimentation and viscosity analysis. Food Hydrocoll 22(8):1435–1442

Morris GA, Adams GG, Harding SE (2014) On hydrodynamic methods for the analysis of the sizes and shapes of polysaccharides in dilute solution: a short review. Food Hydrocoll 42(P3):318–334

Müller-Maatsch J, Bencivenni M, Caligiani A, Tedeschi T, Bruggeman G, Bosch M, Petrusan J, Van Droogenbroeck B, Elst K, Sforza S (2016) Pectin content and composition from different food waste streams. Food Chem 201:37–45

Muthukumaran C, Banupriya L, Harinee S, Sivaranjani S, Sharmila G, Rajasekar V, Kumar NM (2017) Pectin from muskmelon (Cucumis melo var. reticulatus) peels: extraction optimization and physicochemical properties. Biotech 7(1):66

Muthusamy S, Manickam LP, Murugesan V, Muthukumaran C, Pugazhendhi A (2019) Pectin extraction from Helianthus annuus (sunflower) heads using RSM and ANN modelling by a genetic algorithm approach. Int J Biol Macromol 124:750–758

Nilsson M, Birnbaum S, Wahlund KG (1996) Determination of relative amounts of ribosome and subunits in Escherichia coli using asymmetrical flow field-flow fractionation. J Biochem Biophys Methods 33(1):9–23

Oakenfull DG (1991) Chapter 5 - the chemistry of high-methoxyl pectins. In: Walter RH (ed) The chemistry and technology of pectin. Academic, San Diego, pp 87–108

Ognyanov M, Remoroza C, Schols HA, Georgiev Y, Kratchanova M, Kratchanov C (2016) Isolation and structure elucidation of pectic polysaccharide from rose hip fruits (Rosa canina L.). Carbohydr Polym 151:803–811

Oliveira TÍS, Rosa MF, Cavalcante FL, Pereira PHF, Moates GK, Wellner N, Mazzetto SE, Waldron KW, Azeredo HMC (2016) Optimization of pectin extraction from banana peels with citric acid by using response surface methodology. Food Chem 198:113–118

Owens HS, McCready RK, AShepherd AD, Schultz TH, Pippen EL, Swensen HA, Liers JC, Erlandsen RF, MacLay WD (1952) Methods used at western regional research laboratory for extraction of pectin materials. United States Department of Agriculture, Washington, DC

Pagán J, Ibarz A (1999) Extraction and rheological properties of pectin from fresh peach pomace. J Food Eng 39(2):193–201

Pagán J, Ibarz A, Llorca M, Coll L (1999) Quality of industrial pectin extracted from peach pomace at different pH and temperatures. J Sci Food Agric 79(7):1038–1042

Pagán J, Ibarz A, Llorca M, Pagán A, Barbosa-Cánovas GV (2001) Extraction and characterization of pectin from stored peach pomace. Food Res Int 34(7):605–612

Panouillé M, Thibault J-F, Bonnin E (2006) Cellulase and protease preparations can extract pectins from various plant byproducts. J Agric Food Chem 54(23):8926–8935

Pauck T, Cölfen H (1998) Hydrodynamic analysis of macromolecular conformation. A comparative study of flow field flow fractionation and analytical ultracentrifugation. Anal Chem 70(18):3886–3891

Pereira PHF, Oliveira TÍS, Rosa MF, Cavalcante FL, Moates GK, Wellner N, Waldron KW, Azeredo HMC (2016) Pectin extraction from pomegranate peels with citric acid. Int J Biol Macromol 88:373–379

Prasanna V, Prabha TN, Tharanathan RN (2007) Fruit ripening phenomena-an overview. Crit Rev Food Sci Nutr 47(1):1–19

Qiu WY, Cai WD, Wang M, Yan JK (2019) Effect of ultrasonic intensity on the conformational changes in citrus pectin under ultrasonic processing. Food Chem 297:125021

Robert C, Emaga TH, Wathelet B, Paquot M (2008) Effect of variety and harvest date on pectin extracted from chicory roots (Cichorium intybus L.). Food Chem 108(3):1008–1018

Rodsamran P, Sothornvit R (2019) Microwave heating extraction of pectin from lime peel: characterization and properties compared with the conventional heating method. Food Chem 278:364–372

Rojas R, Contreras-Esquivel JC, Orozco-Esquivel MT, Muñoz C, Aguirre-Joya JA, Aguilar CN (2015) Mango peel as source of antioxidants and pectin: microwave assisted extraction. Waste Biomass Valorization 6(6):1095–1102

Rombouts FM, Thibault JF (1986) Feruloylated pectic substances from sugar-beet pulp. Carbohydr Res 154(1):177–187

Round AN, MacDougall AJ, Ring SG, Morris VJ (1997) Unexpected branching in pectin observed by atomic force microscopy. Carbohydr Res 303(3):251–253

Runyon JR, Ulmius M, Nilsson L (2014) A perspective on the characterization of colloids and macromolecules using asymmetrical flow field-flow fractionation. Colloids Surf A 442:25–33

Sabater C, Sabater V, Olano A, Montilla A, Corzo N (2020) Ultrasound-assisted extraction of pectin from artichoke by-products. An artificial neural network approach to pectin characterisation. Food Hydrocoll 98:105238

Samavati V (2013) Central composite rotatable design for investigation of microwave-assisted extraction of okra pod hydrocolloid. Int J Biol Macromol 61:142–149

Sari AM, Ishartani D, Dewanty PS (2018) Effects of microwave power and irradiation time on pectin extraction from watermelon rinds (Citrullus lanatus) with acetic acid using microwave assisted extraction method. IOP Conf Ser 102(1):012085

Seixas FL, Fukuda DL, Turbiani FRB, Garcia PS, Petkowicz CLDO, Jagadevan S, Gimenes ML (2014) Extraction of pectin from passion fruit peel (Passiflora edulis f. flavicarpa) by microwave-induced heating. Food Hydrocoll 38:186–192

Seymour GB, Harding SE, Taylor AJ, Hobson GE, Tucker GA (1987) Polyuronide solubilization during ripening of normal and mutant tomato fruit. Phytochemistry 26(7):1871–1875

Simpson R, Morris GA (2014) The anti-diabetic potential of polysaccharides extracted from members of the cucurbit family: a review. Bioact Carbohydr Diet Fibre 3(2):106–114

Sims IM, Carnachan SM, Bell TJ, Hinkley SFR (2018) Methylation analysis of polysaccharides: technical advice. Carbohydr Polym 188:1–7

Solomon OF, Ciută IZ (1962) Détermination de la viscosité intrinsèque de solutions de polymères par une simple détermination de la viscosité. J Appl Polym Sci 6(24):683–686
Solomon OF, Gottesman BS (1968) Correct determination of intrinsic viscosity. J Appl Polym Sci 12(4):971–972
Sparkman OD, Penton Z, Kitson FG (2011) Gas chromatography and mass spectrometry: a practical guide. In: Gas chromatography and mass spectrometry: a practical guide. Elsevier, Alpharetta
Sudhakar DV, Maini SB (2000) Isolation and characterization of mango peel pectins. J Food Process Preserv 24(3):209–227
Sukhenko Y, Mushtruk M, Vasyliv V, Sukhenko V, Dudchenko V (2020) Production of pumpkin pectin paste. Lecture notes in mechanical engineering
Sun L, Ropartz D, Cui L, Shi H, Ralet MC, Zhou Y (2019) Structural characterization of rhamnogalacturonan domains from Panax ginseng C. A. Meyer. Carbohydr Polym 203:119–127
Szymańska-Chargot M, Chylińska M, Pieczywek PM, Rösch P, Schmitt M, Popp J, Zdunek A (2016) Raman imaging of changes in the polysaccharides distribution in the cell wall during apple fruit development and senescence. Planta 243(4):935–945
Tanford C (1961) Physical chemistry of macromolecules. Wiley, New York
Thielking H, Kulicke WM (1998) Determination of the structural parameters of aqueous polymer solutions in the molecular, partially aggregated, and particulate states by means of FFFF/MALLS. J Microcolumn Sep 10(1):51–56
Urias-Orona V, Rascón-Chu A, Lizardi-Mendoza J, Carvajal-Millán E, Gardea AA, Ramírez-Wong B (2010) A novel pectin material: extraction, characterization and gelling properties. Int J Mol Sci 11(10):3686–3695
Vasco-Correa J, Zapata Zapata AD (2017) Enzymatic extraction of pectin from passion fruit peel (Passiflora edulis f. flavicarpa) at laboratory and bench scale. LWT 80:280–285
Wahlund KG, Giddings JC (1987) Properties of an asymmetrical flow field-flow fractionation channel having one permeable wall. Anal Chem 59(9):1332–1339
Wang S, Chen F, Wu J, Wang Z, Liao X, Hu X (2007) Optimization of pectin extraction assisted by microwave from apple pomace using response surface methodology. J Food Eng 78(2):693–700
Wang M, Huang B, Fan C, Zhao K, Hu H, Xu X, Pan S, Liu F (2016) Characterization and functional properties of mango peel pectin extracted by ultrasound assisted citric acid. Int J Biol Macromol 91:794–803
Wikiera A, Mika M, Grabacka M (2015) Multicatalytic enzyme preparations as effective alternative to acid in pectin extraction. Food Hydrocoll 44:156–161
Wyatt PJ (1993) Light scattering and the absolute characterization of macromolecules. Anal Chim Acta 272(1):1–40
Yapo BM (2009a) Biochemical characteristics and gelling capacity of pectin from yellow passion fruit rind as affected by acid extractant nature. J Agric Food Chem 57(4):1572–1578
Yapo BM (2009b) Pineapple and banana pectins comprise fewer homogalacturonan building blocks with a smaller degree of polymerization as compared with yellow passion fruit and lemon pectins: implication for gelling properties. Biomacromolecules 10(4):717–721
Zarei M, Ahmadi Zenouz A, Saari N, Ghanbari R, Nikkhah M, Vaziri M (2017) Effect of microwave-assisted extraction on the yield and quality of apple pomace and lemon peel pectins. Int Food Res J 24(6):2402–2407
Zhang W, Xie F, Liu X, Luo J, Wu J, Wang Z (2019) Pectin from black tomato pomace: characterization, interaction with gallotannin, and emulsifying stability properties. Starch 71(3-4)
Zhang C, Zhu X, Zhang F, Yang X, Ni L, Zhang W, Liu Z, Zhang Y (2020) Improving viscosity and gelling properties of leaf pectin by comparing five pectin extraction methods using green tea leaf as a model material. Food Hydrocoll 98
Zimm BH (1948) The scattering of light and the radial distribution function of high polymer solutions. J Chem Phys 16(12):1093–1099

Chapter 5
Emulsification Properties of Pectin

Katerina Alba and Vassilis Kontogiorgos

5.1 Introduction

Food and pharmaceutical industry frequently design their formulations aiming to improve human health (e.g., foods that lower cholesterol), produce products with consumer-tailored specifications (e.g., products for vegetarians) or deliver bioactives to the required site of uptake (e.g., colon). Among other biopolymers, pectin may be also used as a carrier for the protection and targeted delivery of bioactive compounds and for increasing their shelf life and stability (Rehman et al. 2019). The challenges arise from the increasing public interest in the availability of "natural" food ingredients where only naturally available materials such as carbohydrates or proteins should be used in the formulations. In addition, complexities also arise from the gastric environment that usually the product needs to bypass before reaching the desired location in the gastrointestinal tract.

Polysaccharides, in general, are routinely used in food and pharmaceutical industries, mostly as thickeners, dispersion stabilisers or water structuring agents. These functional properties are employed to create structures with reproducible physical properties. In recent years, however, the need to create advanced formulations that bypass gastric environment, delay lipid digestion to prolong satiety, and deliver bioactives in the gastrointestinal tract at the site of interest has boosted research on the fundamental properties of polysaccharides at interfaces (McClements and Jafari 2018; Araiza-Calahorra et al. 2018; Kontogiorgos 2019). The main reason is that polysaccharide-based structures may resist attack from proteases as well as the acidic environment of stomach that frequently impair the performance of protein

K. Alba
Department of Biological Sciences, University of Huddersfield, Huddersfield, UK

V. Kontogiorgos (✉)
School of Agriculture and Food Sciences, The University of Queensland,
Brisbane, QLD, Australia
e-mail: v.kontogiorgos@uq.edu.au

V. Kontogiorgos (ed.), *Pectin: Technological and Physiological Properties*,
https://doi.org/10.1007/978-3-030-53421-9_5

and surfactant-based formulations (McClements and Gumus 2016). In addition, surface active compounds are used in acidic drinks to emulsify flavour oils, prevent their oxidation and deliver them in a sustained manner (e.g., in the oral cavity) (Matalanis et al. 2011).

Technological performance of polysaccharides as emulsifiers is controlled by their macromolecular properties (e.g., conformation, surface charge density, molecular weight etc.) and intra- and inter-chain interactions that act cooperatively to determine adsorption strength (Kontogiorgos 2019). Pectin is obtained from natural sources using suitable extraction methodologies and may be tailored with chemical or physical modifications to improve the functionality of the extracted material. Depending on the source of extraction (Chap. 4) pectin has the ability to rapidly adsorb at the interface, reduce interfacial tension to facilitate droplet disruption, and impede droplet aggregation. This is typically attributed to the presence of hydrophobic elements in the structure such as proteins, ferulic acids, methyl or acetyl groups (Alba and Kontogiorgos 2017). The objective of this chapter is to identify the role of the surface active functional groups and provide a mechanistic understanding of the phenomenology of pectin adsorption at the oil-water interface.

5.2 Role of Structural Elements on the Interfacial Activity

The emulsifying capacity of pectin is typically associated with the chemical structure of its backbone such as the degree of methylation (DM) and acetylation (DA), the macromolecular characteristics of pectin chains (molecular weight (M_w), branching, hydrodynamic volume etc.) and the presence of functional units such as protein or ferulic acids. The evaluation of the contribution to the emulsification capacity of pectin of each of these structural parameters is still in progress and a matter of debate. However, some general principles may be drawn that may form the basis for further investigations and greater understanding of pectin functionality at the oil-water interface. In this section, we identify the most important structural elements that contribute to its interfacial activity.

5.2.1 The Role of Protein

The protein content in pectin varies depending on the source, isolation conditions and detection methods with higher values typically reported for sugar beet (up to ~9%) and okra (~5%), in contrast to citrus or apple pectin (e.g., ~3% and ~1%, respectively) (Funami et al. 2011; Yapo et al. 2007a; Chen et al. 2016a, 2018; Alba et al. 2015; Schmidt et al. 2015). Proteins are either present as contaminants that are co-extracted during the isolation process or associated with pectin structure through covalent linkages usually attached on the side chains. This association has been also probed by atomic force microscopy describing the protein-pectin complexes as "tadpoles" or as a network of "rods and spheres" (Fishman et al. 2015; Kirby et al. 2008).

Sugar beet pectin stabilised emulsions require about 3% protein for optimum surface activity (Chen et al. 2016a, 2018) whereas enzymic removal of protein results in reduction of interfacial activity and increase of droplet size compared to emulsions fabricated with non-enzymically modified pectin (Funami et al. 2007). The enzymatic treatment also reduces its molecular weight and radius of gyration thus restricting its steric stabilisation efficiency. It has been also shown that adsorbed pectin fractions at the oil-water interface have high protein concentration hinting at the importance of the protein component on emulsion stability (Leroux et al. 2003; Akhtar et al. 2002; Yapo et al. 2007a; Siew and Williams 2008b; Nakamura et al. 2004). Some pectins, as for instance those from pomegranate peel, show limited capacity to lower the surface tension, and its emulsifying properties are mostly attributed to the presence of protein and ester groups (Yang et al. 2018). In contrast, protein-rich and protein-depleted sugar beet pectin fractions have shown a range of emulsion stabilisation properties with protein playing a secondary role (Karnik and Wicker 2018; Chen et al. 2018). Further complications may also arise from the fact that in some pectins covalently-linked ferulic acid-arabinogalactan-protein complex has more notable impact on the interfacial activity and emulsifying capacity than protein alone (Chen et al. 2016b, 2019; Siew and Williams 2008b).

Another school of thought proposes that the accessibility and chemical nature of protein (e.g., amino acid composition and conformation) is more important determinant of emulsification capacity than its overall concentration. For instance, sugar beet pectin fractions with different protein amount ranging between 0.8% and 5.9% result in formation of emulsions of comparable droplet sizes and stability (Williams et al. 2005). In addition, extensin, a hydroxyproline-rich glycoprotein associated with the plant cell walls, was reported to be the main protein-type in pectin isolated from a range of botanical sources (Karnik et al. 2016; Nuñez et al. 2009). However, similar to total protein content, hydroxyproline-rich fractions did not show good emulsifying capacity and could not be directly associated with the emulsifying activity of sugar beet pectin. This is in general agreement with other investigations that have not identified a direct relationship between protein content and emulsifying capacity (Yapo et al. 2007a; Alba et al. 2016) suggesting that protein accessibility to the interface may be hindered by the bulky carbohydrate chains thus restricting interfacial arrangement (Castellani et al. 2010). A mechanistic description of the complex relationships between protein and pectin at the interface is presented in Sect. 6.3 where the different modes of adsorption are detailed.

5.2.2 *The Role of Acetyl and Methyl Groups*

Acetyl groups, similarly to ferulic groups, enhance interfacial activity of pectin resulting in smaller droplets during emulsification (Akhtar et al. 2002; Dea and Madden 1986; Leroux et al. 2003; Siew and Williams 2008a). Early studies using de-acetylated pectin revealed that the presence of acetyl groups does not contribute to a great extent to emulsification capacity (Leroux et al. 2003). However, the samples had different protein content making difficult to decouple the role of protein and acetyl groups on the overall emulsification performance. For instance, recent studies demonstrate that acetyl groups with a minimum degree of acetylation of ~10% improve considerably the emulsifying properties of pectin, particularly at low protein contents (Chen et al. 2016b; Schmidt et al. 2014).

In addition to the acetyl groups, the presence of methyl groups also contributes to interfacial activity of pectin although the results are sometimes contradicting. Some authors have demonstrated a direct relationship between the DM and emulsifying capacity of citrus pectin with increments of DM from ~70% to ~80% (Schmidt et al. 2014). Interestingly, it has been also shown that increase of DM beyond 80% did not result in further reduction of droplet size due to the self-association of citrus pectin thus restricting the accessibility of hydrophobic groups to the interface. However, recent studies using ultra-high methylated pectin (DM > 90%) of low molecular weight resulted in formation of stable nano-emulsions demonstrating that the importance of methyl group may manifest only at very high degrees of esterification (Hua et al. 2019). Block-wise distribution of carboxylic acid groups at comparable degree of methylation (~63%) showed negligible differences on interfacial tensions of apple pectin also supporting that the overall DM rather than other structural details plays critical role on the interfacial activity (Lutz et al. 2009). In contrast, other authors investigated citrus pectin with DM ranging from 22 to 73% and concluded that the content of methyl esters is of minor importance for the emulsifying properties pectin (Akhtar et al. 2002). The de-methylesterification of sugar beet (Chen et al. 2016b) or citrus pectin (Wan et al. 2019) also resulted in particularly stable dispersions showing that it is possible to create stable emulsions with LM pectin. Other hydrophobic groups may also be attached on the pectin backbone to confer hydrophobicity on the structure. To that end, alkylated citrus pectins with different alkyl chain lengths and degree of alkyl substitution demonstrated improved emulsifying activity, as evidenced by smaller droplet diameters than those stabilised with non-alkylated pectin (Liang et al. 2015).

5.2.3 The Role of Molecular Weight and Side Chains

The accessibility of protein and the other surface active components may be linked to pectin molecular weight although its impact on emulsification is currently inconsistent. Early reports suggested that low molecular weight (e.g., 35–90 × 10^3 g mol^{-1}) favours emulsifying activity of pectin, possibly due to better accessibility of interfacially active groups. However, pectin fractions of very low molecular weight result in lower interfacial activity and coarser emulsions because of the inability of short chains to provide efficient steric stabilisation (Yapo et al. 2007a, b; Akhtar et al. 2002; Leroux et al. 2003). On the contrary, very low Mw (15,000 g mol^{-1}) but also ultra-high methoxylated pectin spontaneously emulsifies oil (Hua et al. 2019) arguing that the influence of chain size should be viewed in conjunction with its group functionalisation. Similarly, sugar beet pectin of low M_w may form emulsions with smaller droplet diameters than those stabilised with its high M_w counterparts (Williams et al. 2005). However, other studies have not demonstrated a direct relationship between M_w of citrus pectin and its emulsifying capacity, particularly after adjusting the viscosity of emulsions (Schmidt et al. 2014). It has been also shown that reduction of M_w from 76×10^3 to 47×10^3 g mol^{-1} did not improve emulsifying properties of citrus pectin. In contrast, increase of Mw of sugar beet pectin via cross-linking of ferulic acid groups has shown that emulsions fabricated with cross-linked pectin (Mw ~ 1860×10^3 g mol^{-1}) have smaller droplet mean diameters and improved long term stability compared to those stabilised with non-cross-linked pectin (Mw ~ 780×10^3 g mol^{-1}) (Zhang et al. 2015). The lack of consensus on the impact of molecular weight on the emulsifying capacity of pectin also suggests that the other structural characteristics discussed earlier (acetyl and methyl groups or ferulic acids) cannot be disregarded.

Pectin fractions adsorbed at the oil-water interface are enriched in neutral sugars (e.g., arabinose and galactose) suggesting that RG-I containing pectins could have better emulsifying properties than those with linear backbone (Siew and Williams 2008a). These results were further supported by the enzymatic degradation of sugar beet pectin side chains revealing a reduction in its interfacial and stabilising capacity (Chen et al. 2016b). The impact of side-chains on emulsion-forming properties of sugar beet pectin is attributed to the interfacial activity of protein and presence of ferulic acid that are attached to the side-chains and act as anchors for the attachment of the entire pectin chain. In addition, the presence of neutral sugar side-chains contributes to the long-term emulsion stability due to the formation of thick interfacial layers thus providing effective steric stabilisation that impedes emulsion coarsening (Funami et al. 2011). Results using highly branched okra pectin also confirmed that the prevalence of RG-I segments and the length of their branches influence emulsion stability (Kpodo et al. 2018). It has been also reported that multilayer adsorption of sugar beet pectin at the interface is possible and originates from electrostatic interactions between positively charged protein moieties and the negatively charged galacturonic acid residues (Chee et al. 2008). Generally, emulsions stabilised with pectin are pH- and ionic strength- sensitive and changes in these factors result in alterations in its emulsifying capacity (Table 5.1). At pH values greater than ~3.5 carboxyl groups of pectin are ionised and the biopolymer chains are extended due to the electrostatic repulsions between the carboxylate anions. The number and distribution of negative charges is determined by the degree of methyl esterification and degree of blockiness (DB) of methyl groups. The ionisation of carboxylic groups decreases with pH ($pH < pKa$) and consequently promotes self-association of the chains. It has been shown that pectin stabilises oil-water interfaces at low pH values, where chains adopt highly compact conformations resulting in the formation of thick interfacial layers thus providing effective steric stabilisation (Alba et al. 2016, 2018; Castellani et al. 2010; Kpodo et al. 2018; Zhao et al. 2018). It becomes evident that modification of conformational characteristics of pectin with the aid of environmental conditions (e.g., pH, ionic strength, type of cation etc.) modulates

Table 5.1 Parameters that influence emulsification capacity of pectin

Parameter	Comments	Effect on emulsification capacity
Branching length	Short ($R_3 < 2$)[a]	Unfavourable
	Intermediate ($2 < R_3 < 3$)	Favourable
	Long ($R_3 > 3$)	Unfavourable
pH	pH < ~3.5	Favourable
	pH > ~3.5	Unfavourable
Salts	Addition of mono- or di-valent cations	Unfavourable
Degree of methylesterification (%)	High (>~70)	Generally favourable
	Low (<~40)	Generally favourable
	Intermediate (~40 < DM < ~70)	Unfavourable
Degree of acetylation (%)	>~10	Favourable
Protein content (%)	>~3	Favourable
Ferulic esters	In sugar beet pectin	Favourable
Molecular weight ($\times 10^3$ g/mol)	High (>200)	Generally unfavourable
	Intermediate (100–200)	Generally favourable
	Low (<100)	Unfavourable

The word "generally" in the last column is used to indicate that there is no straightforward relationship and other parameters also need to be taken into consideration
[a]R_3 values indicate the length of branching of RG-I segments (R_3 = (Ara + Gal)/Rha). Higher values indicate longer branches

emulsifying capacity and may enhance steric stabilisation and long-term emulsion stability. From the above discussion it becomes apparent that pectin has multiple elements that cooperatively influence its emulsification properties. This is partially due to the large number of protocols and botanical sources that may be used to extract pectin and control the outcome of the structure. Synthesing information from the discussion that has been presented so far, it emerges that some critical parameters may be successfully manipulated and general guidelines may be drawn in an effort to rationally design pectin-stabilised emulsions with desirable physical properties (Table 5.1).

5.3 Phenomenology of Pectin Adsorption at the Oil-Water Interface

In synthetic polymer chemistry, the result of linking two different monomers to form a polymer chain is termed "copolymer" (Carraher 2014). There are different classes of copolymers depending on the arrangement of monomers along the polymer chain. In alternating copolymers the two monomers are arranged in an alternating fashion whereas in random copolymers are arranged randomly along the

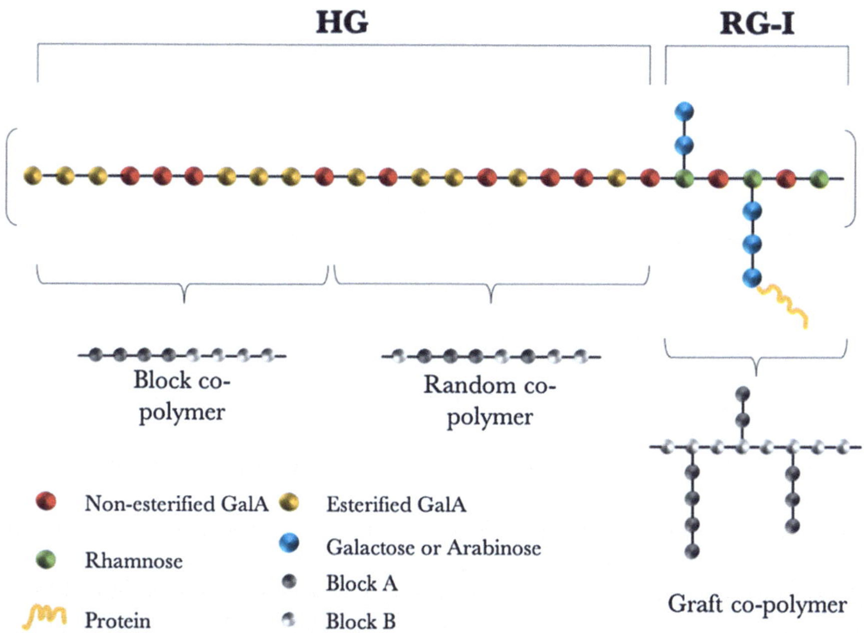

Fig. 5.1 Idealised description of pectin structure as copolymer. Homogalacturonan (HG) sections may have blockwise or random distribution of esterified groups. The former case may be described as block co-polymer whereas the latter as random co-polymer (grey chains). In pectins where the presence of rhamnogalacturonan segments (RG-I) are prevalent the term graft copolymer may be used as branches made of arabinose, galactose and protein may be present

polymer chain (Fig. 5.1). A block copolymer contains two or more polymer chains attached at their ends and is termed di-block, tri-block or multi-block for two, three or more than three chains linked together, respectively. Finally, when side chains are attached on the main backbone then it is termed graft copolymer. Pectin consists of more than one sugar monomers thus making it a heteropolysaccharide (Chap. 2). It becomes evident that an idealised pectin structure (Fig. 5.1) may be viewed using synthetic polymer terminology. Specifically, the HG part of the chain may be described as a random or block copolymer depending on the source or post-extraction modifications of pectin. To that end, methyl esterified and non-methyl esterified galacturonic acid residues may arrange randomly or in a blocky pattern along the chain (Voragen et al. 2009; Chan et al. 2017; Mohnen 2008). In addition, the RG-I segment may be described as a graft copolymer due to the presence of lateral arabinose or galactose branches that may also be decorated with covalently linked protein moieties. To further complicate the landscape, the presence of pH-dependant charges due to the presence of carboxylic groups makes pectin a suitable polymer to be described as polyelectrolyte. A crucial distinction between pectin and synthetic copolymers is that in the latter the architecture is controlled by precise polymerisation reactions yielding well-defined chains. Pectin biosynthesis, however, yields complex structures that are further modified during plant maturation or during extraction procedures (Alba and Kontogiorgos 2017).

The aforementioned structure description provides new avenues for greater understanding of pectin interfacial activity, as we may now use well-established theoretical approaches to describe its adsorption at the oil-water interface. Although such a remarkably complex structure is not normally observed in synthetic copolymers, the principles of polymer adsorption at interfaces still apply (Fleer et al. 1998). Adsorption of copolymers at interfaces may be described as the accumulation of chains at the interface that depends strongly on the chemical nature of the chains and the solvent quality. In the case of copolymers, two cases may be distinguished were the aqueous phase is non-selective or selective. In the former case, both blocks (both A and B, Fig. 5.1) are soluble whereas in the latter only one (either A or B) is soluble in the aqueous buffer. The relationship between buffer composition (e.g., pH, ionic strength, type of cation etc.) and pectin will dictate the strength of pectin-buffer interactions, the conformation of the chains, and the amount that is adsorbed thus controlling the overall stability of the dispersion. It starts becoming evident that precise control of buffer composition is one of the first steps towards successful fabrication of pectin-stabilised emulsions.

Block copolymers are adsorbed when one of the blocks has a high affinity for the interface (frequently termed "anchor") while the other for the continuous phase ("buoy") (Fleer et al. 1998). Adsorption of copolymers from non-selective solvents occurs into two stages where initially the polymer diffuses to the interface forming a monolayer (Motschmann et al. 1991). It should be mentioned that upon initial adsorption at low interfacial coverage the conformation of buoy blocks is not particularly different than of those in the bulk solution. This may be even more relevant when emulsification proceeds via a covalently-linked protein-assisted mechanisms (see below), as pectin conformation will not be affected by its adsorption at the interface. In the second stage, the adsorbed layer will grow by diffusion of pectin to the surface from the bulk that results in chain overlap and formation of multilayers thus leading to conformational rearrangements. At this stage, chain conformations will be described by the correlation length ξ of the chains. This may also require penetration of chains through the barrier created by those already attached to the droplet (Motschmann et al. 1991). Pectin, depending on the source of extraction, may have multiple segments that are able to adsorb at the oil-water interface, however, adsorption may also take place into two steps (Fig. 5.2a). Initially, the transport of pectin from the bulk to the oil interface is due to diffusion towards the droplets that occurs instantaneously during emulsification. After the initial adsorption, pectin conformation at the interface strongly depends on the pH of the continuous phase and the degree of methylation (Alba et al. 2018) that will in turn determine the ability of pectin to stabilise the dispersion. Specifically, at pH $< \sim 3.5$ that is below the dissociation constant of galacturonic acid, pectin conformation is relatively unaffected by the degree of methylation and space occupancy is efficient at these conditions (Fig. 5.2b). In contrast, at pH $> \sim 3.5$ chains of low methoxylated pectins attain extended conformation (Fig. 5.2c) but the effect is suppressed with increase of degree of methylation (HM pectin) due to the decrease in charge density

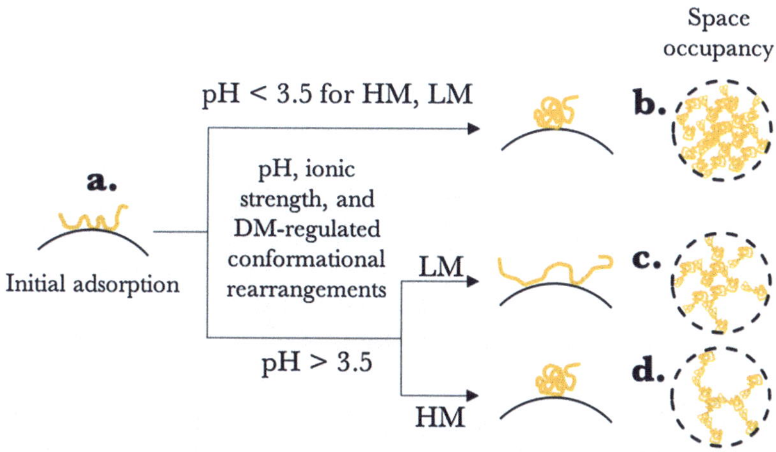

Fig. 5.2 Adsorption of pectin at the oil-water interface takes place into two steps. (**a**) Initially, pectin diffuses from the bulk to the oil-water interface and occurs instantaneously during emulsification. In the second step, pectin rearranges at the interface depending on the pH of the continuous phase and the degree of methylation (**b**) At pH < ~3.5 pectin conformation is relatively unaffected by the degree of methylation, and space occupancy is efficient at these conditions (**c**) At pH > ~3.5 chains of low methoxylated pectins attain extended conformations with space occupancy being less efficient than in (**b**), (**d**) Increase of degree of methylation (HM pectin) leads to compact conformations due to the decrease in charge density and steric hindrance because of the presence of methyl groups. Space occupancy is the least efficient compared to (**b**) and (**c**)

and steric hindrance because of the presence of methyl groups (Fig. 5.3d) (Alba et al. 2017, 2018; Cros et al. 1996). This behaviour is preserved in the semi-dilute regime, or in other words, at concentrations where most likely pectin will be used as emulsifier. Consequently, the space filling capacity of pectin in solution, as controlled by pH and degree of methylation as well as branching, has consequences for the thickness of interfacial layer and the effectiveness of steric stabilisation. Indeed these parameters have been extensively investigated revealing that pectin emulsification is responsive to buffer and chain architecture (Schmidt et al. 2015, 2017; Alba et al. 2016; Kpodo et al. 2018; Verkempinck et al. 2018; Hua et al. 2019; Liu et al. 2019; Chen et al. 2016c). This description proposes that the most efficient steric stabilisation capacity would be at acidic environments (Fig. 5.2b) whereas the least efficient with HM pectin at high pH (Fig. 5.2d).

Pectin may anchor at the interface *via* several mechanisms that act concurrently. The prevalence of one over another depends on the molecular weight and sugar composition of pectin, the strength of interactions between pectin and continuous phase, the chemical properties of interface (e.g., triglyceride or terpene) and the amount of protein present. Specifically, pectin may adsorb unassisted at the interface only with the aid of the hydrophobic groups that are present along the backbone (e.g., methyl, acetyl or ferulic, Fig. 5.3a-I). This mechanism is particularly important in highly methylated (e.g., HM-citrus) or highly acetylated pectin (e.g., from sugar beet or okra). Another dominant mechanism is through anchoring of the

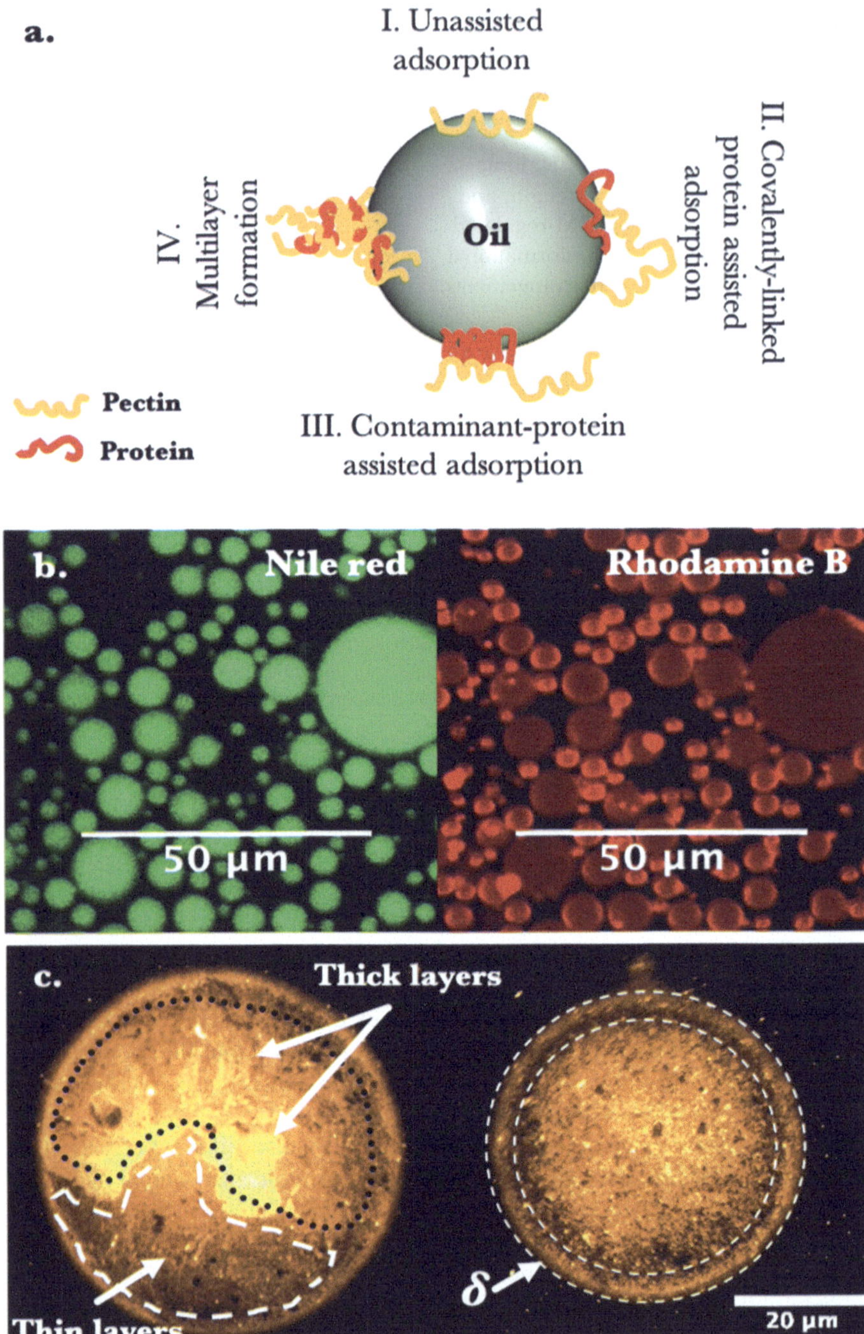

Fig 5.3 (**a**) Four different anchoring mechanisms of pectin at the oil-water interface (see text) (**b**) interfacial protein localisation in emulsions formed with pectin with substantial amounts of contaminant proteins. Rhodamine B stained protein may be observed forming layers (red layers, right)

chains with the aid of covalently-linked proteins that are found in RG-I units of some pectins (Fig. 5.3a-II). Protein may be also present as contaminant, particularly in pectin isolated in laboratory settings or from novel sources with inherently high protein content (Fig. 5.3a-III). In these cases, contaminant protein may adsorb first with pectin following, resulting in formation of multilayers without pectin having any interaction with the interface. For instance, thick interfacial protein layers may be observed at the interface (Fig. 5.3b) in emulsions formed with pectin containing substantial amounts of contaminant proteins (Alba et al. 2013). Protein-assisted polysaccharide adsorption has been described extensively in the literature either through covalently-linked Maillard conjugates or bilayer formation through intentional protein addition (Dickinson 2008, 2009; Evans et al. 2013; Rodríguez Patino and Pilosof 2011). Despite of these distinct mechanisms of adsorption, multilayer formation would be expected in most cases depending on pectin architecture and solvent composition with some role always given to the protein fraction (Fig. 5.3a-IV). It should be noted that the layer thickness is not necessarily uniform along the interface of the droplet thus resulting in quite complex mixed interfacial layers (Alba et al. 2016) or presence of pectin microgels at the interface (Schmidt et al. 2017) (Fig. 5.3c, left) with intricate interfacial rheology (Sagis and Fischer 2014; Fischer 2013). The thickness, δ, of this layer (Fig. 5.3c, right), is normally responsible for the extent and effectiveness of stabilisation, as it protrudes laterally from the droplets thus conferring stabilisation through steric mechanisms. As a result, some areas in the droplet may be covered by thick multiple mixed layers of pectin and protein whereas other areas may present a thinner interfacial coverage. The latter regions may act as destabilisation centres particularly when adsorption strength is not sufficient and desorption may occur during long term storage. Desorption will expose the oil interface that may lead to coarsening through, for example, coalescence or bridging flocculation.

5.4 Conclusions

The structural components that influence the emulsification properties of pectin and its mechanisms of interfacial arrangement have been discussed with the aim to design pectin that may be used as emulsifier. It is challenging to assign a straightforward structure and function relationships owing to the structural complexities of pectin architecture. It is possible, however, to suggest that in order for pectin to effectively arrange at the interface and to provide efficient emulsification and long-term stabilisation it requires protein content of around 3% with a minimum 10% degree of acetylation. Although higher degree of methylation supports interfacial

Fig. 5.3 (continued) around Nile red stained oil droplets (green droplets, left)) (**c**) multilayer formation in pectin-stabilised emulsions forming thin or thick layers depending on pectin architecture and aqueous phase composition (left). Lateral protrusion of the interfacial layer with thickness δ, is responsible for the extent and effectiveness of steric stabilisation (right)

arrangement it does not seem to be a critical factor. Side chains support steric stabilisation and RG-I rich pectins are generally more efficient emulsifiers. Side chains are important as both protein and ferulic acids are located on the side chains. High molecular weight restricts accessibility of protein at the interface and the particularly high viscosity impedes fast adsorption and reorganisation at the interface. Intermediate molecular weight ($\sim 150 \times 10^3$ g mol^{-1}) pectins are preferred as lower values do not confer efficient steric stabilisation. In addition, efficient interfacial functionality of pectin requires a certain degree of repetitive structure similar to that of copolymers. Pectin is a typical block co-polymer that depending on the source may be di-block, triblock, or grafted. Accordingly, theories that have been developed for co-polymer adsorption at interfaces are better suited to theoretically analyse and treat experimental data.

References

Akhtar M, Dickinson E, Mazoyer J, Langendorff V (2002) Emulsion stabilizing properties of depolymerized pectin. Food Hydrocoll 16(16):249–256

Alba K, Kontogiorgos V (2017) Pectin at the oil-water interface: Relationship of molecular composition and structure to functionality. Food Hydrocoll 68:211–218

Alba K, Ritzoulis C, Georgiadis N, Kontogiorgos V (2013) Okra extracts as emulsifiers for acidic emulsions. Food Res Int 54:1730–1737

Alba K, Laws AP, Kontogiorgos V (2015) Isolation and characterization of acetylated LM-pectins extracted from okra pods. Food Hydrocoll 43:726–735

Alba K, Sagis LMC, Kontogiorgos V (2016) Engineering of acidic o/w emulsions with pectin. Colloids Surf B Biointerfaces 145:301–308

Alba K, Bingham RJ, Kontogiorgos V (2017) Mesoscopic structure of pectin in solution. Biopolymers 107(6):1–8

Alba K, Bingham RJ, Gunning PA, Wilde PJ, Kontogiorgos V (2018) Pectin conformation in solution. J Phys Chem B122(29):7286–7294

Araiza-Calahorra A, Akhtar M, Sarkar A (2018) Recent advances in emulsion-based delivery approaches for curcumin: From encapsulation to bioaccessibility. Trends Food Sci Technol 71:155–169

Carraher CE (2014) Carraher's polymer chemistry. In: Carraher E (ed) Polymer chemistry, 9th edn. CRC Press, Boca Raton

Castellani O, Al-Assaf S, Axelos M, Phillips GO, Anton M (2010) Hydrocolloids with emulsifying capacity. Part 2 – adsorption properties at the n-hexadecane–water interface. Food Hydrocoll 24(2):121–130

Chan SY, Choo WS, Young DJ, Loh XJ (2017) Pectin as a rheology modifier: origin, structure, commercial production and rheology. Carbohydr Polym 161:118–139

Chee KS, Williams PA, Cui SW, Wang Q (2008) Characterization of the surface-active components of sugar beet pectin and the hydrodynamic thickness of the adsorbed pectin layer. J Agric Food Chem 56(17):8111–8120

Chen H, Qiu S, Gan J, Liu Y, Zhu Q, Yin L (2016a) New insights into the functionality of protein to the emulsifying properties of sugar beet pectin. Food Hydrocoll 57:262–270

Chen H-M, Fu X, Luo Z-G (2016b) Effect of molecular structure on emulsifying properties of sugar beet pulp pectin. Food Hydrocoll 54:99–106

Chen HM, Fu X, Luo ZG (2016c) Effect of molecular structure on emulsifying properties of sugar beet pulp pectin. Food Hydrocoll 54:99–106

Chen H, Qiu S, Liu Y, Zhu Q, Yin L (2018) Emulsifying properties and functional compositions of sugar beet pectins extracted under different conditions. J Dispers Sci Technol 39(4):484–490
Chen H, Niu H, Zhang H, Yun Y, Chen W, Zhong Q, Chen W, Fu X (2019) Preparation and properties of ferulic acid-sugar beet pulp pectin ester and its application as a physical and antioxidative stabilizer in a fish oil-water emulsion. Int J Biol Macromol 139:290–297
Cros S, Garnier C, Axelos MAV, Imberty A, Perez S (1996) Solution conformations of pectin polysaccharides: determination of chain characteristics by small angle neutron scattering, viscometry, and molecular modeling. Biopolymers 39(3):339–352
Dea ICM, Madden JK (1986) Acetylated pectic polysaccharides of sugar beet. Food Hydrocoll 1(1):71–88
Dickinson E (2008) Interfacial structure and stability of food emulsions as affected by protein–polysaccharide interactions. Soft Matter 4(5):932–942
Dickinson E (2009) Hydrocolloids as emulsifiers and emulsion stabilizers. Food Hydrocoll 23(6):1473–1482
Evans M, Ratcliffe I, Williams PA (2013) Emulsion stabilisation using polysaccharide-protein complexes. Curr Opin Colloid Interface Sci 18(4):272–282
Fischer P (2013) Rheology of interfacial protein-polysaccharide composites. Eur Phys J Spec Top 222(1):73–81
Fishman ML, Chau HK, Qi PX, Hotchkiss AT, Garcia RA, Cooke PH (2015) Characterization of the global structure of low methoxyl pectin in solution. Food Hydrocoll 46:153–159
Fleer GJ, Cohen-Stuart MA, Scheutjens JMHM, Cosgrove T, Vincent B (1998) Polymers at interfaces. Chapman and Hall, London
Funami T, Zhang G, Hiroe M, Noda S, Nakauma M, Asai I, Cowman MK, Al-Assaf S, Phillips GO (2007) Effects of the proteinaceous moiety on the emulsifying properties of sugar beet pectin. Food Hydrocoll 21(8):1319–1329
Funami T, Nakauma M, Ishihara S, Tanaka R, Inoue T, Phillips GO (2011) Structural modifications of sugar beet pectin and the relationship of structure to functionality. Food Hydrocoll 25(2):221–229
Hua X, Ding P, Wang M, Chi K, Yang R, Cao Y (2019) Emulsions prepared by ultrahigh methoxylated pectin through the phase inversion method. Int J Biol Macromol 128:167–175
Karnik D, Wicker L (2018) Emulsion stability of sugar beet pectin fractions obtained by isopropanol fractionation. Food Hydrocoll 74:249–254
Karnik D, Jung J, Hawking S, Wicker L (2016) Sugar beet pectin fractionated using isopropanol differs in galacturonic acid, protein, ferulic acid and surface hydrophobicity. Food Hydrocoll 60:179–185
Kirby A, Macdougall A, Morris V (2008) Atomic force microscopy of tomato and sugar beet pectin molecules. Carbohydr Polym 71(4):640–647
Kontogiorgos V (2019) Polysaccharides at fluid interfaces of food systems. Adv Colloid Interf Sci 270:28–37
Kpodo FM, Agbenorhevi JK, Alba K, Oduro IN, Morris GA, Kontogiorgos V (2018) Structure-function relationships in pectin emulsification. Food Biophys 13:71–79
Leroux J, Langendorff V, Schick G, Vaishnav V, Mazoyer J (2003) Emulsion stabilizing properties of pectin. Food Hydrocoll 17(4):455–462
Liang R-H, Wang L-H, Chen J, Liu W, Liu C-M (2015) Alkylated pectin: synthesis, characterization, viscosity and emulsifying properties. Food Hydrocoll 50:65–73
Liu Z, Pi F, Guo X, Guo X, Yu S (2019) Characterization of the structural and emulsifying properties of sugar beet pectins obtained by sequential extraction. Food Hydrocoll 88:31–42
Lutz R, Aserin A, Wicker L, Garti N (2009) Structure and physical properties of pectins with block-wise distribution of carboxylic acid groups. Food Hydrocoll 23(3):786–794
Matalanis A, Jones OG, McClements DJ (2011) Structured biopolymer-based delivery systems for encapsulation, protection, and release of lipophilic compounds. Food Hydrocoll 25(8): 1865–1880

McClements DJ, Gumus CE (2016) Natural emulsifiers — biosurfactants, phospholipids, biopolymers, and colloidal particles: molecular and physicochemical basis of functional performance. Adv Colloid Interf Sci 234:3–26
McClements DJ, Jafari SM (2018) Improving emulsion formation, stability and performance using mixed emulsifiers: a review. Adv Colloid Interf Sci 251:55–79
Mohnen D (2008) Pectin structure and biosynthesis. Curr Opin Plant Biol 11(3):266–277
Motschmann H, Stamm M, Toprakcioglu C (1991) Adsorption kinetics of block copolymers from a good solvent: a two-stage process. Macromolecules 24(12):3681–3688
Nakamura A, Yoshida R, Maeda H, Furuta H, Corredig M (2004) Study of the role of the carbohydrate and protein moieties of soy soluble polysaccharides in their emulsifying properties. J Agric Food Chem 52(17):5506–5512
Nuñez A, Fishman ML, Fortis LL, Cooke PH, Hotchkiss AT (2009) Identification of extensin protein associated with sugar beet pectin. J Agric Food Chem 57(22):10951–10958
Rehman A, Ahmad T, Aadil RM, Spotti MJ, Bakry AM, Khan IM, Zhao L, Riaz T, Tong Q (2019) Pectin polymers as wall materials for the nano-encapsulation of bioactive compounds. Trends Food Sci Technol 90:35–46
Rodríguez Patino JM, Pilosof AMR (2011) Protein–polysaccharide interactions at fluid interfaces. Food Hydrocoll 25(8):1925–1937
Sagis LMC, Fischer P (2014) Nonlinear rheology of complex fluid–fluid interfaces. Curr Opin Colloid Interface Sci 19(6):520–529
Schmidt US, Koch L, Rentschler C, Kurz T, Endreß HU, Schuchmann HP (2014) Effect of molecular weight reduction, acetylation and esterification on the emulsification properties of citrus pectin. Food Biophys 10(2):217–227
Schmidt US, Schmidt K, Kurz T, Endreß HU, Schuchmann HP (2015) Pectins of different origin and their performance in forming and stabilizing oil-in-water-emulsions. Food Hydrocoll 46:59–66
Schmidt US, Schütz L, Schuchmann HP (2017) Interfacial and emulsifying properties of citrus pectin: Interaction of pH, ionic strength and degree of esterification. Food Hydrocoll 62:288–298
Siew CK, Williams PA (2008a) Characterization of the surface-active components of sugar beet pectin and the hydrodynamic thickness of the adsorbed pectin layer. J Agric Food Chem 56:8111–8120
Siew CK, Williams PA (2008b) Role of protein and ferulic acid in the emulsification properties of sugar beet pectin. J Agric Food Chem 56:4164–4171
Verkempinck SHE, Kyomugasho C, Salvia-Trujillo L, Denis S, Bourgeois M, Van Loey AM, Hendrickx ME, Grauwet T (2018) Emulsion stabilizing properties of citrus pectin and its interactions with conventional emulsifiers in oil-in-water emulsions. Food Hydrocoll 85:144–157
Voragen AGJ, Coenen G-J, Verhoef RP, Schols HA (2009) Pectin, a versatile polysaccharide present in plant cell walls. Struct Chem 20(2):263
Wan L, Chen Q, Huang M, Liu F, Pan S (2019) Physiochemical, rheological and emulsifying properties of low methoxyl pectin prepared by high hydrostatic pressure-assisted enzymatic, conventional enzymatic, and alkaline de-esterification: a comparison study. Food Hydrocoll 93:146–155
Williams PA, Sayers C, Viebke C, Senan C (2005) Elucidation of the emulsification properties of sugar beet pectin. J Agric Food Chem 53(53):3592–3597
Yang X, Nisar T, Hou Y, Gou X, Sun L, Guo Y (2018) Pomegranate peel pectin can be used as an effective emulsifier. Food Hydrocoll 85:30–38
Yapo BM, Robert C, Etienne I, Wathelet B, Paquot M (2007a) Effect of extraction conditions on the yield, purity and surface properties of sugar beet pulp pectin extracts. Food Chem 100(4):1356–1364
Yapo BM, Wathelet B, Paquot M (2007b) Comparison of alcohol precipitation and membrane filtration effects on sugar beet pulp pectin chemical features and surface properties. Food Hydrocoll 21(2):245–255

Zhang L, Shi Z, Shangguan W, Fang Y, Nishinari K, Phillips GO, Jiang F (2015) Emulsification properties of sugar beet pectin after modification with horseradish peroxidase. Food Hydrocoll 43:107–113

Zhao S, Gao W, Tian G, Zhao C, DiMarco-Crook C, Fan B, Li C, Xiao H, Lian Y, Zheng J (2018) Citrus oil emulsions stabilized by citrus pectin: the influence mechanism of citrus variety and acid treatment. J Agric Food Chem 66(49):12978–12988

Chapter 6
Edible Films and Coatings with Pectin

Athina Lazaridou and Costas G. Biliaderis

6.1 Introduction

Edible films and coatings, based on biopolymers, have received considerable attention in recent years as means to improve and preserve food quality, enhance safety and prolong the shelf-life of perishable products. As raw materials, biopolymers constitute an alternative environmentally friendly and renewable source for packaging development due to their palatability and biodegradability, compared to synthetic films which impose a great burden on the environment as non-biodegradable waste. Moreover, edible film or coating preparations may serve as vehicles to deliver other desirable functionalities by inclusion of (bio)active compounds or even microbial cultures within a composite polymeric matrix, e.g., antimicrobial agents, antioxidants, colourants, flavorings, other fortifying nutrients (Kester and Fennema 1986; Gialamas et al. 2010; Mellinas et al. 2016; Otoni et al. 2017; Hassan et al. 2018; Saha et al. 2017). The main purposes in using edible films and coatings are to provide a barrier against diffusion of gases (water vapors, O_2, CO_2), migration of moisture or volatile components, control chemical reactions (e.g., oxidation) and microbial invasion–growth (both spoilage flora and pathogens), improve appearance of the product as well as maintain separation of different components in multicomponent foods (Lacroix 2009). Edible films, in particular, are primarily composed of polysaccharides, proteins and lipids, alone or in combination as blends or layers with varying ratios of the structuring components, and often include other functionalizing ingredients to modify the physicochemical properties (e.g., film integrity and structural stability as well as barrier properties, mechanical strength and adhesion behaviour). The main challenge for the development of effective edible films

A. Lazaridou · C. G. Biliaderis (✉)
Department of Food Science and Technology, Faculty of Agriculture, Aristotle University of Thessaloniki, Thessaloniki, Greece
e-mail: athlazar@agro.auth.gr; biliader@agro.auth.gr

V. Kontogiorgos (ed.), *Pectin: Technological and Physiological Properties*,
https://doi.org/10.1007/978-3-030-53421-9_6

and food coatings is thus to maintain the desired functionalities and overall acceptability throughout the intended shelf life of the food product.

Among the natural polymers used in edible films and coatings, polysaccharides are often employed as structuring matrices either alone or in combination with other bio-based materials. Polysaccharides, regardless of their origin (plant, seaweeds, microbes or exoskeletons of crustaceans and mollusks), are generally very hydrophilic, unless chemically modified (e.g., cellulose and starch derivatives with hydrophobic groups), and therefore may be susceptible to moisture and have poor water vapor permeability (WVP). As a result, when comparing the WVP of films made from polysaccharides or most proteins with those of conventional synthetic polymers, it becomes clear that the latter are more efficient barriers to water vapor transfer and moisture loss. Lipids, instead, having a low affinity for water, can improve the WVP of biopolymer films, although their incorporation requires emulsification or the use of a lamination process (e.g., formation of bilayer or laminant films) by combining multiple coatings with more effective water barriers, e.g., fatty acids or waxes (Lacroix 2009; Kristo et al. 2007). Bilayer films also require multiple drying steps and often tend to delaminate over time, developing cracks (brittleness) due to development of differential stresses between the successive layers upon storage. In contrast to moisture sensitivity, polysaccharide-based films exhibit good barrier properties to gas diffusion and permeability (O_2, CO_2), volatiles and resistance to oils (Baldwin et al. 1995; Biliaderis et al. 1999; Costa et al. 2015). Migration properties, structural integrity and mechanical properties of polysaccharide films are largely influenced by molecular structure (component monosaccharide residues and linkages, linear, branched, neutral or charged, monosaccharide distribution on the chain backbone) and size of the polysaccharide, presence of plasticisers (including water) and cross-linkers, inclusion of micro- or nano-sized particles as structure modifiers of the composite film matrix as well as solvent composition (ion type–ionic strength, pH) of the medium employed to initially solubilise/disperse all the film components, particularly in the case of charged polymers. All these molecular-compositional factors, solvent, method of preparation and processing parameters can affect chain conformation, polymer-polymer interactions and packing efficiency upon solvent removal. Recent reviews have focused on edible films based on polysaccharides, including hemicelluloses (Hansen and Plackett 2008), chitosan (Dutta et al. 2009), starch (Jiménez et al. 2012) and pectins (Espitia et al. 2014).

Pectins are important anionic heteropolysaccharides which exist in the cell walls of higher plants, consisting mostly of polymeric regions (with average degree of polymerization, DP, 50–100 of α-D-GalpA residues; Voragen et al. 2009) rich in D-galacturonic acid (GalA) units, as a homogalacturonan (HG) backbone segment; these α-(1,4) interlinked residues are partly methylesterified at C-6 and/or *O*-acetylated at *O*-2 or *O*-3 of the GalA units. In some pectins (e.g., sugar beet, spinach) there may be also some other functional units, covalently attached to some sugar moieties, i.e., ferulic acid, protein, etc. Chemical modification is also feasible (e.g., amidation of carboxyl groups by reaction with ammonia in alcoholic media), creating pectin derivatives with varying properties. Depending on the degree of methylesterification (DM) of the carboxyl group, pectins are distinguished as high methoxy pectins (HMP), with DM ≥ 50%, and low methoxy pectins (LMP), with DM < 50%. The functional properties (solution rheology and gelling behaviour) of pectins are affected by all these structural parameters (DM, degree of acetylation, amount and DP of HG cobiopolymer segments). Besides the HG (being ca. 50–65% of most pectin preparations), pectins are structurally more complex, containing several other pectic cobiopolymer components (Voragen et al. 2009; Noreen et al. 2017; Yapo and Gnakri 2015; Gawkowska et al. 2018): (a) rhamnogalacturonan I (RG-I) with up to 100 repeating units of [-2)-α-L-Rha*p*-(1,4)-α-D-Gal*p*A-(1-] as backbone, partly branched at *O*-4 (mainly) and/or *O*-3 positions of the α-L-Rhap residues, with small side chains of various types and sizes, i.e., (1,5)-α-L-arabinan, (1,5)-β-D-galactan, galactoarabinan and arabinogalactans; this material constitutes the second most abundant cobiopolymer structure (depending on the source is between 10 and 40% of complex pectic substances). (b) Rhamnogalacturonan II (RG-II), a block cobiopolymer of more complex structure, with an oligogalacturonan backbone (DP 11-15) without rhamnose residues, which is branched with four well-defined side chains (A, B, C, and D, at C-2 and C-3 of the Gal*p*A) containing six unusual sugars, namely apiose, 2-*O*-methyl-fucose, 2-*O*-methyl-xylose, aceric acid, 3-deoxy-manno-2-octulosonic acid (kdo) and 3-deoxy-D-lyxo-heptulosaric acid (dha); this is a highly conserved structure throughout vascular plants and is believed to exist in the primary cell wall as a minor fraction of ca. 10% in pectic substances. (c) xylogalacturonan (XGA), a partly methylesterified (1,4)-linked α-D-galacturonan backbone, substituted at *O*-3(mainly)/*O*-2 by single nonreducing β-D-Xyl*p* residues or short side chains (DP 2-8) of 1 → 2, 1 → 3, 1 → 4, 1 → 2,3, 1 → 2,4, 1 → 3,4–linked β-D-xylans; moreover, the β-D-xylan side chains of XGA may be substituted with α-L-Ara*f*, α-L-Fuc*p*, and/or β-D-Gal*p* units. The XGA is present in the vicinity of RG-I, thus constituting a part of the so-called hairy regions of pectic substances (Schols and Voragen 1996). (d) Apiogalacturonan (ApGA), a low DM methylesterified (1,4)-linked α-D-galacturonan core backbone, partly substituted at *O*-3(mainly)/*O*-2 by single nonreducing β-D-Api*f*, (1,5)-linked β-D-Api*f* residues and/or by 'apiobioside' [β-D-Api*f*-(1,3′)-β-D-Api*f*-(1-)]. (e) Depending on the source, other minor substituted galacturonans are also present, i.e., galactogalacturonan, arabinogalacturonan, galacturonogalacturonan. The relative amount, distribution and fine structure of all these structural elements in various pectin preparations are expected to play a major role in their water solubility, conformation and functional properties such as thickening–gelling and film forming behaviour, emulsifying capacity, interchain associations and interactions with other biomolecules. For the latter two, hydrogen bonding, hydrophobic interactions, electrostatic and ionic binding allow formation of either self-assembled structures or more complex functional entities (e.g., polyelectrolyte complexes, microgel particles, nanocomposites, nanoliposomes and nanofibers, emulsions, etc.), useful as delivery systems of other active substances or for some applications in the biomedical field and the food industry (Noreen et al. 2017; Gawkowska et al. 2018; Lu et al. 2019; Rehman et al. 2019). In a recent article, Einhorn-Stoll (2018) provided a thorough overview of pectin-water interactions as modulated by many intrinsic (molecular features of the polymer as well as intrinsic material properties, when it is used in a powder form) and extrinsic (pH, ionic environment, temperature, humidity, applied shear, etc.) factors and how they impact on pectin powder wetting-solubilization, thickening-gelation behaviour, water binding and water retention in pectin gels. Overall, the extraordinary diverse functionality of pectins, as biologically active natural polymers (involvement in plant protection mechanisms and as human health promoting bioactives with multiple activities, e.g., biosorbent, immunomodulating, prebiotic, antiproliferative, antioxidant, antimicrobial, hypocholesterolemic, hypoglycemic) as well as technologically important structuring agents and stabilisers of food and pharmaceutical formulations, is certainly related to their molecular structure variability (Voragen et al. 2009; Yapo and Gnakri 2015). To better understand the (bio)functionality of pectic polysaccharides, scientific elucidation of the structural components responsible for the beneficial effects in human nutrition or performance in technological applications of this biopolymer is thus essential.

6.2 Edible Films and Coatings from Pectins: Elaboration and Physical Properties

Edible films and coatings involving different polysaccharides, and pectins in particular, have been proposed in food related applications for preservation purposes, since they provide a barrier to oil, O_2, CO_2 and volatile (metabolites) losses, enhance product quality (reduction of physical damage and improvement of structural integrity, colour intensification, prevention of oxidation and off-flavors development) and inhibition of microbial growth, both for spoilage flora and pathogens (Campos et al. 2011). To differentiate between films and coatings, the former can be defined as a dried, self-standing biopolymer matrix, separately being added on the food product, like any conventional synthetic packaging material (as cover, wrapping, pouch, casing or separation layer in multi-component food products), whereas coatings involve the formation of a thin layer directly on the surface of the product and thus are regarded as part of the final product (Fig. 6.1); nevertheless, in both cases there is formation of a three-dimensional biopolymer structure upon solvent evaporation, involving closely packed chains which interact via non-covalent (mostly H-bonding, hydrophobic interactions, electrostatic and ionic bonding) or covalent linkages, depending on the primary structure of the biopolymer. For pectins, in particular, all these types of interactions are pertinent depending on the plant source (origin of pectin), the isolation procedure and other processing treatments involved. For example, the DM of pectins largely affects their gelling behaviour: LMP can gel in the presence of multivalent ions (Ca^{2+}), acting as bridges between pairs of unesterified carboxylate groups of adjacent polymer chains; i.e., electrostatic interactions between the cations and the negatively charged cavities formed by polysaccharide chains (COO^- groups), a mechanism known as 'egg-box'. The neighboring chains are also stabilised by van der Waals interactions and H-bonding. In contrast,

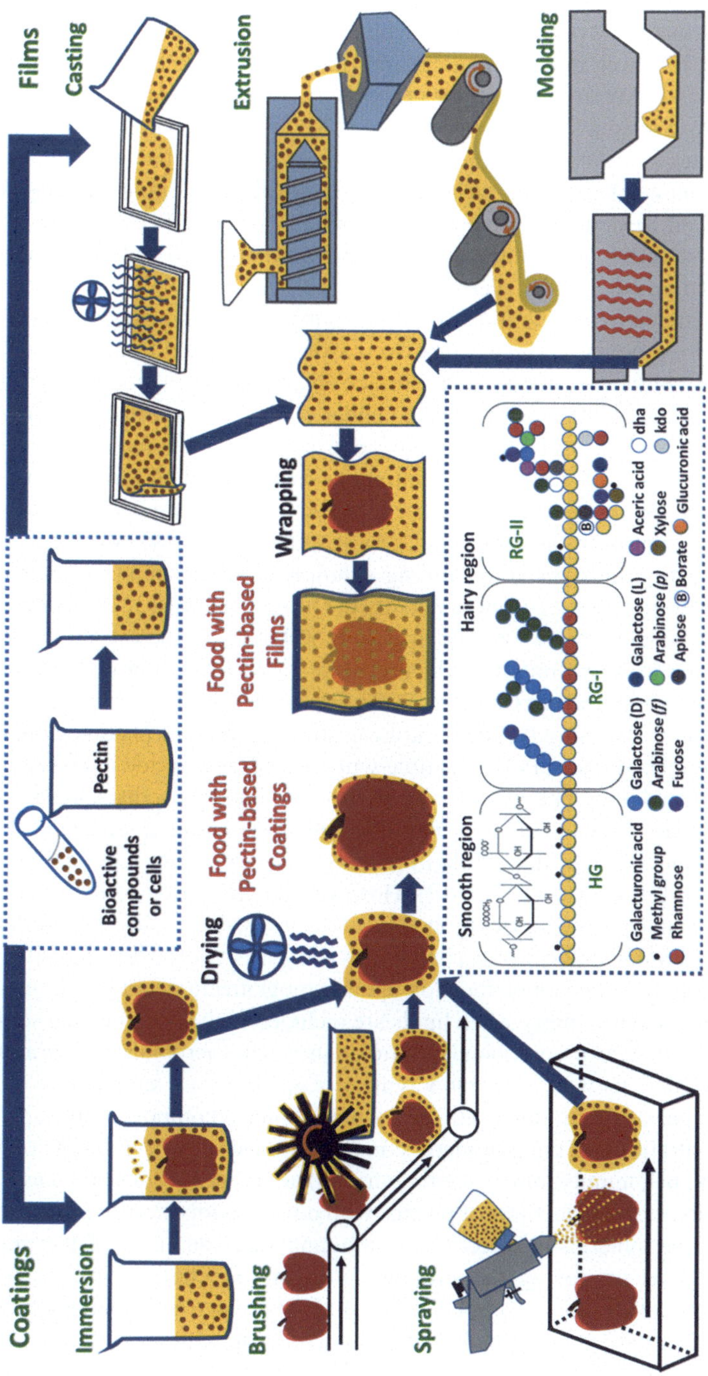

Fig. 6.1 A pictorial representation of methods employed for formation of pectin-based edible films and coatings properly designed to preserve food commodities. (left side) Immersion (dipping), brushing, or spraying of coating solutions to cover perishable products; (right side) casting, extrusion or thermo-compression molding of edible film formulations to develop self-standing film matrices as bio-based packaging materials

HMP form gels in acidic media with the addition of sugars; i.e., electrostatic repulsions are suppressed by lowering the pH, while water activity is reduced by the added co-solutes, resulting in chain aggregation (interchain H-bonding and hydrophobic interactions between methyl ester groups of the HG chain segments). The presence of feruloyl residues on some pectins (e.g., sugar beet and spinach contain ferulic acid residues in the arabinan side chains) also permits interchain associations by oxidative coupling of the phenolic groups (i.e., covalent bridging) leading to cross-linked network structures with high water holding capacity; this type of cross-linking is similar to that found for arabinoxylans present in the cell wall material of cereal grains (Izydorczyk and Biliaderis 1995).

Production methods for edible films from pectin materials can be accomplished by either wet or dry processes. The wet process (casting) involves the application of the polymer solution/dispersion onto a flat surface and removal of the solvent (usually water or mixtures of water with ethanol) under controlled conditions (drying) to form an edible film (Fig. 6.1). This process is convenient at a laboratory scale but not applicable for industrial scale-up. A continuous casting method (knife coating or doctor blade coating technique) can be used instead at an industrial scale, allowing more effective control of the film's thickness (Espitia et al. 2014). Moreover, in the continuous casting unit, the film forming solution (apple pectins) is partially dried first by an infrared heater and subsequently the film is dried at 132 °C (convective heating stage) under a high velocity air flow (1500 m/min); the overall drying process takes place in a few minutes which enables rapid formation of the films without microbial contamination (Du et al. 2008; Rossman 2009).

Other components that may be added in the casting solution of pectins include low molecular weight plasticisers (compatible with the polysaccharide matrix, e.g., glycerol, sorbitol, polyethylene glycol) which increase interchain distances (free volume) and increase chain segmental mobility; all these result in reducing the cohesion and firmness of the film, while improving its extensibility and flexibility (Biliaderis et al. 1999; Espitia et al. 2014). However, incorporation of these additives may bring changes in the barrier properties of the films, decreasing their ability to act as barriers to water vapor, aroma compounds and gases. Moreover, emulsifiers may be also used to facilitate emulsification upon homogenization and proper mixing of all ingredients, particularly when lipids are included in the film forming solution, as well as other active compounds (antioxidants, antimicrobials, colourants, etc.). Blending of pectins with other biopolymers (e.g., chitosan, starch, alginates, proteins) at different weight ratios is also feasible to produce composite films, provided that concentrations of the polymeric components and other conditions (solvent composition, pH, temperature of casting/drying) are properly chosen to ensure good compatibility. Composite films of pectins with other biopolymers or lipids, are designed to achieve synergistic effects by combining functionalities of the pure components and thus improving the structural integrity, mechanical and barrier properties. In this context, polyion complex formation (via electrostatic interactions between oppositely charged macromolecules) of pectins with chitosan (Jindal et al. 2013; Baron et al. 2017; Younis and Zhao 2019; Norcino et al. 2018), soy proteins (Pierro et al. 2005; Liu et al. 2007), gelatin (Liu et al. 2007; Farris et al. 2011) or caseinate (Eghbal et al. 2016, 2017) has been investigated to enhance the mechanical properties (increased tensile strength) and reduce water solubility of the composite pectin films. Greater improvement in these properties has been noted when crosslinking of the protein component in the composite film was carried out by either glutaraldehyde (Liu et al. 2007; Farris et al. 2011) or by transglutaminase treatment (Pierro et al. 2005). The latter approach was also successful in improving mechanical and barrier properties of composite pectin-protein films by cross-linking with microbial transglutaminase (formation of isopeptide bonds between glutamine and lysine residues) with phaseolin (Giosaffato et al. 2014), whey protein (Pierro et al. 2013) and bitter vetch (*Vicia ervilia*) proteins (Porta et al. 2016). Improved tensile strength, better UV/light barrier properties, compared to synthetic polyethylene film, and a lower WVP and solubility in water were also observed for films prepared from blends of low methoxy pectin and gluten (Sartori et al. 2018); these synergistic effects were attributed to electrostatic interactions between the polymeric components of the composite films, as revealed by FTIR spectroscopy. The physical properties of mixed alginate-LMP films, cross-linked with Ca^{2+} ions, have been also examined as a function of the polysaccharide ratio and plasticiser content (Galus and Lenart 2013; Silva et al. 2009; Makaremi et al. 2019). In a recent study by Kalathaki et al. (2019), metal-doped HMP films were prepared by a casting method, using pectin solutions at two pH (2.0 and 7.0) with inclusion of different metals (Na^+, K^+, Ca^{2+}, Mg^{2+}, Al^{3+}, as chloride salts), and examined for their thermal, mechanical and microstructural characteristics. The physical properties of the films were largely dependent on the pH of the casting solution (films made from the low pH solution exhibited higher strength), ion type (Ca^{2+} showed the greatest impact on film strengthening) and moisture content (water effectively plasticises the polymeric film matrix). The observed changes in film microstructure and physical properties were attributed to the modulation of electrostatic forces and interactions among the polymer chains, leading to variation in molecular packing and interchain associations in the film matrix. Cross-linking of polymeric mixtures made from high amylose starch and LMP in alkaline medium with sodium trimetaphosphate, as a preparatory step, allowed the elaboration of films by solvent casting which had improved mechanical properties and reduced WVP (Prezotti et al. 2012); these materials exhibited high resistance to enzymic digestion when exposed to simulated intestinal fluid, an important property for colon-targeted delivery systems. LMP-based composite films with included insoluble corn husk fiber particles (~53 μm) in the continuous pectin matrix, as filler microparticles, exhibited improved mechanical strength and decreased WVP compared to pure pectin films (Bernhardt et al. 2017). Similar findings were observed for HMP films reinforced with spent coffee grounds (SCG), added at 5–20% w/w pectin (Mendes et al. 2019); these films made by continuous casting exhibited improvement in thermal stability and reduction in water vapor transmission rate. Other related studies on mechanical properties, WVP, moisture sensitivity, structure analysis (X-ray diffractometry, SEM) and in vitro enzymatic digestion of cellulose nanobiber reinforced composite films of pectin (LMP) with resistant starch (retrograded high amylose corn starch) have been carried out by Meneguin et al. (2014, 2017); depending on composition of the film forming dispersion (polymeric materials and plasticisers used) such nanocomposite films can become effective vehicles for colonic delivery of drugs or other bioactives.

In the area of nanocomposite materials, the inclusion of chitosan nanoparticles (<100 nm in diameter) into HMP and LMP film matrices made by solution casting improved the mechanical properties (tensile strength) of both composite films, whereas it did not largely affect their WVP (Lorevice et al. 2016). Similar works on composite films of pectin and fruit purees with inclusion of chitosan nanoparticles (~90 to 100 nm) (Martelli et al. 2013; Melo et al. 2019) or nanofibrillated bacterial cellulose (Viana et al. 2018), prepared by casting, also revealed the positive impact of nanoparticles on film strengthening and improvement of barrier properties, e.g., reduced WVP. Preparation methods and characterization (by thermogravimetry, calorimetry, scanning and transmission electron microscopies, atomic force microscopy, X-ray diffraction analysis, and tensile strength responses) of bio-nanocomposite edible films made with crystalline nanocellulose (CNC) particles and HMP (Chaichi et al. 2017) or LMP cross-linked with Ca^{2+} ions and glycerol (Chaichi et al. 2019) were recently reported. According to the findings from these studies, inclusion of 5% CNC into the HMP matrix significantly improves the tensile strength, raises the glass transition temperature (T_g) and enhances the water resistance of the composite film (Chaichi et al. 2017). Moreover, crosslinking of LMP pectin with Ca^{2+} cations, along with incorporation of nanocellulose into the composite film matrix leads to the formation of more flexible and completely water insoluble pectin nanocomposite film in comparison to the crosslinked LMP simply with Ca^{2+} cations and glycerol (Chaichi et al. 2019). The WVP of films made from HMP derived from orange and mango peel was reduced by 30–60%, as measured under various environmental conditions (temperature range 5–35 °C and under different relative humidity gradients), when silicon dioxide (SiO_2) nanoparticles (5–15 nm) were included in the film casting solutions (at 1.3% wt solution solids), compared to pectin films without nanoparticles (Salazar et al. 2019); i.e., SiO_2 nanoparticles interspersed between pectin chains can form a more tortuous path for water molecules to diffuse through the composite film matrix. A similar study by Dash et al. (2019) focused on the production and properties of biodegradable starch-pectin-titanium oxide nanoparticles (0.5–4% w/w TiO_2). Significant decreases in the moisture content, solubility and moisture uptake of the composite films were noted when the TiO_2 particles were incorporated into the mixed polymeric matrix. Moreover, the T_g shifted to higher temperatures, the thermal stability and mechanical strength of the films increased, whereas the moisture barrier properties were slightly improved.

A recent study by Manrich et al. (2017) focused on properties of hydrophobic edible films made from tomato peel-extracted cutin (a biopolyester complex lipid mixture) and a commercial LMP by continuous casting (fast process) to obtain kinetically-stable film forming dispersions; the composite hydrophobic films exhibited increased water resistance. Composite films of pectin with shellac (at different polymer weight ratios) were made by solution casting using two plasticisers at a concentration of 10% w/w film solids, polyethylene glycol, PEG 400 (hydrophilic), and diethyl phthalate, DEP (hydrophobic) (Luangtana-anan et al. 2017); the results indicated that incorporation of both plasticisers could enhance the mechanical characteristics of the composite films, particularly at a pectin/shellac ratio of 50:50, largely reduce the hydrophilicity of single pectin-based films and enhance their moisture protective properties. Overall, in order to develop composite films with pectins and other ingredients, it is necessary to explore synergistic effects among the various components employed in the film formulation and maximise end-use performance (physical properties, quality improvement of the product) via optimization studies as has been recently performed for edible films based on pectin/alginate/whey protein concentrate blends (Chakravartula et al. 2019).

An alternative approach to edible film production involves extrusion processing, similar to what is used in commercial processing of thermoplastic materials (Fig. 6.1); i.e., mixing ingredients in a barrel, with one or multiple screws, in order to forward the ingredients under shear, increased pressure and temperature throughout the different extruder zones towards a die where expansion may take place along with solvent evaporation, and thereby solidification of the composite film matrix. Extrusion processes have been employed by Fishman and co-workers to fabricate edible films from pectin blends with other polymeric materials (e.g., starch, polyvinyl alcohol) including glycerol as plasticiser (Fishman et al. 2000, 2004, 2006); the authors have concluded that the mechanical properties and plasticisation behaviour of the extruded composite films are comparable to those obtained by solution casting. However, as pointed out by these authors as well as in other reports (Liu et al. 2005; Espitia et al. 2014), the successful application of extrusion depends upon several variables that should be controlled during the extrusion process; i.e., material composition (particularly when biopolymer blends are used), plasticiser type and content, feed rates, screw configuration, screw speed, temperatures in different zones of the barrel, inlet/outlet pressures and die configuration. For pectin materials, Liu et al. (2005) have optimised some of these parameters as follows: temperature level two (extruder zone 3 (125 °C) and zone 4 (110 °C)), feeder speed (18.49 g/min), glycerol concentration (50%) and screw speed (225 rpm) in order to produce the best film in terms of tensile strength, elongation, puncture resistance, colour, Young's modulus, tear resistance, turbidity and thickness. An alternative approach to develop biodegradable films from pectin is by thermo-compression molding (Fig. 6.1), using HMP and different plasticisers (alone or in combination to create a deep eutectic mixed solvent, e.g., choline chloride/glycerol), as was recently presented (Gouveia et al. 2019). The effect of processing parameters (120 °C, compression load 196 kN and compression times, 20–25 min) along with the plasticiser content and composition on film properties and microstructure were examined. The mechanical properties of the thermoplastic pectin films were influenced by the compression time and the type of plasticiser used; however, due to their high WVP, applications of these films would be rather limited.

Edible coatings represent a thin layer of edible material which adheres as a protective film/layer/cover on the surface of a product (Fig. 6.1) to improve quality and extend shelf life by controlling microbial, chemical and physicochemical deteriorative processes occurring in foods; i.e., edible coatings create a semi-permeable barrier on the surface of food commodities, restricting gas and moisture transfer rates, chemical reactions (e.g., oxidation) as well as microbial contamination and spoilage of perishable products. Besides the biopolymeric constituents, other additives (plasticisers, emulsifiers, cross-linking agents, antimicrobials, antioxidants, texture modifiers, nutrients, etc.) are often incorporated into edible coating formulations to maximise the beneficial impact these materials have on retention of product quality and nutritional value. Depending on the product (consistency–structure, surface morphology and wettability) and physical properties of the coating solution/dispersion (solvent type, viscosity, surface tension, etc.), three different coating methods can be used (Valdes et al. 2015; Saha et al. 2017; Dhumal and Sarkar 2018): dipping (10 s to 2 min, particularly when the coating solution is highly viscous), spreading or brushing, and spraying (Fig. 6.1). Spraying is carried out in a classical way with a spray gun (controlling nozzle type and temperature, air and liquid flow rates, humidity on incoming air and polymer solution rheology are all critical parameters affecting the coating efficiency for food products); the most usual spraying devices create fine sprays with droplet size ~20 μm. Lately, electrohydrodynamic processes (e.g., electrospraying) have been developed and are finding applications in the food industry for the production of micro and nanosised particles, in which an ultrathin polymeric film (coating wall material) is formed around the core substances to encapsulate them (Drosou et al. 2017); in this technique, electrically charged jets from viscoelastic coating solutions are being forced into a high-voltage electric field to evaporate the solvent without application of a heat treatment. A special variation in the dipping method may be also used, involving layer-by-layer development of multiple coatings on a product surface; i.e., using electrostatic forces between oppositely charged polymeric species (Dhumal and Sarkar 2018). A similar approach can be seen in the case of applying a 'hardening' solution on top of the first coating liquid layer; e.g., when sodium alginate solution is first spread by brushing on the product surface, then a $CaCl_2$ solution is applied by spraying to fix the alginate coating (Gonzales-Forte et al. 2014). As for edible films, optimization of coating formulations (ingredient composition, coating layer thickness, wettability, pleasant taste, etc.) and processing-storage conditions, using appropriate experimental designs and response surface methodologies, is necessary to improve their suitability for specific applications; this is particularly important when coatings are applied on respiring vegetable and fruit tissues to manage water loss, O_2 and CO_2 levels, and thereby control metabolism of the plant cells, while maintaining the integrity of the coating matrix material under the high relative humidity environment of fresh produce.

6.3 Pectin-Based Films and Coatings in Food Preservation

Biopolymer-based edible films and coatings are often fabricated as carriers of active compounds, including different antimicrobials, antioxidants, flavorings, colouring agents, micronutrients and bioactives, etc. (Fig. 6.1) (Coma et al. 2002; Acevedo-Fani et al. 2015; Mellinas et al. 2016; Valdes et al. 2015). Such biomaterials may control microbial contamination and growth, enhance nutritional value and sensory properties, reduce oxidation rates and discolouration, improve barrier properties, modify the structural integrity and texture of the product, and thereby extend its shelf-life and ensure food safety. The primary advantage of 'active' films and coatings is that the incorporated inhibitory or quality promoting agents can be specifically targeted to act on the food surface where most post-processing modifications (deterioration) take place, e.g., oxidation, microbial contamination and growth; for antimicrobials, in particular, the main advantage is that the active compounds will be mostly present at the surface to act on spoilage flora or foodborne pathogens, without the need to use excessive concentrations of the antimicrobial(s) throughout the entire product mass (Campos et al. 2011). The diffusion of the active substances through an edible film or coating is influenced by its composition and manufacturing procedure, intrinsic factors related to the food matrix (pH, water activity, porosity), molecular size and hydrophilic/hydrophobic balance of the active compound(s), as well as storage conditions (temperature, relative humidity environment). Furthermore, the migration rates of the active compound(s) of the film within the product matrix, are controlled by the chemical nature of the compound, the molecular interactions with other food constituents present and the composition-microstructure of the food product itself (Cagri et al. 2004; Campos et al. 2011; Dhumal and Sarkar 2018). Moreover, the efficacy of active films, e.g., those containing antimicrobials, is often tested using laboratory media and not in situ as a means to control microbial activity in actual products; apparently, food constituents might interact with the antimicrobials and thereby modulate their activity. All these issues need to be considered in order to properly design effective edible films and coating systems with desired functionalities for improving food preservation and safety.

Several studies on production and the use of active films based on pectins (HMP) or pectin-containing plant products (e.g., fruit, vegetable and hibiscus purees) with several natural antimicrobials (e.g., carvacrol, cinnamaldehyde, the main ingredients of oregano and cinnamon oil, respectively) have revealed the efficacy of these materials (dose-dependent bactericidal activity) to reduce growth of foodborne pathogens (e.g., *E. coli* O157:H7, *S. enterica*, *Campylobacter jejuni* and *Listeria monocytogenes*) on meat and poultry products (Ravishankar et al. 2009, 2012; Mild et al. 2011). Results showed that edible films containing carvacrol exhibited stronger activity against *S. enterica* and *E. coli* on chicken breast surface at 4 °C than those edible films made with cinnamaldehyde; similarly, edible films with carvacrol exhibited stronger activity against *L. monocytogenes* on ham (Ravishankar et al. 2009). In contrast, for *Campylobacter jejuni*, Mild et al. (2011) showed that the pectin-based films with inclusion of cinnamaldehyde were more effective against growth of different strains of the pathogen on chicken breast compared with the carvacrol containing films. The authors have attributed such differentiation in bactericidal activity to the different mechanisms by which these compounds operate at a cellular level; carvacrol is reported to disrupt bacterial cell membranes by non-covalent interactions, whereas cinnamaldehyde reacts chemically with the membrane surfaces resulting in cell death. These studies have thus confirmed previous findings in vitro (Rojas-Grau et al. 2006, 2007; Du et al. 2008, 2009a, b) for pectin material-based films containing low levels of plant essential oils (cinnamon, lemongrass, oregano). The effect of cinnamaldehyde or carvacrol incorporated in apple or tomato-pectin based edible films on baked chicken was assessed by sensory analysis (Du et al. 2012); adding carvacrol or cinnamaldehyde at low levels (0.5% w/w) in the formulation of tomato-pectin edible films did not seem to affect the preference of baked chicken when compared to the control. In a recent study by Nisar et al. (2018) integration of clove bud essential oil (CEO) into composite films, as assisted by Tween 80 (added at 0.2% v/v of CEO), was elaborated by adding 0.5–1.5% CEO oil dispersion in a citrus pectin film forming solution. The emulsified pectin films with CEO exhibited improved water barrier properties, resistance to breaking and higher thermal stability than control films. Moreover, using the agar disc-diffusion assay, antimicrobial efficiency was demonstrated against common food-borne pathogens, particularly the gram positive *S. aureus* and *L. monocytogenes*. The effectiveness of citrus pectin coatings enriched with clove essential oil was also investigated as preservation strategy for bream (*Megalobrama ambycephala*) fillets during refrigeration over a period of 15 days (Nisar et al. 2019). The inclusion of CEO reduced lipid oxidation and extended the shelf life of fillets by at least 15 days; moreover, the pectin coatings with CEO improved the weight loss, water holding capacity, textural and colour attributes of the beam samples as well as they were effective in decreasing the growth rate of many microbial groups, including psychrophilic bacteria. Orange oil loaded (at 3% w/w) mixed pectin films (LMP commercial product with mango pectin) demonstrated antibacterial activity against *S. aureus* by the agar disc-diffusion test (Chaiwarit et al. 2018). In a related study, enhancement of the antibacterial activity (against *S. aureus* and *Propionibacterium acnes*) was noted when the orange oil was added in the form of microemulsion (Tween 80, propylene glycol as co-surfactants) into the pectin solution (LMP) and thin films were subsequently prepared by casting of the dispersions using ionotropic gelation with Ca^{2+} ions (Jantrawut et al. 2018); moreover, the films loaded with orange oil microemulsion exhibited a dense matrix with negligible micropores, together with high limonene loading content, and showed improved film flexibility and higher percent elongation. Using curcumin nanoemulsions/pectin coatings prepared with cinnamon and garlic essential oils, Abdou et al. (2018) improved the microbiological and sensorial quality, delayed the appearance of microbial spoilage, reduced the oxidation rate and retarded the deterioration of texture of refrigerated chicken fillets, increasing their shelf life up to 12 days. The preparation and characterisation (FT-IR. XRD, Field Emission SEM) of pectin films activated by marjoram (*Origanum majorana* L.) essential oil–loaded nanoemulsions (NE) and Pickering emulsions (PE) were recently described by Almasi et al. (2020). The films containing PE had good mechanical and water barrier properties due to their highly dense and less permeable structure. However, the antioxidant activity of the films containing PE was significantly lower than that of their NE counterparts. Consistent with the antioxidant activity trends, the encapsulation of essential oil with PE provided a slower release profile in the film samples, compared to the NE, implying a greater potential for the use of the former films in active food packaging applications where a prolonged release of the antimicrobial compounds is being targeted. Apparently, there is an increasing number of studies in the area of nanoencapsulation of bioactive compounds, for protection and/or targeted delivery, where pectins have been investigated as wall materials, as has been recently reviewed by Rehman et al. (2019); in this context, pectins are being used for formation of nanocapsules (in the form of nanocomplexes, nanohydrogels, pectin-coated nanoliposomes and nanoemulsions), either alone or in combination with other compounds such as proteins and lipids.

Novel bioactive films based on pectin and cellulose (PC) with different loadings of tea polyphenols and cinnamaldehyde were prepared by Ye et al. (2019). The thermal stability (tested by thermogravimetry) decreased slightly after loading with cinnamaldehyde and tea polyphenols, compared to pure PC films. The composite PC films had good DPPH radical and hydroxyl radical scavenging activities and excellent antibacterial activities against *Escherichia coli*, *Candida albicans* and *Staphylococcus aureus*. High methoxy pectin–konjac glucomannan composite edible films containing tea polyphenols (1–5% w/w of dry film basis) showed improved physical properties (increased tensile strength and decreased WVP for up to 2% fortification level with the tea polyphenols) and enhanced antimicrobial (against *S. aureus* and *E. coli*, using the well diffusion technique) and antioxidant activities (Lei et al. 2019). Other reports demonstrating the antimicrobial activity of essential oils in pectin coating formulations include those for cinnamon leaf oil in preserving quality of fresh-cut peach (decreased bacterial growth, *E. coli* O157:H7, *S. aureus*, *L. monocytogenes*, as well as improved antioxidant status of the fruit tissue; Ayala-Zavala et al. 2013) and grapes (significant inhibition of fungal decay and increased antioxidant activity; Melgarejo-Flores et al. 2013). Similar antifungal activity (against *Alternaria alternata* growth) has been reported for pectin edible coatings carrying oregano essential oil when applied to tomatoes (Rodriguez-Garcia et al. 2015); moreover, the treated tomatoes had acceptable flavor and other sensorial characteristics. The topic of pectin/cinnamon oil antimicrobial packaging has been recently reviewed by Gutierrez-Pacheco et al. (2016), whereas that of edible antimicrobial films and coatings by Raybaudi-Massilia et al. (2016).

The preparation and antimicrobial activity of an extruded composite film containing pectin/polylactic acid and nisin were studied by Jin et al. (2009a). The nisin containing composite films (500 IU nisin/mg of film) were fairly effective in controlling the growth of *L. monocytogenes* in a typical growth medium as well as in liquid food products (e.g., orange juice and liquid egg). Polysaccharide composite films (pullulan/pectin) with inclusion of silver nanoparticles exhibited strong antimicrobial activity against foodborne pathogens, especially *S. typhimurium*, *E. coli*, *L. monocytogenes*. Moreover, the presence of the silver nanoparticles (~400 to 500 nm) slightly improved the mechanical properties of the composite films and increased their WVP. The physicochemical, microbiological and sensorial qualities of cooked pork patties coated with pectin-based material containing green tea leaf extract powder were studied upon storage for 2 weeks at 10 °C (Kang et al. 2007). Cooked pork patties were separated into three groups: uncoated control (C), coated with pectin-based materials (CP), and coated with pectin-based materials containing 0.5% green tea powder (CGP), and were all irradiated (0 and 3 kGy using cobalt-60 gamma rays). Coated patties contained higher moisture contents than the controls (in both air- and under vacuum packaging). The lipid oxidation decreased ($p < 0.05$) and radical scavenging increased ($p < 0.05$) in the CGP or CP pork patties, relative to the controls, when the samples were vacuum packaged; moreover, the CGP coated patties exhibited the lowest growth of bacteria upon storage. The antimicrobial activity of pectin edible films containing nisin and its combination with treatment of ionizing radiation was studied to control *Listeria monocytogenes* on a ready-to eat (RTE) turkey meat product (Jin et al. 2009b). The combined treatments of pectin/nisin films with the irradiation (1–2 kGy) were more effective in reducing the *L. monocytogenes* counts, than the pectin/nisin film alone, implying a synergistic effect for *Listeria* growth inhibition on the surface of RTE turkey meat.

Edible active coatings based on pectins or pectin containing plant materials (purees, pomace, juice), either alone or in combination with other biobased ingredients or treatments (technologies), are employed to retain freshness and extend the shelf life of other highly perishable products, such as fruit and vegetables, as reviewed by Lin and Zhao (2007), Valdes et al. (2015) and Otoni et al. (2017). For these respiring commodities, particularly their fresh-cut products, the traditional methods of low temperature preservation to reduce microbial load and growth, tissue respiration rate and other metabolic activities (enzymic deteriorative processes) are often combined with modified atmosphere packaging (MAP), short thermal treatments, UV or pulsed light, high pressure and other emerging technologies, irradiation and coatings, as a holistic preservation approach in the context of multiple hurdle technology (Rico et al. 2007; Nieto 2009; Zhang et al. 2014; Valdes et al. 2015). For pectin-based coatings, LMP preparations are often employed because of their ability to form stronger network structures in the presence of Ca^{2+} at the low pH of the fresh-cut produce, promoting higher firmness and structural integrity of the product, while reducing at least in part WVP. Coatings, being semi-permeable layers, in addition to providing barrier properties (O_2 and CO_2 exchange, moisture and solute migration) they can be also used as carriers of important preservation additives (antimicrobials, antioxidants) or vitamins and other bioactives; i.e., 'active' coatings. However, coatings can sometimes cause anaerobic respiration and off-flavor development, making necessary the optimization of the formulation to get a balance between gas and water vapor transmission rates. Examples of more recently published works on preservation-extension of shelf life of highly perishable fruit include the following applications: a HMP/beeswax/monoglyceride coating on mango reduced the rate of colour development, weight loss, acid production, texture softening and CO_2 evolution, extending the shelf life from 1–2 weeks (Moalemiyan et al. 2011); a HMP/sorbitol/beeswax/monoglyceride coating applied on avocado restricted the loss of moisture during storage, decreased the respiration rate, delayed the occurrence of the climacteric peak, reduced the intensity of the respiratory peak and delayed the changes in texture and colour, with an overall extension in shelf life to over a month on storage at 10 °C (Maftoonazad and Ramaswamy 2008); application of an edible coating based on HMP, chitosan or pullulan, containing Na-benzoate and K-sorbate as antimicrobials, on strawberries, improved sensory quality, reduced weight loss, delayed alteration of colour, reduced microbial load-growth and decay rate as well as maintained the ascorbic acid content of strawberries throughout storage (Trevino-Garza et al. 2015); pectin or alginate coating solutions, enriched with essential oil constituents, at different concentrations of the polysaccharides (1–2%), citral or eugenol (0.1–0.2%) and ascorbic acid (1%), used as an anti-browning agent, were applied on raspberries and proved to be effective in reducing microbial spoilage and preserving the general sensory attributes of the fruit (the 1% pectin coating solution with 0.1% citral was the optimum formulation for treating fresh fruit and storage for at least 7 days at 0.5 °C, Guerreiro et al. 2015); application of composite edible coatings based on HMP, corn flour and with or without a beetroot powder preparation on tomatoes reduced respiration rate, weight loss and post-harvest decay (Sucheta et al. 2019). Examples for similar approaches in extending shelf life and preserving quality attributes of fresh-cut fruit and vegetables are the following: for fresh-cut papaya, a multilayer active coating prepared by a layer-by-layer deposition process, applying first a chitosan layer by dipping in a chitosan solution containing a complex of β-cyclodextrin with trans-cinnamaldehyde, 2 g/100 g, as antimicrobial, then followed by a LMP pectin layer and finally with immersion into a $CaCl_2$ solution, reduced the microbial contamination and the losses of vitamin C and total carotenoids (overall, the multilayer coating doubled the shelf life of the fresh fruit cuts up to 15 days on storage at 4 °C, Brasil et al. 2012); for fresh-cut cantaloupe, the multilayer active coating from the previous study was equally effective for shelf life extension of the product stored at 4 °C (Martinon et al. 2014); for fresh-cut melon, a LMP based coating, cross-linked with Ca^{2+} ions, improved water vapor resistance, prevented dehydration and maintained the initial firmness during storage (15 days at 4 °C). However, this coating did not reduce the microbiological growth and neither affected the CO_2 and O_2 diffusion rates, but exhibited an inhibitory effect on ethylene production, while maintaining the antioxidant properties and the sensory attributes of the product (Oms-Oliu et al. 2008a); in a similar study, for pieces of pears, LMP coatings solutions, 2% w/v, were first applied and then dipped into a $CaCl_2$ solution (2% w/v), containing N-acetylcysteine, 0.75% w/v, and glutathione, 0.75% w/v, as antibrowning agents. This coating helped to control enzymatic browning and maintain a high antioxidant potential as well as retard the microbial growth, without affecting firmness for 14 days at 4 °C (Oms-Oliu et al. 2008b); for fresh-cut melon, the combination of osmotic dehydration first, in 40° Bx sucrose solution containing 0.5% calcium lactate, followed by coating with LMP pectin (1%) was a good preservation alternative for this product. Apparently, the combined treatment reduced the fruit respiration rate, while maintaining the organoleptic properties and other quality parameters for 14 days of storage at 5 °C and 80% RH (Ferrari et al. 2013); for watermelon, multilayer coatings were successfully prepared with the layer-by-layer technique by using a solution of sodium alginate containing one antimicrobial agent, trans-cinnamaldehyde encapsulated in β-cyclodextrin, in combination with LMP pectin and calcium lactate solutions to obtain stable and homogeneous coatings (Sipahi et al. 2013). The multilayer coatings with 1% alginate, 2% β-CD-cinnamaldehyde and 2% pectins maintained the quality and organoleptic attributes of fresh-cut watermelon, while extending the shelf life at 4 °C from 7 (control) to 12 days; for fresh-cut apples, LMP pectin-based edible coatings, enriched with apple fiber, as well as the application of pulsed light treatments, as a 'hurdle' approach, have been effective to minimize browning and softening of apples, along with reduction of microbial counts, ~2 log CFU/g. However, in this case, with all coating formulations there were off-flavor notes, limiting the acceptability of the pectin-coated samples (Moreira et al. 2017); in a similar study by the same group of authors (Moreira et al. 2015), fresh-cut apples were coated with a LMP preparation enriched with apple fiber together with $CaCl_2$ and ascorbic acid. This coating positively influenced the sensory characteristics, colour and firmness of the product, thereby increasing the nutritional value and maintaining the antioxidant potential for 7 days of storage at 4 °C, without compromising their fresh-like quality attributes; for fresh-cut apples, potatoes, carrots, a composite edible coating was applied which was based on a mixed solution of pectin (LMP)/whey protein, 1/4 w/w, cross-linked by transglutaminase (Marquez et al. 2017). This coating material was effective to avoid spoilage of the fresh-cut samples during ten days of storage (4 °C), as demonstrated by reduction of weight loss, microbial growth prevention, antioxidant activity and no change in fruit and vegetable hardness and chewiness.

Polysaccharide based edible coatings, including those with pectin, because of their hydrophilic nature they tend to reduce lipid uptake in fried-food products as well as to lessen the moisture loss of products during frying, while maintaining acceptable organoleptic characteristics of the product (Albert and Mittal 2002). Edible coatings from low-methoxyl sunflower head pectin (SFHP) were applied to reduce lipid uptake in fried potato chips; formation of pectin coatings was induced by calcium cations within several seconds prior to frying. The results indicated that the coating prepared with 1.0% (w/v) SFHP and 0.05 mol/l $CaCl_2$ can reduce the lipid uptake by about 30% in comparison with the uncoated chips (Hua et al. 2015). In another study, among several polysaccharide coating formulations, a mixture of CMC and pectin (0.5 and 1% w/w) produced fried french-fries with a low oil uptake, while the moisture content of the final products was higher than the uncoated samples (Garmakhany et al. 2014). The use of hydrocolloids (pectin, CMC, alginate) to decrease oil absorption in fried banana chips was also studied (Singthong and Thongkaew 2009); among the hydrocolloids tested, the best results were obtained from the banana chips blanched in 0.5 g $CaCl_2$/100 ml distilled water and treated with a solution of 1 g pectin/100 ml distilled water. With this coating treatment, there was a reduction of oil content from 40.22 g/100 g sample (untreated) to 22.89 g/100 g. The pectin coated chips also exhibited the highest sensory scores in all attributes compared to the untreated product. Scanning electron microscopy photographs indicated that coating banana chips with calcium chloride solution and the solution of pectin, was quite effective in protecting the cellular structure of banana tissue from damage during deep-fat frying. In a similar study, hydrocolloid- based materials (pectin, whey protein and soy protein isolate) were used as coatings to reduce fat uptake during deep-frying of a pre-fried, frozen product preformed from cassava flour or cassava puree (Freitas et al. 2009); the whey proteins showed the best results with respect to fat absorption, presenting a reduction of ~27% for the cassava puree and flour products, whereas the pectin coated products exhibited the highest moisture content after the final frying. In another study, the application of pectin-based edible coating was shown to reduce to around 33.0% the acrylamide content in the processed products, along with a slight improvement of the sensory properties of the fried banana chips (Suyatma et al. 2015); moreover, a greater acrylamide reduction was noted up to 91.9 and 90.8%, for the samples pre-treated by edible coating and then followed by blanching at 90 and 100 °C, respectively. All these findings clearly point to the effectiveness of coatings based on pectin and other hydrocolloids to reduce the lipid uptake and acrylamide content in deep fried potato and banana chips. A novel LMP (2% w/v)–honey (10% w/w added to the pectin solution) based coating was recently developed and tested as a method for dehydrating fruit cuts (apple, cantaloupe, melon, mango), while preserving their quality and safety by exploiting the antimicrobial activity of honey (Santagata et al. 2018); a higher dehydration efficiency (at 4 °C and RH of 40%) was noted with application of the coating solution, if compared to control. Following dehydration, the coated samples exhibited higher polyphenol concentration, antioxidant activity, vitamin C content and showed improved sensorial characteristics (flavour) and microbiological quality, compared to the uncoated dehydrated fruit samples.

There is an increasing interest for the incorporation of plant extracts from agricultural by-products, as sources of antioxidants and antimicrobials, into bio-based edible films for improving their physical properties and bioactivities. Antioxidant films were recently prepared by integrading mature coconut water and lime peel extract, as natural antioxidants, into lime peel pectin films (Rodsamran and Sothornvit 2019). The film water barrier was improved by the combination of coconut water added as a plasticiser and the lime peel extract; as more lime peel extract was included, the films also showed increased total phenolics content and antioxidant activities. Moreover, the bioactive lime pectin films, with or without the lime peel extract, formed as sachets, were quite effective in delaying lipid oxidation of soybean oil during storage for 30 days (at 25 °C, 50% RH). This study indicated that lime peel residues are a good source for pectin production and a natural antioxidant extract; these materials can be combined to develop active packaging films or coatings for various food applications to extend shelf life of food products. Another example of biopolymer-based composite films and coatings, with incorporation of natural antioxidants, are edible pectin (citrus HMP)-fish gelatin films containing the phenols hydroxytyrosol (HT) and 3,4-dihydrophenylglycol (DHPG), two naturally present antioxidants in the olive fruit and leaves (Bermudez-Oria et al. 2017, 2019). The bioactive films exhibited good barrier properties (WVP, O_2), while the mechanical properties of the films were not affected by the inclusion of the two phenols. The respective bioactive edible coating was effective in delaying mold growth on strawberries during storage at room temperature and thus extending the shelf-life of fresh strawberries (Bermudez-Oria et al. 2017). In a follow-up study (Bermudez-Oria et al. 2019), the same group of researchers showed that meat samples wrapped with a film containing HT and DHPG reduced the formation of oxidation products in the form of thiobarbituric acid reaction substances (TBARS) compared with the control film without antioxidants. The HT added at 0.5% in the film with beeswax suppressed the lipid oxidation of beef meat during 7 days of storage at 4 °C, possibly by the combined effect of acting as an oxygen barrier and a specific antioxidant activity due to the HT. Active films made with these natural phenolic antioxidants might have a double positive impact on treated products; one on improving the oxidation stability of the meat fat itself, and a second related to several health benefits for the consumer. For the latter, it is well known that the HT is conferring many beneficial physiological activities on plasma lipoproteins, platelet and cellular function, as well as on bone health. As a result, EFSA has authorised a health claim for olive oil polyphenols, standardised by their content of hydroxytyrosol and its derivatives, on maintenance of normal blood HDL-cholesterol concentrations (European Commission Regulation EC No. 432/2012 2012).

6.4 Conclusions

Pectins are multifunctional polysaccharides widely used in the food industry as thickening and gelling agents, stabilisers and emulsifiers as well as in edible films or coatings as matrix structuring agents. Depending on the plant origin and processing treatments employed during their isolation from the raw plant materials, pectin chains exhibit rather diverse chemical features and therefore they are able to associate with similar (self-interacting) or other biopolymer molecules via covalent, ionic or cation-mediated, and non-ionic interactions (H-bonding, hydrophobic interactions, van der Waals forces), forming network structures, such as gels or films. As film former biopolymers, pectins, being primarily hydrophilic as many other polysaccharides, are excellent barriers of gases exchange (O_2, CO_2, ethylene), but display rather poor water barrier properties which can be improved by the addition of hydrophobic additives (e.g., lipids), in the form of emulsion systems or as bilayers. Moreover, to further resemble the physical properties of conventional packaging films, pectin-based composite materials can be elaborated by inclusion of other biopolymers (proteins, polysaccharides) in the film forming solution/dispersion either as co-solubilised ingredients or in the form of micro-/nanoparticles (cellulose fibrils, chitosan nanoparticles, etc.). Incorporation of other active ingredients (antimicrobials, antioxidants, etc.) into the pectin-based films or coatings, i.e., create 'active' packaging systems, provide further advantages for control release of the (bio) active(s) and a more efficient preservation or formulation strategy for enhancement of the nutritional quality and shelf life of fresh and processed food items. Nevertheless, further research and development is needed in the area of biopolymer-based films and coatings, when including pectins or other polysaccharides as structuring components, to produce these materials at an industrial level and transfer up-scale laboratory results to actual industrial practice. At an even more fundamental level, there is a need to better understand the underlying mechanisms of film composition-structure-processing-property relationships of these functional materials in order to improve their effectiveness and broaden their applicability as quality improving edible ingredients in the food processing and preservation arena.

References

Abdou ES, Galhoum GF, Mohamed EN (2018) Curcumin loaded nanoemulsions/pectin coatings for refrigerated chicken fillets. Food Hydrocoll 83:445–453

Acevedo-Fani A, Salvia-Trujillo L, Rojas-Grau MA, Martin-Belloso O (2015) Edible films from essential-oil-loaded nanoemulsions: physicochemical characterization and antimicrobial properties. Food Hydrocoll 47:168–177

Albert S, Mittal GS (2002) Comparative evaluation of edible coatings to reduce fat uptake in a deep-fried cereal product. Food Res Int 35:445–458

Almasi H, Azizi S, Amjadi S (2020) Development and characterization of pectin films activated by nanoemulsion and Pickering emulsion stabilized marjoram (Origanum majorana L) essential oil. Food Hydrocoll 99:105338

Ayala-Zavala JF, Silva-Espinoza BA, Cruz-Valenzuela MR, Leyva JM, Ortega-Ramirez LA, Carrazco-Lugo DK, Perez-Carlon JJ, Melgarejo-Flores BG, Gonzalez-Aguilar GA, Miranda MRA (2013) Pectin-cinnamon leaf oil coatings add antioxidant and antibacterial properties to fresh-cut peach. Flavour Fragance J 28:39–45

Baldwin EA, Nisperos MO, Baker RA (1995) Use of edible coatings to preserve quality of lightly and slightly processed products. Crit Rev Food Sci Nutr 35:509–524

Baron RD, Perez LL, Salcedo JM, Cordoba LP, Sobral PJDA (2017) Production and characterization of films based on blends of chitosan from blue crab (Callinectes sapidus) waste and pectin from orange (Citrus sinensis Osbeck) peel. Int J Biol Macromol 98:676–683

Bermudez-Oria A, Gutierrez GR, Vioque B, Rubio-Senent F, Fernandez-Bolanos J (2017) Physical and functional properties of pectin-fish gelatin films containing the olive phenols hydroxytyrosol and 3,4-dihydroxyphenylglycol. Carbohydr Polym 178:368–377

Bermudez-Oria A, Gutierrez GR, Rubio-Senent F, Fernandez-Prior A, Fernandez-Bolanos J (2019) Effect of edible pectin-fish gelatin films containing the olive antioxidants hydroxytyrosol and 3,4-dihydroxyphenylglycol on beef meat during refrigerated storage. Meat Sci 148:213–218

Bernhardt DC, Perez CD, Fissore EN, De'Nobili MD, Rojas AM (2017) Pectin-based composite film: effect of corn husk fiber concentration on their properties. Carbohydr Polym 164:13–22

Biliaderis CG, Lazaridou A, Arvanitoyannis I (1999) Glass transition and physical properties of polyol-plasticized pullulan-starch blends at low moisture. Carbohydr Polym 40:29–47

Brasil IM, Gomes C, Puerta-Gomez A, Castell-Perez MRG (2012) Polysaccharide-based multilayered antimicrobial edible coating enhances quality of fresh-cut papaya. LWT Food Sci Technol 47:39–45

Cagri A, Ustunol Z, Ryser E (2004) Antimicrobial edible films and coatings. Review. J Food Prot 67(4):833–848

Campos CA, Gerschenson LN, Flores SK (2011) Development of edible films and coatings with antimicrobial activity. Food Bioprocess Technol 4(6):849–875

Chaichi M, Hashemi M, Badii F, Mohammadi A (2017) Preparation and characterization of a novel bionanocomposite edible film based on pectin and crystalline nanocellulose. Carbohydr Polym 157:167–175

Chaichi M, Badii F, Mohammadi A, Hashemi M (2019) Water resistance and mechanical properties of low methoxy-pectin nanocomposite film responses to interactions of Ca2+ ions and glycerol concentrations as crosslinking agents. Food Chem 293:429–437

Chaiwarit T, Ruksiriwanich W, Jantanasakulwong K, Jantrawut P (2018) Use of orange oil loaded pectin films as antibacterial material for food packaging. Polymers 10:1144

Chakravartula SSN, Soccio M, Lotti N, Balestra F, Rosa MD, Siracusa V (2019) Characterization of composite edible films based on pectin/alginate/whey protein concentrate. Materials 12:2454

Coma V, Martial-Gros A, Garreau S, Copinet A, Salin F, Deschamps A (2002) Edible antimicrobial films based on chitosan matrix. J Food Sci 67(3):1162–1169

Costa MJ, Cerqueira MA, Ruiz HA, Fougnies C, Richel A, Vicente AV, Teixeira JA, Aguedo M (2015) Use of wheat bran arabinoxylans in chitosan-based films: Effect on physicochemical properties. Ind Crop Prod 66:305–311

Dash KK, Ali NA, Das D, Mohanta D (2019) Thorough evaluation of sweet potato starch and lemon-waste pectin based-edible films with nano-titania inclusions for food packaging applications. Int J Biol Macromol 139:449–458
Dhumal CV, Sarkar P (2018) Composite edible films and coating from food-grade biopolymers. J Food Sci Technol 55(11):4369–4383
Drosou CG, Krokida MK, Biliaderis CG (2017) Encapsulation of bioactive compounds through electrospinning/electrospraying and spray drying: a comparative assessment of food-related applications. Dry Technol 35(2):139–162
Du W-X, Olsen CW, Avena-Bustillos RJ, McHugh TH, Levin CE, Friedman M (2008) Storage stability and antibacterial activity against Escherichia coli O157:H7 of carvacrol in edible apple films made by two different casting methods. J Agric Food Chem 56:3082–3088
Du W-X, Olsen CW, Avena-Bustillos RJ, McHugh TH, Levin CE, Friedman M (2009a) Effects of allspice, cinnamon, and clove bud essential oils in edible apple films on physical properties and antimicrobial activities. J Food Sci 74(7):373–378
Du W-X, Olsen CW, Avena-Bustillos RJ, McHugh TH, Levin CE, Mandrell R, Friedman M (2009b) Antibacterial effects of allspice, garlic, and oregano essential oils in tomato films determined by overlay and vapor-phase methods. J Food Sci 74(7):390–397
Du W-X, Avena-Bustillos RJ, Woods R, Breksa AP, McHugh TH, Friedman M, Levin CE, Mandrell R (2012) Sensory evaluation of baked chicken wrapped with antimicrobial apple and tomato edible films formulated with cinnamaldehyde and carvacrol. J Agric Food Chem 60(32):7799–7804
Dutta PK, Tripathi S, Mehrotra GK, Dutta J (2009) Perspectives for chitosan based antimicrobial films in food applications. Food Chem 114(4):1173–1182
Eghbal N, Yarmand MS, Mousavi M, Degraeve P, Oulahal N, Gharsallaoui A (2016) Complex coacervation for the development of composite edible films based on LM pectin and sodium caseinate. Carbohydr Polym 151:947–956
Eghbal N, Degraeve P, Oulahal N, Yarmand MS, Mousavi M, Gharsallaoui A (2017) Low methoxyl pectin/sodium caseinate interactions and composite film formation at neutral pH. Food Hydrocoll 69:132–140
Einhorn-Stoll U (2018) Pectin-water interactions in foods - from powder to gel. Food Hydrocoll 78:109–119
Espitia PJP, Du W-X, Avena-Bustillos RJ, Soares NFF, McHugh TH (2014) Edible films from pectin: physical-mechanical and antimicrobial properties - a review. Food Hydrocoll 35:287–296
European Commission Regulation EC No. 432/2012 (2012) Establishing a list of permitted health claims made on foods, other than those referring to the reduction of disease risk and to children's development and health. J Eur Union 136:1–40
Farris S, Schaich KM, Liu L, Cooke PH, Piergiovanni L, Yam KL (2011) Gelatin-pectin composite films from polyion-complex hydrogels. Food Hydrocoll 25:61–70
Ferrari CC, Sarantópoulos CIGL, Carmello-Guerreiro SM, Hubinger MD (2013) Effect of osmotic dehydration and pectin edible coatings on quality and shelf life of fresh-cut melon. Food Bioprocess Technol 6:80–91
Fishman ML, Coffin DR, Konstance RP, Onwulata CI (2000) Extrusion of pectin/starch blends plasticized with glycerol. Carbohydr Polym 41:317–325
Fishman ML, Coffin DR, Onwulata CI, Konstance RP (2004) Extrusion of pectin and glycerol with various combinations of orange albedo and starch. Carbohydr Polym 57:401–413
Fishman ML, Coffin DR, Onwulata CI, Willet JL (2006) Two stage extrusion of plasticized pectin/poly(vinyl alcohol) blends. Carbohydr Polym 65:421–429
Freitas DGC, Berbari SAG, Prati P, Fakhouri FM, Queiroz FPC, Vicente E (2009) Reducing of fat uptake in cassava product during deep-fat frying. J Food Eng 94:390–394
Galus S, Lenart A (2013) Development and characterization of composite edible films based on sodium alginate and pectin. J Food Eng 115:459–465

Garmakhany AD, Mirzaei HO, Maghsudlo Y, Kashaninejad M, Jafari SM (2014) Production of low fat french-fries with single and multi-layer hydrocolloid coatings. J Food Sci Technol 51(7):1334–1341

Gawkowska D, Cybulska J, Zdunek A (2018) Structure-related gelling of pectins and linking with other natural compounds: a review. Polymers 10:762

Gialamas H, Zinoviadou KG, Biliaderis CG, Koutsoumanis KP (2010) Development of a novel bioactive packaging based on the incorporation of Lactobacillus sakei into sodium caseinate films for controlling Listeria monocytogenes in foods. Food Res Int 43(10):2402–2408

Giosaffato CVL, Pierro PD, Gunning P, Mackie A, Porta R, Mariniello L (2014) Characterization of Citrus pectin edible films containing transglutaminase-modified phaseolin. Carbohydr Polym 106:200–208

Gonzales-Forte L, Bruno E, Martino M (2014) Application of coating on dog biscuits for extended survival of probiotic bacteria. Anim Feed Sci Technol 195:76–84

Gouveia TIA, Biernacki K, Castro MCR, Concalves MP, Souza HKS (2019) A new approach to develop biodegradable films based on thermoplastic pectin. Food Hydrocoll 97:105175

Guerreiro AC, Gago CML, Faleiro ML, Miguel MGC, Antunes MDC (2015) Raspberry fresh fruit quality as affected by pectin- and alginate-based edible coatings enriched with essential oils. Sci Hortic 194:138–146

Gutierrez-Pacheco MM, Ortega-Ramirez LA, Cruz-Valenzuela MR, Silva-Espinoza BA, Gonzalez-Aguilar GA, Ayala-Zavala JF (2016) In antimicrobial food packaging. Academic, New York, pp 609–617

Hansen NML, Plackett D (2008) Sustainable films and coatings from hemicelluloses: a review. Biomacromolecules 9(6):1493–1505

Hassan B, Chatha SAS, Hussain AI, Zia KM, Akhtar N (2018) Recent advances on polysaccharides, lipids and protein based edible films and coating: a review. Int J Biol Macromol 109:1095–1107

Hua X, Wang K, Yang R, Kang J, Yang H (2015) Edible coatings from sunflower head pectin to reduce lipid uptake in fried potato chips. LWT- Food Sci Technol 62:1220–1225

Izydorczyk MS, Biliaderis CG (1995) Cereal arabinoxylans: advances in structure and physicochemical properties. Carbohydr Polym 28:33–48

Jantrawut P, Boonsermsukcharoen K, Thipnan K, Chaiwarit T, Hwang K-M, Park E-S (2018) Enhancement of antibacterial activity of orange oil in pectin thin film by microemulsion. Nanomaterials 8:545

Jiménez A, Fabra M, Talens P, Chiralt A (2012) Edible and biodegradable starch films: a review. Food Bioprocess Technol 5(6):2058–2076

Jin T, Liu L, Zhang H, Hicks K (2009a) Antimicrobial activity of nisin incorporated in pectin and polylactic acid composite films against Listeria monocytogenes. Int J Food Sci Technol 44:322–329

Jin T, Liu L, Sommers CH, Boyd G, Zhang H (2009b) Radiation sensitization and post irradiation proliferation of Listeria monocytogenes on ready-to-eat deli meat in the presence of pectin-nisin films. J Food Prot 72(3):644–649

Jindal M, Kumar V, Rana V, Tiwary AK (2013) An insight into properties of Aegle marmelos pectin-chitosan cross-linked films. Int J Biol Macromol 52:77–84

Kalathaki I, Alba K, Muhamedsalih H, Kontogiorgos V (2019) Fabrication and characterisation of metal-doped pectin films. Food Hydrocoll 92:259–266

Kang HJ, Jo C, Kwon JH, Kim JH, Chung HJ, Byun MW (2007) Effect of a pectin-based edible coating containing green tea powder on the quality of irradiated pork patty. Food Control 18:430–435

Kester JJ, Fennema OR (1986) Edible films and coatings: a review. Food Technol 40(12):47–59

Kristo E, Biliaderis CG, Zampraka A (2007) Water vapour barrier and tensile properties of composite caseinate-pullulan films: Biopolymer composition effects and impact of beeswax lamination. Food Chem 101:753–764

Lacroix M (2009) Mechanical and permeability properties of edible films and coatings for food and pharmaceutical applications. In: Embuscado ME, Huber KC (eds) Edible films and coatings for food applications. Springer, New York, pp 347–366
Lei Y, Wu H, Jiao C, Jing Y, Liu R, Xiao D, Lu J, Zhang Z, Shen G, Li S (2019) Investigation of the structural and physical properties, antioxidant and antimicrobial activity of pectin-konjac glucomannan composite edible films incorporated with tea polyphenol. Food Hydrocoll 94:128–135
Lin D, Zhao Y (2007) Innovations in the development and application of edible coatings for fresh and minimally processed fruits and vegetables. Compr Rev Food Sci Food Saf 6(3):60–75
Liu L, Kerry JF, Kerry JP (2005) Selection of optimum extrusion technology parameters in the manufacture of edible/biodegradable packaging films derived from food-based polymers. J Food Agric Environ 3:51–58
Liu L, Liu C-K, Fishman ML, Hicks KB (2007) Composite films from pectin and fish skin gelatin or soybean flour protein. J Agric Food Chem 55:2349–2355
Lorevice MV, Otoni CG, Moura MR, Mattoso LHC (2016) Chitosan nanoparticles on the improvement of thermal, barrier, and mechanical properties of high- and low-methyl pectin films. Food Hydrocoll 52:732–740
Lu X, Chen J, Guo Z, Zheng Y, Rea MC, Su H, Zheng X, Zheng B, Miao S (2019) Using polysaccharides for the enhancement of functionality of foods: a review. Trends Food Sci Technol 86:311–327
Luangtana-Anan M, Soradech S, Saengsod S, Nunthanid J, Limmatvapirat S (2017) Enhancement of moisture protective properties and stability of pectin through formation of a composite film: effect of shellac and plasticizer. J Food Sci 82:2915–2925
Maftoonazad M, Ramaswamy HS (2008) Effect of pectin-based coating on the kinetics of quality change associated with stored avocados. J Food Process Preserv 32:621–643
Makaremi M, Yousefi H, Cavallaro G, Lazzara G, Goh CBS, Lee SM, Solouk A, Pasbakhsh P (2019) Safely dissolvable and healable active packaging films based on alginate and pectin. Polymers 11:1594
Manrich A, Moreira FKV, Otoni CG, Lorevice MV, Martins MA, Mattoso LHC (2017) Hydrophobic edible films made up of tomato cutin and pectin. Carbohydr Polym 164:83–91
Marquez GR, Di Piero P, Mariniello L, Esposito M, Giosafatto CVL, Porta R (2017) Fresh-cut fruit and vegetable coatings by transglutaminase-crosslinked whey protein/pectin edible films. LWT- Food Sci Technol 75:124–130
Martelli MR, Barros TT, Moura MR, Mattoso LHC, Assis OBG (2013) Effect of chitosan nanoparticles and pectin content on mechanical properties and water vapor permeability of banana puree films. J Food Sci 78(1):98–104
Martinon ME, Moreira RG, Castell-Perez EC, Gomes C (2014) Development of a multilayered antimicrobial edible coating for shelf-life extension of fresh-cut cantaloupe (*Cucumis melo L.*) stored at 4oC. LWT- Food Sci Technol 56(2):341–350
Melgarejo-Flores BG, Ortega-Ramirez LA, Silva-Espinoza BA, Gonzalez-Aguilar GA, Miranda MRA, Ayala-Zavala JF (2013) Antifungal protection and antioxidant enhancement of table grapes treated with emulsions, vapors, and coatings of cinnamon leaf oil. Postharvest Biol Biotechnol 86:321–328
Mellinas C, Valdes A, Ramos M, Burgos N, Garrigos MC, Jimenez A (2016) Active edible films: current state and future trends. J Appl Polym Sci 133:42631
Melo PTS, Nunes JC, Otoni CG, Aouada FA, Moura MR (2019) Combining cupuassu (Theobroma grandiflorum) puree, pectin, and chitosan nanoparticles into novel edible films for food packaging applications. J Food Sci 84(8):2228–2233
Mendes JF, Martins JT, Manrich A, Sena Neto AR, Pinheiro ACM, Mattoso LHC, Martins MA (2019) Development and physical-chemical properties of pectin film reinforced with spent coffee grounds by continuous casting. Carbohydr Polym 210:92–99
Meneguin AB, Cury BSF, Evangelista RC (2014) Films from resistant starch-pectin dispersions intended for colonic drug delivery. Carbohydr Polym 99:140–149

Meneguin AB, Cury BSF, Santos AM, FrancoDF BHS, Filho EC (2017) Resistant starch/pectinfre-standing films reinforced with nanocellulose intended for colonic methotrexate release. Carbohydr Polym 157:1013–1023

Mild R, Joens LA, Friedman M, Olsen CW, McHugh TH, Law B, Ravishankar S (2011) Antimicrobial sdible apple films inactivate antibiotic resistant and susceptible Campylobacter jejuni strains on chicken breast. J Food Sci 76:163–168

Moalemiyan M, Ramaswamy HS, Maftoonazad N (2011) Pectin-based edible coating for shelf-life extension of Ataulfo Mango. J Food Process Eng 35:572–600

Moreira MR, Cassani L, Martín-Belloso O, Soliva-Fortuny R (2015) Effects of polysaccharide-based edible coatings enriched with dietary fiber on quality attributes of fresh-cut apples. J Food Sci Technol 52(12):7795–7805

Moreira MR, Alvarez MV, Martin-Belloso O, Soliva-Fortuny R (2017) Effects of pulsed light treatments and pectin edible coatings on the quality of fresh-cut apples: a hurdle technology approach. J Sci Food Agric 97:261–268

Nieto MB (2009) Structure and function of polysaccharide gum-based edible films and coatings. In: Embuscado ME, Huber KC (eds) Edible films and coatings for food applications. Springer, New York, pp 57–112

Nisar T, Wang Z-C, Yang X, Tian Y, Igbal M, Guo Y (2018) Characterization of citrus pectin films integrated with clove bud essential oil: physical, thermal, barrier, antioxidant and antibacterial properties. Int J Biol Macromol 124:1156–1168

Nisar T, Yang X, Alim A, Igbal M, Wang Z-C, Guo Y (2019) Physicochemical responses and microbiological changes of bream (Megalobrama ambycephala) to pectin based coatings enriched with clove essential oil during refrigeration. Int J Biol Macromol 106:670–680

Norcino LB, Oliveira JE, FKV M, Marconcini JM, Mattoso LHC (2018) Rheological and thermo-mechanical evaluation of bio-based chitosan/pectin blends with tunable ionic cross-linking. Int J Biol Macromol 118:1817–1823

Noreen A, Nazli Z-I-H, Akram J, Rasul I, Mansha A, Yaqoob N, Igbal R, Tabasum S, Zuber M, Zia KM (2017) Pectins functionalized biomaterials: a new viable approach for biomedical applications: a review. Int J Biol Macromol 101:254–272

Oms-Oliu G, Soliva-Fortuny R, Martín-Belloso O (2008a) Using polysaccharide-based edible coatings to enhance quality and antioxidant properties of fresh-cut melon. LWT- Food Sci Technol 41:1862–1870

Oms-Oliu G, Soliva-Fortuny R, Martín-Belloso O (2008b) Edible coatings with antibrowning agents to maintain sensory quality and antioxidant properties of fresh-cut pears. Postharvest Biol Biotechnol 50:87–94

Otoni CG, Avena-Bustillos RJ, Azeredo HMC, Lorevice MV, Moura MR, Mattoso LHC, McHugh TH (2017) Recent advances on on edible films based on fruits and vegetables – a review. Compr Rev Food Sci Food Saf 16:1151–1169

Pierro PD, Mariniello L, Glosafatto CVL, Masi P, Porta R (2005) Solubility and permeability properties of edible pectin-soy flour films obtained in the absence or presence of transglutaminase. Food Biotechnol 19:37–49

Pierro PD, Marquez GR, Mariniello L, Sorrentino A, Villalonga R, Porta R (2013) Effect of transglutaminase on the mechanical and barrier properties of whey protein/pectin films prepared at complexation pH. J Agric Food Chem 61:4593–4598

Porta R, Pierro PD, Sabbah M, Regalado-Gonzales C, Mariniello L, Kadivar M, Arabestani A (2016) Blend films of pectin and bitter vetch (*Vicia ervilia*) proteins: properties and effect of transglutaminase. Innovative Food Sci Emerg Technol 36:245–251

Prezotti FG, Meneguin AB, Evangelista RC, Cury BSF (2012) Preparation and characterization of free films of high amylose/pectin mixtures cross-linked with sodium trimetaphoshate. Drug Dev Ind Pharm 38(11):1354–1359

Ravishankar S, Zhu L, Olsen C, McHugh T, Friedman M (2009) Edible apple film wraps containing plant antimicrobials inactivate foodborne pathogens on meat and poultry products. J Food Sci 74(8):440–445

Ravishankar S, Jaroni D, Zhu L, Olsen C, McHugh T, Friedman M (2012) Inactivation of Listeria monocytogenes on ham and bologna using pectin-based apple, carrot, and hibiscus edible films containing carvacrol and cinnamaldehyde. J Food Sci 77(7):377–382

Raybaudi-Massilia R, Mosqueda-Melgar J, Soliva-Fortuny R, Martin-Belloso O (2016) Combinational edible films and coatings. In: Barros-Velazquez J (ed) Antimicrobial food packaging. Academic, New York, pp 633–646

Rehman A, Ahmad T, Aadil RM, Spotti MJ, Bakry AM, Khan IM, Zhao L, Riaz T, Tong Q (2019) Pectin polymers as wall materials for the nano-encapsulation of bioactive compounds. Trends Food Sci Technol 90:35–46

Rico D, Martin-Diana AB, Barat JM, Barry-Ryan C (2007) Extending and measuring the quality of fresh-cut fruit and vegetables: a review. Trends Food Sci Technol 18:373–386

Rodriguez-Garcia I, Cruz-Valenzuela MR, Silva-Espinoza BA, Gonzalez-Aguilar GA, Moctezuma E, Gutierrez-Pacheco MM, Tapia-Rodriguez MR, Ortega-Ramirez LA, Ayala-Zavala JF (2015) Oregano (Lippia graveolens) essential oil added within pectin edible coatings prevents fungal decay and increases the antioxidant capacity of treated tomatoes. J Sci Food Agric 96(11):3772–3778

Rodsamran P, Sothornvit R (2019) Lime peel pectin integrated with coconut water and lime peel extract as a new bioactive film sachet to retard soybean oil oxidation. Food Hydrocoll 97:105173

Rojas-Grau MA, Avena-Bustillos RJ, Friedman M, Henika PR, Martin-Belloso O, McHugh TH (2006) Mechanical barrier and antimicrobial properties of apple puree edible films containing plant essential oils. J Agric Food Chem 54(24):9262–9267

Rojas-Grau MA, Avena-Bustillos RJ, Olsen C, Friedman M, Henika PR, Martin-Belloso O, Pan Z, McHugh TH (2007) Effect of plant essential oils and oil compounds on mechanical, barrier and antimicrobial properties of alginate-apple puree edible films. J Food Eng 81(3):634–641

Rossman JM (2009) Commercial manufacture of edible films. In: Embuscado ME, Huber KC (eds) Edible films and coatings for food applications. Springer, New York, pp 367–390

Saha A, Tyagi S, Gupta RK, Tyagi YK (2017) Natural gums of plant origins edible coatings for food industry applications. Crit Rev Biotechnol 37(8):959–973

Salazar ASS, Cavazos PAS, Paz HM, Fragoso AV (2019) External factors and nanoparticles effect on water vapor permeability of pectin-based films. J Food Eng 245:73–79

Santagata G, Mallardo S, Fasulo G, Lavermicocca P, Valerio F, Di Biase M, Di Stasio M, Malinconico M, Volpe MG (2018) Pectin-honey coating as novel dehydrating bioactive agent for cut fruit. Food Chem 258:104–110

Sartori T, Feltre G, Sobral PJDA, Cunha RLD, Menegalli FC (2018) Properties of films produced from blends of pectin and gluten. Food Packag Shelf Life 18:221–229

Schols HA, Voragen AGJ (1996) Complex pectins: structure elucidation using enzymes. In: Pectins and pectinases. Elsevier, London, pp 3–19

Silva MA, Bierhalz AC, Kieckbusch TG (2009) Alginate and pectin composite films crosslinked with Ca^{2+} ions: effect of the plasticizer concentration. Carbohydr Polym 77:736–742

Singthong J, Thongkaew C (2009) Using hydrocolloids to decrease oil absorption in banana chips. LWT- Food Sci Technol 42:1199–1203

Sipahi RE, Castell-Perez ME, Moreira RG, Gomes C, Castillo A (2013) Improved multilayered antimicrobial alginate-based edible coating extends the shelf life of fresh-cut watermelon (*Citrullus lanatus*). LWT- Food Sci Technol 51:9–15

Sucheta CK, Sharma N, Yadav SK (2019) Composite edible coatings from commercial pectin, corn flour and beetroot powder minimize post-harvest decay, reduces ripening and improves sensory liking of tomatoes. Int J Biol Macromol 133:284–293

Suyatma NE, Ulfah K, Prangdimurti E, Ishikawa Y (2015) Effect of blanching and pectin coating as pre-frying treatments to reduce acrylamide formation in banana chips. Int Food Res J 22(3):936–942

Trevino-Garza MZ, Garcia S, Flores-Gonzalez MS, Arevalo-Nino K (2015) Edible active coatings based on pectin, pullulan, and chitosan increase quality and shelf life of strawberries (Fragaria ananassa). J Food Sci 80(8):1823–1829

Valdes A, Burgos N, Jimenez A, Garrigos MC (2015) Natural pectin polysaccharides as edible coatings. CoatingsTech 5:865–886

Viana RM, Sa NM, Barros MO, Borges MF, Azeredo HMC (2018) Nanofibrillated bacterial cellulose and pectin edible films added with fruit purees. Carbohydr Polym 196:27–32

Voragen AGJ, Coenen GJ, Verhoef RP, Schols HA (2009) Pectin, a versatile polysaccharide present in plant cell walls. Struct Chem 20:263–275

Yapo BM, Gnakri D (2015) Pectic polysaccharides and their functional properties. In: Ramawat KG, Merillon J-M (eds) Polysaccharides. Springer, New York, pp 1729–1749

Ye S, Zhu Z, Wen Y, Su C, Jiang L, He S, Shao W (2019) Facile and green preparation of pectin/cellulose composite films with enhanced antibacterial and antioxidant behaviors. Polymers 11:57

Younis HGR, Zhao G (2019) Physicochemical properties of the edible films from the blends of high methoxyl apple pectin and chitosan. Int J Biol Macromol 131:1057–1066

Zhang Y, Rempel C, McLaren D (2014) Edible coating and film materials: carbohydrates. In: Han JH (ed) Innovations in food packaging, 2nd edn. Academic, New York, pp 305–323

Chapter 7
Pectin Gelation and Its Assembly into Functional Materials

M. A. K. Williams

7.1 Introduction

7.1.1 A Soft Matter Exemplar

Over the last few decades the appreciation of the importance of soft condensed matter, both in the biological and physical sciences, has continued to grow (Witten 1999; Poon and Andelman 2006; van der Gucht 2018). Pectin gels offer a fascinating soft matter exemplar for study (Owens et al. 1954; Thakur et al. 1997; Willats et al. 2006; Ström et al. 2007; Munarin et al. 2012a, b). From the physical science perspective the decoration of the pectic components yields multiple different interactions that can be tuned spatially and temporally through fine structure modifications and the physiochemical environment in order to manipulate the properties of assembled macroscopic matrices. From a biological perspective pectin networks in the plant cell wall (Jarvis 1984; Levesque-Tremblay et al. 2015; Braybrook and Peaucelle 2013) co-evolved with other life on Earth so that not only plants but also bacteria, fungi and insects offer a host of enzymatic tools for pectin modification and processing (Rexová-Benková and Markoviĉ 1976; Pilnik and Voragen 1993; Schols and Voragen 1996; Alkorta et al. 1998; Kent et al. 2016). From a materials engineering perspective, learning how to make materials and devices from pectic substrates holds the promise of harnessing their biocompatible, reconfigurable and sustainable nature, as we turn our sights toward a zero-waste society (Di Giacomo

M. A. K. Williams (✉)
School of Fundamental Sciences, Massey University, Palmerston North, New Zealand

The MacDiarmid Institute for Advanced Materials and Nanotechnology, Wellington, New Zealand

The Riddet Institute, Massey University, Palmerston North, New Zealand
e-mail: m.williams@massey.ac.nz

V. Kontogiorgos (ed.), *Pectin: Technological and Physiological Properties*,
https://doi.org/10.1007/978-3-030-53421-9_7

et al. 2015; Cifarelli et al. 2016; Sun et al. 2017, 2018; Mohammadinejad et al. 2019; Vanitha and Khan 2019; Thakur et al. 2019).

7.1.2 Evolved Pectin Networks

Pectin networks pervade the biosphere on planet Earth. They are the epitome of smart responsive matrices whose properties are controlled in order to fulfil vital physiological roles, in particular in the plant kingdom (Jarvis 1984; Levesque-Tremblay et al. 2015; Braybrook and Peaucelle 2013). The evolution of the fine structures of these key biological polyelectrolytes has led to a variety of enzymatically modifiable components, (predominantly homo- and rhamno-galacturonans), that can occur in different ratios and with various patterns of substituent groups and sidechains, and can be linked chemically and physically to themselves and to other cell wall components (Thibault and Ralet 2008; Mohnen 2008; Sila et al. 2009; Round et al. 2010; Voragen et al. 2013; Anderson 2019). The sheer amount of genes annotated as coding for pectin-modifying enzymes in plant genomes (Henrissat et al. 2001), and their presence in the genomes of other organisms from every branch of the tree of life (fungi, insects and bacteria) speaks volumes for the importance of pectin in the natural world (Pelloux et al. 2007; Kent et al. 2016). Its key role in controlling the mechanics and porosity of the plant cell wall also makes it a key player to understand in attempts to develop solutions to our burgeoning energy crisis based on the exploitation of biomass (Edwards and Doran-Peterson 2012; Xiao and Anderson 2013).

7.1.3 Current Human Exploitation

While undoubtedly the pragmatic human exploitation of pectin dates back into the mists of time, with notable mentions of jam-recipes in Roman cookery books from the fourth century (Apicius and Alföldi-Rosenbaum 1958), its modern use, primarily as a gelling agent, dates to the beginning of the nineteenth century (Braconnot 1825). Today the extraction, modification and supply of pectin as a food ingredient, as a stabiliser or a gelling agent, is commonplace. However, only a small amount of the potential of pectin for producing smart controllable matrices, as demonstrated in the natural world, has been exploited to date. With pectin assembly traditionally taking place in acidic high-sugar content environments or in biomimetic calcium-binding systems, where the carboxyl groups of the galacturonic acid moieties play a major role, homogalacturonan-rich fractions naturally dominate the current supply for use in technologies exploiting pectin gelation.

7.1.4 Unlocking the Potential

As alluded to above, Nature uses pectin in a more sophisticated manner than has been exploited to date in technological settings, performing 'bottom-up' processing with aplomb, assembling structures at the molecular level in order to yield materials with desired macroscopic attributes. Naturally occurring pectin biomaterials are 'smart', changing their properties in response to external stimuli, and are processed at ambient temperatures from sustainable resources, before being seamlessly recycled. Eager to construct high-tech materials and devices that also display these properties, the first demonstrations of the use of pectin networks in a number of somewhat less-than-traditional applications have recently appeared, including the manufacture of memresistive components and high performance temperature sensors (Di Giacomo et al. 2015; Cifarelli et al. 2016; Sun et al. 2017, 2018). It is clear however that in order to be able to imitate Nature's molecular wizardry in constructing optimised materials and devices, and truly exploit pectin's potential as a state-of-the-art materials-engineering substrate, the structure-function relationships it exploits must first be understood in more detail. Firstly, the measurement of the molecular structure of extracted pectin polymers, in particular how the different pectic polysaccharides are connected and the patterning of substituent groups is itself a substantial challenge. Furthermore, the importance of long-range electrostatic repulsions between chains in governing the complex solution state structure and dynamics in systems of these biological polyelectrolytes must be appreciated. Finally, once assembly is initiated, the relevance of phenomena as varied as counterion condensation, the non-trivial role of specific ion interactions, the polymer physics of viscoelasticity and glassy worm-like chains, and the response of non-equilibrium systems to thermal or stress-driven perturbations, means that the mission to illuminate structure-function relationships in pectin gels is both exciting and challenging in equal measure.

7.2 Making Pectin Gels

7.2.1 Different Types of Junction Zones

As alluded to in the introduction, pectic polymers can be assembled into networks in many different ways, using interactions between different substituent groups, that respond differently to changes in the physiochemical environment, yielding a complex responsive web of interactions. The connections formed between different chains can be chemical or physical in nature, with the former more often of the point-like type and the latter often relying on a minimum number of concerted groups to be involved in order to overcome thermal fluctuations that typically overwhelm physical bonds consisting of single ion-mediated groups or hydrogen bonds. Such junction zones might be seen as local crystallites of limited extent that tie the separate chains together into a single space-spanning macromolecule that can support system-wide stress transmission pathways and control the dynamics of entrained solvent. The formation of large length-scale junction-zone arrays, potentially exhibiting a propensity to precipitate, is typically prevented either by the way substituent groups are patterned, generating contiguous runs of interacting and inert residues, or simply by the topological trapping and dynamical frustration instigated by previously formed connections preventing potentially cross-linkable sections of chain subsequently encountering each other.

In a biological context having different accessible interactions with differing dependencies on environmental conditions provides enough redundancy to ensure mechanical homeostasis where required, while in a technological setting it provides a host of opportunities to select specific gel types (described below) suitable for applications in different physiochemical environments.

7.2.1.1 Ionotropic Gels

Ionotropic gels are routinely formed by exploiting the interactions that exist between divalent cations and pectins that consist predominantly of homogalacturonan, either of a low degree of methylesterification (DM) or of a higher DM but where the unmethylesterifed (charged) regions are arranged in a blockwise pattern along the polymer backbone, at pH values where the carboxyl acid moiety of the galacturonic acid residues are charged (Morris et al. 1982; Powell et al. 1982; Willats et al. 2006; Ström et al. 2007; Mansel and Williams 2015). Calcium is typically the ion exploited in the natural world, and the interplay between calcium concentration and enzyme-induced demethylesterification observed in the cell wall of the growing pollen tube provides a beautiful example of the mechanical control of growth using this kind of network (Chebli and Geitmann 2007; Zerzour et al. 2009). It is worth noting that while pectin-methylesterases (PMEs) are multitudinous and varied (Kent et al. 2016), most plant or bacterial PMEs produce blocky and thereby calcium sensitive regions of contiguous charged residues upon removing the methylesters that are carried by something of the order of eighty percent of the HG carboxyl groups when biosynthesised (Owen et al. 2017). From a biophysical point of view, bacterial and plant PMEs are of particular interest for this ability to catalyse consecutive reactions along the HG chain before dissociating from it. Such processivity requires the upgrade of thermal energy available in the form of Brownian fluctuations into mechanical work. Such upgrades are of fundamental interest and have been widely studied, for example in the cases of the canonical ATP-powered actin- and microtubule-associated linear molecular motors or the ion-gradient-powered rotary motors Fo-F1-ATP synthase. Of particular interest here is that bacterial and plant PMEs do not require ATP hydrolysis or the relaxation of ion gradients to function; with the chemical free energy that rectifies the Brownian fluctuations carried instead by removable substituents on the substrate itself (Mercadante et al. 2014).

The formation of junction-zones with divalent ions has been modelled in silico with increasing precision, with models progressively revealing improvements in the original "egg-box" model (first suggested for alginate), and capturing ion-specific and solvent-mediated effects (Grant et al. 1973; Braccini and Pérez 2001; Li et al. 2007; Sikorski et al. 2007; Plazinski 2011; Borgogna et al. 2013; Assifaoui et al. 2015). The amount of ion-binding groups on the polymer, the ion type, and their relative concentrations are crucial parameters affecting the outcome of the gel assembly process. The latter has traditionally been characterised by a so-called "reactivity ratio" or R-value, defined as the concentration of divalent ions (+2) to the concentration of pairs of carboxyl acid groups ($2 \times -1 = -2$), so that R = 1 corresponds to every ion having the potential to bind two sugar residues and gives a first back-of-the-envelope estimate of the amount of ions that saturates the system. In reality, the fact that a minimum number of contiguous charged (unmethylesterifed) residues are required to form stable junctions zones (the order of ten at room temperature) (Powell et al. 1982; Liners et al. 1992), means that the intramolecular patterning of the polymer substrate is crucial and that R = 1 is always going to be an overestimate of the amount of ions required for system saturation—particularly for higher DM samples with randomly distributed unmethylesterified residues. For lower DM random substrates or for substrates where the majority of charged residues occur in blocks longer than those required to form the minimum stable junction zone, then the R value is more likely to be a parameter that is at least roughly proportional to the number of junction-zones formed, and thereby proportional to the elastic modulus in the simplest of mechanical models. But even in these cases the kinetic trapping and slowing dynamics concurrent with the development of the network means that many potential junction zones cannot form and many ion-binding groups counted in the R ratio will never form stress-bearing connections in the gel. While some attempts to formulate an effective reactivity ratio that takes into account the intramolecular distribution of different lengths of contiguous charged residues in a typical chain ensemble have been made (Vincent and Williams 2009), this requires prior knowledge of the statistics of the patterning of methylesterification. Estimating further the amount of potential ion-binding runs that do not form connections owing to the topological frustration is even more difficult to quantify and likely depends on the particular way the ions and ion-binding group are introduced to one another (discussed below). In any case, the divalent ion-pectin interactions can clearly be modulated by modifying the ionic strength of the solution (with for example added monovalent ions) (Yoo et al. 2009; Fraeye et al. 2010; Ström et al. 2014), and it seems unclear that using R values approaching, or even in excess of one, can benefit an understanding of ionotropic network assembly.

7.2.1.2 Acidic Gels

Acidic gels formed with pectin are traditionally best known in high sugar systems such as jams and confectionaries, where the interactions have been described to originate from hydrogen bonding between hydroxyl and carboxyl protons, and some degree of hydrophobicity, enhanced by reduced water activity (Oakenfull and Scott 1984; Evageliou et al. 2000; Tsoga et al. 2004; Löfgren et al. 2005). Pectin gels formed in this way are typically made with high DM pectin substrates. In the last few decades, however, it has been conclusively demonstrated that, in low DM systems, and particularly for pectins with blockwise charge distributions, interacting networks of chains can also be assembled simply by reducing the pH, without the addition of sugar (Gilsenan et al. 2000; Ström et al. 2007). Computational modelling of the exact nature of these junction-zones, certainly in the latter case, has not been carried out to the same extent as with the egg-box model (Walkinshaw and Arnott 1981a, b). However, several things about the mechanism are clear, including the role of the pH in modifying the propensity for interchain H-bond formation as changes in the polymer charge density yield changes in the average local conformation of the HG sections (Pérez et al. 2000; Mansel et al. 2019a, b). In particular, techniques such as optical rotation show a change with pH that has been postulated to reflect a change in the preferred helical twist along the backbone (Gilsenan et al. 2000). Figure 7.1a shows the relative population of states in a set of HG chains with bond angles characterising the orientation between neighbouring sugar rings, as calculated by molecular dynamics (MD) simulations for (red) charged and (black) uncharged (protonated) residues (Irani 2018). Regardless of the starting state, the preferred most populated states after tens of ns of MD, are a 2(1) helix for charged HG and a 3(1) helix for the protonated species. Figure 7.1b shows the formation of several interchain hydrogen bonds between 3(1) chains that are favoured in this configuration and are likely play a role in the gel formation.

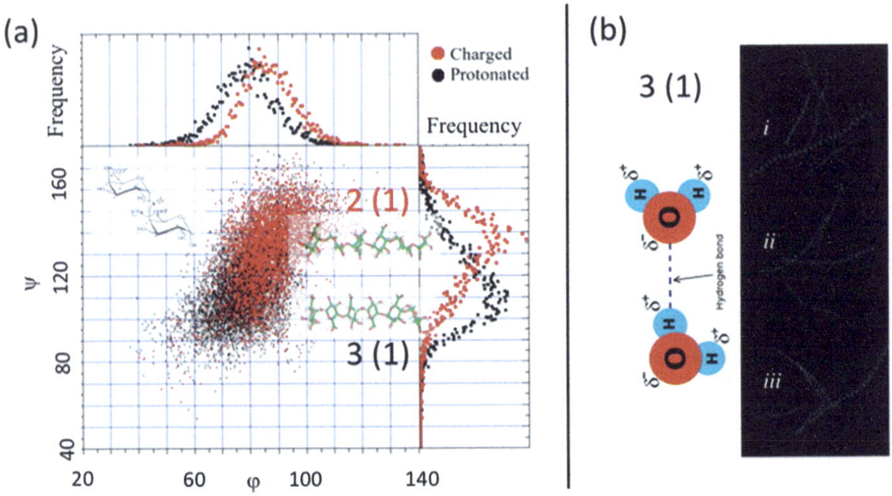

Fig. 7.1 (**a**) The relative population of HG chains in states characterised by bond angles defining the orientation between neighbouring sugar rings (Inset), for charged (Red) and uncharged (protonated) residues (Black) as calculated by MD (Irani 2018). 2(1) helices (Inset) are favoured for charged HG, 3(1) helices are favoured for the protonated species, (**b**) The formation of several interchain hydrogen bonds between 3(1) chains—favoured in the protonated configuration—are seen to lead to chain association (*i, ii* and *iii*) are 10, 50 and 100 ns after the beginning of the simulation)

In addition, experimental measurements of the mechanical properties of these gels have found a large effect of the addition of urea as would be expected where hydrogen bonding is important (Vincent et al. 2013), and perhaps more interestingly, the basic modulus is found to be significantly increased when water is substituted for D_2O.

7.2.1.3 Other Pectin Gels

Other pectin gel-types can be made, with notable mentions including: (1) those where the more traditional physically-linked junction zones are replaced with covalent linkages formed through the chemical coupling of groups that have either been pre-incorporated into pectic components (Munarin et al. 2011, 2012a, b; Pereira et al. 2018) or are naturally occurring, for example ferulic acid substituents, particularly common in sugar beet (Rombouts and Thibault 1986; Norsker et al. 2000), (2) those where the assembly of the polymer is largely driven by changes in the solute composition, such as introducing alcohol (while avoiding precipitation) or large amounts of monovalent salt (Yoo et al. 2003; Tho et al. 2006); (3) those where specific interactions with other biological polyelectrolytes, in particular alginate, are exploited (Morris and Chilvers 1984; Toft et al. 1986) and (4) those where more exotic pectic components are used, in particular the propensity of RG-I to self-associate and bare potentially interacting sidechains (Mikshina et al. 2015, 2017), and RG-II's ability to assemble with boron (Kobayashi et al. 1996; Ishii and Matsunaga 1996).

7.2.2 *Different Assembly Procedures*

That networks created using different types of junction-zones, for example those expounded above, can have different properties that makes them the optimal choice for fulfilling a particular function in a particular spatiotemporal location, either in the plant cell wall or in an industrial application, is not surprising. The process of assembling the gel, however, adds yet another level of complexity. Taking the canonical ionotropic gels as an example, the interaction between divalent ions and ion-binding groups quickly forms connections, with the burgeoning assembly influencing the mechanical properties of the system and thereby affecting the ability to homogenously further mix the polymer and the ions. Simply mixing free calcium ions into pectin solutions at ambient temperature yields a complex, spatially-inhomogeneous, kinetically-trapped system that is history and shear dependent to the extent that revealing simplistic structure-function relationships in such systems becomes problematic.

Broadly speaking two preparation methodologies have been developed in an attempt to distribute both polymer and ions as homogeneously as possible throughout solutions before the interactions that knit together the pectic chains are manifest.

The first of these is simply to heat solutions of free calcium ions (typically calcium chloride) and pectin separately, mix them vigorously at the highest temperature possible (for as short a time as possible, to avoid instigating chemical changes to the polymer) and then cool the resulting mixed solution, whereupon gelation occurs. While this procedure is relatively straightforward it is not uncommon to still observe a small amount of precipitation or microgel formation, even at higher temperatures, presumably from the smallest (and thereby the least dynamically encumbered) chains with the most blockwise (calcium sensitive) charge distributions. In addition it requires that not only the mixing, but also the transfer of the post-mixed but pregelled solution into sample cells must be carried out in a timely manner, before cooling-induced gelation proceeds inside transfer pipettes. Largely as an attempt to circumvent these issues, so-called controlled-release protocols were developed.

Controlled release protocols attempt to introduce a slower rate-limiting step into the initiation of interactions between ions and ion-binding groups so that the homogeneous mixing of ingredients can happen before the interactions are "activated". Such "activation" typically involves releasing free calcium ions from an inert form, using a release of protons (Ström and Williams 2003; Farrés and Norton 2014) or photoactivation (Cui et al. 2013; Heymann et al. 2017). The former typically uses glucono-delta-lactone (GDL), a compound that hydrolyses with time and releases protons. GDL has been used extensively, particularly in food systems—for example to mimic the pH drop caused by bacteria during fermentation. The rate of proton release is determined by the temperature, concentration, and physical properties of the GDL powder sourced. In controlled release applications using GDL the liberated protons interact with calcium salts (typically calcium carbonate) instead of lowering the pH, releasing water, carbon dioxide and free calcium ions. By ensuring the correct stoichiometry of the GDL to the initially-insoluble calcium-salt is selected all the protons released are sequestered by the salt and the pH of the solution does not change with time, guaranteeing that effects observed in the assembly of the network result only from the introduction of the free ions. Although the dispersion, dissolution and mixing of ingredients must still be performed in a timely manner (quickly with respect to the GDL hydrolysis and the sedimentation of insoluble ion-releasing salts), using this methodology does circumvent issues with the control of heating and after the dissolution of all the salt particles produces qualitatively more limpid gels. This particle dissolution can take many hours, and can be rate limited by the GDL hydrolysis or by the calcium release depending on the size of the salt particles (Ström and Williams 2003).

However, despite their homogeneous appearance and the less fractious nature of their preparation, controlled-release gels might be argued to be formed in a more precarious non-equilibrium state than those where the components are mixed at high temperature and slowly cooled. Given that typical samples contain examples of chains exhibiting a variety of lengths of consecutive runs of charged galacturonic acid residues (set according to biosynthesis and processing in the cell wall or in industrial extraction and modification processes) and that these different motifs will have different binding energies that will control whether or not they can form network nodes at different temperatures it would not be surprising if the melting

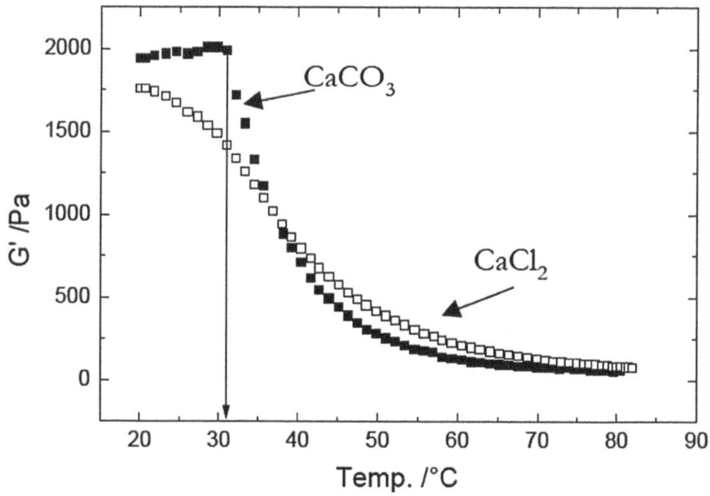

Fig. 7.2 Melting of 2% pectin (DM ~40%) networks, with R = 0.4 (and a background concentration of monovalent ions, 0.2M NaCl). G′ is shown as a function of temperature during heating at 1 °C min^{-1}. Filled symbols show the results obtained using a gel formed using the controlled release of calcium ions, over 12 h at room temperature; while open circles show the results obtained when a temperature set gel was studied (Ström 2006)

behaviour depended on the preparation history. Indeed Fig. 7.2 demonstrates exactly that (Ström and Williams 2003). It should be stressed that the two systems whose rheological behaviour upon heating is displayed in the figure are very close in terms of composition: in particular the initial gels have the same concentration of the same pectin, and the same R value (simply in one case the calcium ions are introduced as calcium chloride into heated solutions that are cooled, while in the other the ions are released from calcium carbonate using GDL at room temperature). It can be hypothesised that in the controlled release gel, where all junction-zones are formed at room temperature, longer runs of charged residues that exist are bound less-optimally (in the sense of junction-zones being formed with smaller spatial extents and lower melting temperatures) by interacting with smaller lengths competing for binding. On the other hand by initiating the interactions at a high temperature, where only the longest of the charged runs might have a chance to form stable connections, and then slowly reducing the temperature, mismatches between the length of the charged regions contributed to the junction-zone formation by each of the partners are minimised and the gelling and subsequent melting curves of the gel would be expected to be similar.

The data shown in Fig. 7.2 is consistent with this hypothesis, with the temperature cycling of the gel that was formed by controlled release essentially annealing the system: after melting and then setting again by further reducing the temperature the subsequent response to heating mimics that of the entirely 'temperature set' gel (Ström and Williams 2003). This highlights the fact that not only the polymeric fine structure, crosslinker type and concentration can be used as levers in system design but also that, the assembly process itself should be carefully considered, and may play an important role in affecting the final network structure and properties.

In a similar vein it has been shown that ionotropic gels made from low DM pectin, where the ions are liberated by controlled release, and gels with the same final composition where ions are initially mixed with impotent high DM pectin and ion-binding groups are released over time by the action of a PME do not have identical mechanical properties (Slavov et al. 2009; O'Brien et al. 2009; Vincent et al. 2009). This potential dependence of network properties on the assembly history adds complexity but also flexibility; makes understanding and comparing systems more difficult, but promises yet more control once they can be tamed. It has particular relevance for the networks formed in the plant cell wall, where the spatiotemporal control of the concentration of polymer, active enzymes, and ions is able to provide a rich landscape of functional metastable states.

7.2.3 Practical Top-Tips for Making Gels

Whichever one of the myriad of gel-species in the pectin network zoo is being studied, the initial proper dissolution of the pectic substrate is crucial to ensuring reproducible experiments. Typically, after extraction from the cell wall (and some chemical or enzymatic processing, most commonly carried out to modify the degree of methylesterification) pectin is precipitated with alcohol and ground to a fine powder. Commercially-available samples will most commonly arrive at the research scientist's desk in this form, and while the fluffy white samples produced by freeze-drying are more commonly used in research labs for storing and sharing samples and often appear a little easier to dissolve, presumably owing to their more porous particulate morphology, the comments below still generally apply.

Biopolymers are notoriously difficult to molecularly re-dissolve after they have been encouraged to associate. Indeed using the freezing and subsequent thawing of solutions in order to form gels (Zhang et al. 2013), and the differences attainable in the viscosity of polysaccharide solutions prepared in different conditions attest to this fact (Wang et al. 2001; Picout et al. 2001, 2002). Each powder grain is essentially a glassy particle and its dissolution involves many fascinating processes: from the initial wetting of the grain by the solvent, through solvent ingress and the concurrent glass to gel transition of the outer layers, to eventual molecular separation (Parker et al. 2000). The rate of escape of polymer chains from the gel layer formed by the glass-gel transition occurring at the grain periphery upon imbibing solvent can be slow, ultimately being determined by reptation dynamics (that depend on molecular weight cubed). During this period the maximum dispersion of the grains of polymer powder is crucial in order to avoid the agglomeration of sticky, partially-hydrated grains into solvent-impenetrable lumps. Vain attempts to increase the polymer dissolution rate by, for example, using hot water will typically dramatically fail by producing sticky particles before dispersion can separate the grains from each other. Dispersing the grains well in cold water and then subsequently heating the solution is generally much more successful. With gentle heating to ~50 to 60 °C pectin solutions at concentrations below or moderately above the overlap concentration are typically well-dissolved within half an hour. Heating solutions for extended periods, particularly at pH values above ~6 to 7 does increase the risk of inducing chemical changes to the fine structure, such as demethylesterification or depolymerisation (Renard and Thibault 1996), and for the more cautious it is worth noting that solutions with indistinguishable light scattering and low resolution NMR behaviour (proxies for the state of dissolution) can be formed simply at room temperature (~20 °C) by overnight stirring. Gentle and patient generally works best.

The pH of the solutions can also be tweaked to aid dissolution, with low DM or blocky pectins often reluctant to dissolve at the low pH values typically achieved when the powder is added to unbuffered water (pectins are generally supplied in their protonated, acidic form, as opposed to other anionic polysaccharides such as alginate and carrageenan that are generally supplied in their sodium form for example). As described previously such conditions are exactly those that favour the formation of acid-gels through pH induced changes to the conformation, and it should not be expected that glassy polymer grains will easily dissolve under the same conditions that favour the interaction of their components. In such cases raising the pH slightly can aid dissolution but again patience is required: substantially overshooting the target pH on titration is not difficult and pH-induced fine structure modifications can occur rapidly even at room temperature at pH values above 11.

Industrially-developed tricks that aid dispersion, such as adding the polymer powder to solvent after dry-mixing it with other (non-polymeric) particulates can also help prepare good quality solutions at ambient temperatures at the expense of adding another component to the system. In situations where such an added component is also an ingredient of the system being prepared (for example sugar, or surfactants (whose Marangoni-induced flows can aid dispersion significantly)) these grain-spacers naturally dissolve themselves, while insoluble particles can be removed after sample preparation is complete (being mindful of the possible of some of the polymer interacting with these particles).

Once a pectin solution has been prepared gelation is typically induced via the addition and of a second component, such as free ions, ion-sources that are insoluble at the starting pH of the solutions and proton-providing GDL, or enzymes. As described in some detail above, achieving a spatially homogeneous distribution of this second component is not trivial, especially if immediate interactions with the pectin polymers change the mechanical properties of the solution. It is generally preferable, rather than adding the second component to the pre-prepared pectin solution as solid powder, to pre-disperse such components in a small amount of the solvent, and add this solution to the former, mixing as quickly as possible under conditions relevant to the assembly methodology selected. Of course the concentrations of the initial solutions should be selected based on how they will be diluted according to the ratio of their mixing and the required amounts of the different components in the final mixture. While time is of the essence when mixing such solutions, it is important to try and minimise the production of any bubbles. While some gel properties may only minimally depend on the presence of bubbles, light scattering experiments, for example, would be drastically affected, as potentially would be internal-stress redistributions generated during gel formation and the macroscopic fracture behaviour (Liu et al. 2017). Keeping prepared samples in the fridge increases the amount of bubbles that form after removal, owing to the increased solubility of gases at lower temperature and their expulsion on warming to ambient temperature, so that while such storage may increase the useful lifetime of samples that may otherwise develop bacterial growth more quickly, it is not always to be favoured. If long term storage is required then adding a small amount of sodium azide may provide a better solution. Finally, when using an initially insoluble calcium ion source it is important to ascertain that these particles do not substantially sediment during the viscosification and gelation period if gradients in mechanical properties are to be avoided. Problems of this type can typically be ameliorated simply by experimenting with the size of the particles.

7.3 Measuring Gel Properties

Gels, famously easier to recognise that to define, have a range of properties that are exploited by evolved natural systems and technological products. Among their physical attributes the mechanical and internal transport properties are perhaps the most important, and how these are determined by the fine structure of the constituent polymers and the way in which they have been assembled is a continuing focus for scientists pursuing real bottom-up soft matter engineering. Here we focus on describing techniques that can measure: (1) the spatial organisation of systems; from the molecular details of the junction-zones, through the network architecture, to inhomogeneities that exist at scales many times larger than the individual polymers and (2) the mechanical properties of pectin gels over a range of applied forces, and on length and time scales over many orders of magnitude.

Structural characterisation of pectin networks revolves around applying techniques to understand the position and connectivity of pectic polymers within gels. Broadly, these techniques can be divided into real-space imaging using microscopies of various flavours (Gierlinger et al. 2012; Lopez-Sanchez et al. 2015), most notably with electron microscopy (EM) techniques (Löfgren et al. 2002; Bernin et al. 2011) being used to successfully map network architectures (Mansel et al. 2015; Hernandez-Cerdan 2018; Hernandez-Cerdan et al. 2018); and scattering techniques where X-ray studies aim to illuminate not only the size and structure of junction zones (Stokke et al. 2000; Schuster et al. 2011), but also, to report on the organisation of networks on larger length-scales (Alba et al. 2017; Mansel et al. 2019a, b). Indeed by using different distances between the sample and the detector at synchrotron radiation facilities a range of wavevectors corresponding to equivalent real space dimensions of up to around one micron can be accessed, after which experiments using small angle light scattering (SALS) can be applied to extend this range. Ultra-small angle neutron scattering (USANS) can also access larger length-scale information, although it should be noted that light scattering experiments are considerably more accessible. While neutron scattering experiments excel in mixed systems where contrast matching can be carried out, for water-rich biopolymer gels the requirement for the use of heavy water is generally a non-trivial substitution (Mansel et al. 2015). USAXS experiments, while not commonly possible at all beamlines at present, do have the potential to access this lengthscale without the necessity for the use of heavy water and may become more popular in the future as synchrotron facilities are continually upgraded. While generally EM experiments on pectin gels require detailed sample preparation to be carried out, predominantly owing to their large water content and the naturally low electron density of carbohydrates, they can, when carried out by those skilled in the art, provide spectacular three dimensional images of networks allowing network connectivities to be obtained and biomimetic webs to be constructed in silico (Hernandez-Cerdan 2018; Hernandez-Cerdan et al. 2018). In contrast, scattering experiments are intrinsically spatially averaged, but do have the advantage that gel samples can be examined without adulterations resulting from invasive sample preparation. In modelling the structure and size of junction zones X-ray scattering experiments excel whereas, although techniques are constantly improving, direct imaging at this scale is still a substantial challenge at present.

Figure 7.3 shows the scattering behaviour of visible light and X-rays from a pectin solution, contrasted with results obtained from the same solution after being gelled using a controlled-release of calcium ions. A number of points are worth highlighting: (1) the high q (small length-scale) data is consistent with calculations based on the proposed atomic co-ordinates obtained from fibre X-ray diffraction data (Mansel et al. 2019a, b); (2) the pre-gelled solution shows organisation on mul-

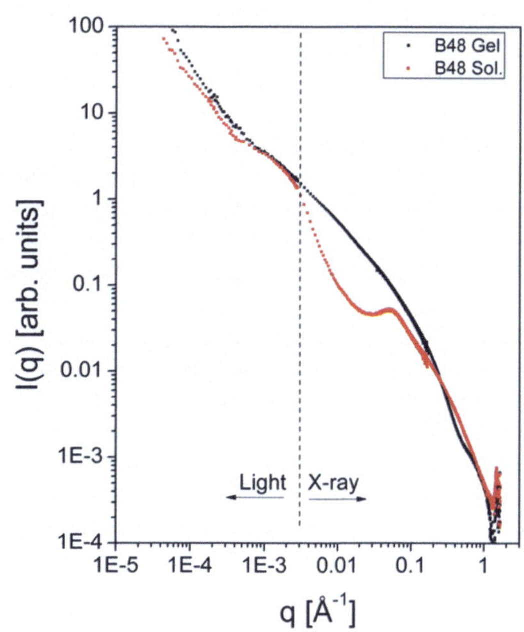

Fig. 7.3 The results of scattering experiments (SAXS, SLS, and SALS) carried out using a 1% pectin solution (blockwise DM = 48%) (red), and the solution after gelation was induced using a controlled release of calcium ions (R = 0.4) (Williams et al. Unpublished results)

tiple scales, in particular yielding a so-called polyelectrolyte peak that reflects how the electrostatic interaction leads to the organisation of the charged polymers in solution (Alba et al. 2017; Mansel et al. 2019a, b), and a low *q* (large length-scale) upturn reflecting the spatial structure of density fluctuations generated by correlated motions, the length-scale of which is fundamentally related to the compressibility of the system; (3) perhaps counterintuitively, in this system the gel is more homogeneous at intermediate lengthscales than its pre-gelled counterpart, with crosslinking-ions physically connecting the chains and attendant internal stresses now dominating the relevant distribution of the connected polymers, restricting the ability of the electrostatic repulsions between chains to freely organise them as in solution, which manifests in the disappearance of the polyelectrolyte peak; (4) the large length-scale density fluctuations in the structure of the gel (microns and above) appear to originate in the pre-gelled solution and are trapped by gelling rather than being a consequence of the assembly process per se.

Mechanical properties of different sorts of pectin-related networks most likely started being measured by humans with the squeezing of fruit, the assessment of whether, during heating, extracted plant liquors stuck to the back of a spoon, observing what happened on tilting and ultimately inverting a volume of the fluid, or feeling how they spread on the skin or were deformed in the mouth. Other sections of the animal kingdom such as aphids have been prodding plant material with their proboscises for eons (Wattier et al. 2019) and moderating their feeding patterns accordingly.

Today, scientists have an expanding arsenal of techniques available to prod and poke pectin gels on a range of timescales and with a variety of applied forces: microscopic hammers in atomic force microscopy (AFM) (Peaucelle et al. 2011; Milani et al. 2013; Peaucelle 2014; Braybrook 2015), using optical and magnetic fields to shake probe particles embedded in gel networks (Wilson and Poon 2011), grasping and wobbling samples at different rates and under different pre-stresses (Endress et al. 1996; Vincent et al. 2013), and using large mechanical racks that stretch and break macroscopic samples (Vliet 1995). Thermal fluctuations or internal stresses also drive observable dynamics of embedded probe particles and of the network itself (Cicuta and Donald 2007; Vincent et al. 2007; Williams et al. 2008; Mansel et al. 2013). Microrheological analysis uses the essence of the fluctuation-dissipation theorem to provide a clear link between such dynamics and the mechanical properties of the system, as embodied by the so-called Generalised Stokes-Einstein relation (Mason et al. 1997; Valentine et al. 2001; Nishi et al. 2018). Local dynamical measurements using microrheology are able to assess mechanical heterogeneity in a way that bulk rheological measurements cannot (Chen et al. 2003). Small amplitude oscillatory spectroscopy (SAOS) was the first technique widely used for the research measurement of the bulk mechanical properties of soft materials (Beveridge and Timbers 1989; Clark et al. 1994), and to some extent is still regarded as the gold standard. Typical SAOS and microrheology experiments generally report on the linear viscoelastic materials properties, but the non-linear properties of pectin gels are also fascinating, being specifically examined in large amplitude oscillatory (LAOS) experiments (John et al. 2019), and manifest in such behaviour as strain-stiffening (Schuster et al. 2012; Vincent et al. 2013) and failure at large deformation (Kim et al. 1978; Ström et al. 2007).

Despite the outward placid appearance of prepared pectin gels, deep within the material fascinating internal dynamics are playing out; polymer cliques battle for supremacy, gathering followers, strengthening alliances and tearing apart neighbouring associates, in an orgy of stress re-distribution and dissipation. Several localities hold an uneasy truce for hours until mounting tensions are eventually released in kingdom-wide climactic bursts of activity, and the present local behaviour becomes inextricably entangled with the history of the whole system. Simply by shining a laser through a pectin gel and directing the emerging speckle pattern onto a CCD camera hypnotising movies can be recorded that contain signatures of these internal dynamics. A surprising multi-scale dance takes place that reveals the complex internal struggles of the scattering constituent biopolymers. By placing a pinhole in front of the camera to spatially filter the light, experiments can be designed to measure fluctuations going on in different parts of the same gel over different time-scales—and the behaviour is surprisingly rich (Duri et al. 2009; Lieleg et al. 2011; Secchi et al. 2013, 2014).

In recent experiments the effects of increasing the strength of intermolecular associations on such internal dynamical processes taking place in pectin gels was studied (Mansel et al. 2015). Pectin samples were engineered to have longer and longer blocks of calcium-binding cross-link-forming regions, and while initially the internal dynamics of their assembled gels slowed, as might be expected, they then increased considerably, and exhibited increasingly complex behaviour. In fact "overtightening" internal connections leads to catastrophic re-distributions of stress, resulting in systems exhibiting faster, spatially-correlated dynamics, and the appearance of gel-quakes in close analogy with seismic events. Further multi-speckle light scattering experiments carried out as a function of wavevector revealed that indeed such fluctuations appeared "ballistic" in nature rather than "diffusive"; showing that they were driven by stress, not by thermal fluctuations. The battle was raging, with large internal stresses generated by the association of stronger bundles ripping weaker junctions apart. One might imagine the response of such a material to external stresses would include creeping flow; and it is tempting to speculate that in biological materials (such as the plant cell wall) that have to maintain mechanical integrity between periods of growth, that tuning the polymer association using structure modifying enzymes might be involved in placing the system in different dynamical regimes.

7.4 Modelling Mechanical Properties

As the different techniques available for measuring mechanical properties have expanded so several paradigms have been explored for understanding and modelling the results of different rheological measurements. Results from bulk SAOS experiments carried out on gels are generally reported in terms of the growth of a storage (elastic) shear modulus, G′, during gel formation and subsequently measurements of the frequency dependence of the storage and loss moduli. The simplest interpretation of the elastic modulus appeals to entropic-elasticity theory (Everaers and Kremer 1995), where the restoring force generated in response to an externally-applied stress that displaces each polymer chain is provided by random thermal fluctuations maximising the available configurations of the chain. In this framework the modulus is calculated by counting the number of elastically active network chains (EANCs) that constitute the network, each of which contributes kT to the restoring potential. Early attempts to relate the number of EANCs to concentration relied on estimating the number of non-contributing loops using branching theories (Gordon 1975; Clark 1994). The modelling of the reduction of the storage modulus of pectin gels during heating and melting has been examined within this framework and mapped to the reduction in the number of chains attached to the network with increasing temperature, thereby obtaining estimates of the thermodynamic parameters relating to the junction-zones (Clark et al. 1994).

More recently using EM tomography, image analysis, and computational tools, attempts have been made to directly image the chains in pectin networks, and map their connected webs to networks that can then be generated in silico (Hernandez-Cerdan 2018; Hernandez-Cerdan et al. 2018). These representations of the network architecture have the potential to replace those that have typically been generated entirely in silico with faithful network representations firmly embedded in the results of physical measurement. Once the relative position of junction-zones and their connectivities can be mapped to graphs of nodes and edges then the basic mechanical behaviour of the system can be modelled and compared with the predictions of analytical theories and with the results obtained from experimental rheological inquisitions of the same gel (Schuster et al. 2012; Carrillo et al. 2013). In order to perform such computations, the potentials between the junction-zones (nodes) are needed so that restoring forces can be calculated as the nodes move relative to one another under the influence of applied external stresses. Where gels have been fabricated specifically in order to manifest short junction-zones connected by longer sections of single polymer chains (Schuster et al. 2012), the basic gel moduli (and the characteristic strain-stiffening behaviour of the gels) has been shown to be surprising well captured simply by using the single-pectin-chain force extension behaviour as measured experimentally using AFM (Marszalek et al. 2002; Williams et al. 2007) to derive inter-node potentials. While the application of these kinds of models to pectin gels is still in its infancy they do have the potential to be extended in order to calculate the frequency dependent mechanical properties and even simulate large deformation, failure, and non-affine behaviour of such systems. The challenge for the future is to make the network architecture mapping sufficiently available and facile, and to characterise the force response behaviours of "single multichain network strands" and junction zones to the same degree as has been carried out with single chains.

Whilst the idea of microrheology dates back at least to Einstein's work on Brownian motion, improvements in both the experimental techniques available to monitor the thermally driven microscopic movements of embedded tracer particles, and in the mathematical methodologies for turning data on the motion of particles with time into the dependence of environmental mechanical moduli with frequency, have made it increasingly popular. In particular the use of techniques like diffusing wave spectroscopy (DWS) to measure nm displacements of microscopic tracers occurring in 10s to 100s of ns has provided access to high frequency rheological data that is difficult to obtain in any bulk experiment but uniquely provides information on the nature of the filaments comprising the networks examined (Schuster et al. 2011).

In the spirit of the mesoscopic approaches that have proved so illuminating in the study of cytoskeletal networks, the linear microrheological and strain-stiffening responses of biopolymeric networks reconstituted from pectin have been examined and found to be well-captured by the glassy worm-like chain (GWLC) model of self-assembled semi-flexible filaments (Vincent et al. 2013). Rather than focus on percolation theories and viewing the gel transition as a first order phase change, the GWLC is an example of an alternative paradigm where the internal dynamics of systems continually slow and correlated length-scales slowly diverge, in close analogy to a glass transition. In a somewhat parallel effort to polymer network theory, poroelastic theories have also been developed to explain some aspects of the mechanical response. In this framework the main role of the network is to control the rate at which internal fluid transport is possible, as opposed to generating the restoring forced per se (Lopez-Sanchez et al. 2016; Kaplan et al. 2019).

7.5 Conclusions

In 2016 the Food and Agriculture Organization (FAO) noted a global volume of food wastage of approximately 1.6 billion tons/year, in which fruit waste accounted for the lion's share. As society increasingly turns its attentions to sustainable and recyclable systems there is a vast untapped resource of pectin in fruit wastes and byproducts, most of which currently pose disposal problems without causing environmental pollution. At the same time consumer's perceptions of the use of pectin in processed food are favourable, and its use in non-traditional, even high-tech areas is beginning to be taken seriously. In 2017 it was reported that only two hydrocolloid categories exceeded US $1.0 billion in value for food applications alone. In 2018, the global value of pectin catapulted it into the same league. At current growth rates and barring any upsets, pectin will become the most valuable hydrocolloid used in food applications by 2023.

To reiterate the central theme of this review, scientists and technologists have the potential to create versatile, smart, soft materials and devices from this ubiquitous, recyclable, sustainable resource, in the way that Nature has done, by understanding the structure-function relationships exploited over millennia of evolutionary history and applying biomimetic principles to our own manufacturing processes. There is still much to be done to achieve this goal. Access to pectic components with pure fine structures of known substituent patterns (through controlled extraction, or chemical or enzymatic modification) remains a major challenge. Polysaccharides have fallen into the too-hard basket for the majority of biophysicists for too long, with neither facile chemistry nor biological tools analogous to those found in molecular biology available. In attempts to fill such gaps alginate researchers have trial-blazed some excellent approaches (Sletmoen et al. 2004, 2005; Bowman et al. 2016), with systematic and simultaneous study of enzymatic tools and use of single molecule techniques, the spirit of which could be applied to pectic systems. Further, understanding how heterogeneity is exploited in network construction (certainly antibody-binding studies seem to suggest that even before extraction from the cell wall pectins with different fine structures are prevalent in different spatial locations (Willats et al. 2001)) is a ripe area for investigation. What benefits this may confer to materials through, for example, increasing the robustness of networks in the light of the random failure of nodes remains largely unstudied. Finally, understanding these physical gels as inherently non-equilibrium materials with sacrificial bonds (Hartmann and Fratzl 2009; McKee et al. 2014; Neal et al. 2015), where internal stresses and stress redistribution provides a mechanism of quiescent creeping flow, and as networks that can be rewired in the light of external stresses and enzymatic tinkering in order to optimise their behaviour promises to provide a realistic biomimetic paradigm for pectin gels. The future is bright. The future in an orange.

Acknowledgements The author would like to thank all pectin friends and colleagues for many interesting discussions; and past and present members of the Biophysics and Soft Matter Group at Massey. Amir Irani, Anna Ström, Brad Mansel and Jessie Owen, are specifically acknowledged for their contributions to the work portrayed in Figs. 7.1, 7.2 and 7.3, and Brad Mansel and Ben Westberry are thanked for a critical reading of the manuscript.

References

Alba K, Bingham RJ, Kontogiorgos V (2017) Mesoscopic structure of pectin in solution. Biopolymers 107(6):e23016

Alkorta I, Garbisu C, Llama MJ, Serra JL (1998) Industrial applications of pectic enzymes: a review. Process Biochem 33(1):21–28

Anderson CT (2019) Pectic polysaccharides in plants: structure, biosynthesis, functions, and applications. In: Extracellular sugar-based biopolymers matrices. Springer, Cham, pp 487–514

Apicius FB, Alföldi-Rosenbaum E (1958) The Roman cookery book: a critical translation of the art of cooking by apicius: for use in the study and the kitchen. Harrap, London

Assifaoui A, Lerbret A, Uyen HT, Neiers F, Chambin O, Loupiac C, Cousin F (2015) Structural behaviour differences in low methoxy pectin solutions in the presence of divalent cations (Ca^{2+} and Zn^{2+}): a process driven by the binding mechanism of the cation with the galacturonate unit. Soft Matter 11(3):551–560

Bernin D, Goudappel GJ, van Ruijven M, Altskär A, Ström A, Rudemo M, Nydén M (2011) Microstructure of polymer hydrogels studied by pulsed field gradient NMR diffusion and TEM methods. Soft Matter 7(12):5711–5716

Beveridge T, Timbers GE (1989) Small amplitude oscillatory testing (SAOT). Application to pectin gelation. J Texture Stud 20(3):317–324

Borgogna M, Skjåk-Bræk G, Paoletti S, Donati I (2013) On the initial binding of alginate by calcium ions. The tilted egg-box hypothesis. J Phys Chem B 117(24):7277–7282
Bowman KA, Aarstad OA, Nakamura M, Stokke BT, Skjåk-Bræk G, Round AN (2016) Single molecule investigation of the onset and minimum size of the calcium-mediated junction zone in alginate. Carbohydr Polym 148:52–60
Braccini I, Pérez S (2001) Molecular basis of Ca2+-induced gelation in alginates and pectins: the egg-box model revisited. Biomacromolecules 2(4):1089–1096
Braconnot H (1825) Recherches sur un nouvel acide universellement répandu dans tous les végétaux. Ann Chim Phys 28(2):173–178
Braybrook SA (2015) Measuring the elasticity of plant cells with atomic force microscopy. In: Methods in cell biology, vol 125. Academic, New York, pp 237–254
Braybrook SA, Peaucelle A (2013) Mechano-chemical aspects of organ formation in *Arabidopsis thaliana*: the relationship between auxin and pectin. PLoS One 8(3):e57813
Carrillo JMY, MacKintosh FC, Dobrynin AV (2013) Nonlinear elasticity: from single chain to networks and gels. Macromolecules 46(9):3679–3692
Chebli Y, Geitmann A (2007) Mechanical principles governing pollen tube growth. Funct Plant Sci Biotechnol 1(2):232–245
Chen DT, Weeks ER, Crocker JC, Islam MF, Verma R, Gruber J, Yodh AG (2003) Rheological microscopy: local mechanical properties from microrheology. Phys Rev Lett 90(10):108301
Cicuta P, Donald AM (2007) Microrheology: a review of the method and applications. Soft Matter 3(12):1449–1455
Cifarelli A, Berzina T, Parisini A, Erokhin V, Iannotta S (2016) Polysaccarides-based gels and solid-state electronic devices with memresistive properties: synergy between polyaniline electrochemistry and biology. AIP Adv 6(11):111302
Clark AH (1994) Rationalisation of the elastic modulus-molecular weight relationship for kappa-carrageenan gels using cascade theory. Carbohydr Polym 23(4):247–251
Clark AH, Evans KT, Farrer DB (1994) Shear modulus-temperature meltdown profiles of gelatin and pectin gels. A cascade theory description. Int J Biol Macromol 16(3):125–130
Cui JX, Wang M, Zheng YJ, Muniz GMR, delCampo A (2013) Light-triggered cross-linking of alginates with caged Ca2+. Biomacromolecules 14(5):1251–1256
Di Giacomo R, Daraio C, Maresca B (2015) Plant nanobionic materials with a giant temperature response mediated by pectin-Ca^{2+}. Proc Natl Acad Sci U S A 112(15):4541–4545
Duri A, Sessoms DA, Trappe V, Cipelletti L (2009) Resolving long-range spatial correlations in jammed colloidal systems using photon correlation imaging. Phys Rev Lett 102(8):085702
Edwards MC, Doran-Peterson J (2012) Pectin-rich biomass as feedstock for fuel ethanol production. Appl Microbiol Biotechnol 95(3):565–575
Endress HU, Döschl-Volle C, Dengler K (1996) Rheological methods to characterize pectins in solutions and gels. In: Progress in biotechnology, vol 14. Elsevier, New York, pp 407–423
Evageliou V, Richardson RK, Morris ER (2000) Effect of pH, sugar type and thermal annealing on high-methoxy pectin gels. Carbohydr Polym 42(3):245–259
Everaers R, Kremer K (1995) Test of the foundations of classical rubber elasticity. Macromolecules 28(21):7291–7294
Farrés IF, Norton IT (2014) Formation kinetics and rheology of alginate fluid gels produced by in-situ calcium release. Food Hydrocoll 40:76–84
Fraeye I, Duvetter T, Doungla E, Van Loey A, Hendrickx M (2010) Fine-tuning the properties of pectin–calcium gels by control of pectin fine structure, gel composition and environmental conditions. Trends Food Sci Technol 21(5):219–228
Gierlinger N, Keplinger T, Harrington M (2012) Imaging of plant cell walls by confocal Raman microscopy. Nat Protoc 7(9):1694
Gilsenan PM, Richardson RK, Morris ER (2000) Thermally reversible acid-induced gelation of low-methoxy pectin. Carbohydr Polym 41(4):339–349
Gordon M (1975) Rubber elasticity. Flaws in the theory of networks. Macromolecules 8(2):247–248

Grant GT, Morris ER, Rees DA, Smith PJ, Thom D (1973) Biological interactions between polysaccharides and divalent cations: the egg-box model. FEBS Lett 32(1):195–198
Hartmann MA, Fratzl P (2009) Sacrificial ionic bonds need to be randomly distributed to provide shear deformability. Nano Lett 9(10):3603–3607
Henrissat B, Coutinho PM, Davies GJ (2001) A census of carbohydrate-active enzymes in the genome of Arabidopsis thaliana. Plant Mol Biol 47(1-2):55–72
Hernandez-Cerdan P (2018) Biopolymer networks: image analysis, reconstruction and modeling, PhD Thesis, Massey University, New Zealand
Hernandez-Cerdan P, Mansel BW, Leis A, Lundin L, Williams MA (2018) Structural analysis of polysaccharide networks by transmission electron microscopy: comparison with small-angle X-ray scattering. Biomacromolecules 19(3):989–995
Heymann RR, Thum MD, Hardee AL, Falvey DE (2017) Visible light initiated release of calcium ions through photochemical electron transfer reactions. Photochem Photobiol Sci 16(6):1003–1008
Ishii T, Matsunaga T (1996) Isolation and characterization of a boron-rhamnogalacturon-II complex from cell walls of sugar beet pulp. Carohydr Res 284(1):1–9
Irani A (2018) Charged polysaccharides as model polyelectrolytes: computational studies of transport and conformational properties, PhD Thesis, Massey University, New Zealand
Jarvis MC (1984) Structure and properties of pectin gels in plant cell walls. Plant Cell Environ 7(3):153–164
John J, Ray D, Aswal VK, Deshpande AP, Varughese S (2019) Dissipation and strain-stiffening behavior of pectin–Ca gels under LAOS. Soft Matter 15(34):6852–6866
Kaplan JL, Torode TA, Daher FB, Braybrook SA. (2019) On pectin methyl-esterification: implications for in vitro and in vivo viscoelasticity. BioRxiv 565614
Kent LM, Loo TS, Melton LD, Mercadante D, Williams MA, Jameson GB (2016) Structure and properties of a non-processive, salt-requiring, and acidophilic pectin methylesterase from Aspergillus niger provide insights into the key determinants of processivity control. J Biol Chem 291(3):1289–1306
Kim WJ, Rao VNM, Smit CJB (1978) Effect of chemical composition on compressive mechanical properties of low ester pectin gels. J Food Sci 43(2):572–575
Kobayashi M, Matoh T, Azuma JI (1996) Two chains of rhamnogalacturonan II are cross-linked by borate-diol ester bonds in higher plant cell walls. Plant Physiol 110(3):1017–1020
Levesque-Tremblay G, Pelloux J, Braybrook SA, Müller K (2015) Tuning of pectin methylesterification: consequences for cell wall biomechanics and development. Planta 242(4):791–811
Li L, Fang Y, Vreeker R, Appelqvist I, Mendes E (2007) Re-examining the egg-box model in calcium−alginate gels with X-ray diffraction. Biomacromolecules 8(2):464–468
Lieleg O, Kayser J, Brambilla G, Cipelletti L, Bausch AR (2011) Slow dynamics and internal stress relaxation in bundled cytoskeletal networks. Nat Mater 10(3):236
Liners F, Thibault JF, Van Cutsem P (1992) Influence of the degree of polymerization of oligogalacturonates and of esterification pattern of pectin on their recognition by monoclonal antibodies. Plant Physiol 99(3):1099–1104
Liu D, Šavija B, Smith GE, Flewitt PE, Lowe T, Schlangen E (2017) Towards understanding the influence of porosity on mechanical and fracture behaviour of quasi-brittle materials: experiments and modelling. Int J Fract 205(1):57–72
Löfgren C, Walkenström P, Hermansson AM (2002) Microstructure and rheological behavior of pure and mixed pectin gels. Biomacromolecules 3(6):1144–1153
Löfgren C, Guillotin S, Evenbratt H, Schols H, Hermansson AM (2005) Effects of calcium, pH, and blockiness on kinetic rheological behavior and microstructure of HM pectin gels. Biomacromolecules 6(2):646–652
Lopez-Sanchez P, Schuster E, Wang D, Gidley MJ, Strom A (2015) Diffusion of macromolecules in self-assembled cellulose/hemicellulose hydrogels. Soft Matter 11(20):4002–4010

Lopez-Sanchez P, Martinez-Sanz M, Bonilla MR, Wang D, Walsh CT, Gilbert EP, Gidley MJ (2016) Pectin impacts cellulose fibre architecture and hydrogel mechanics in the absence of calcium. Carbohydr Polym 153:236–245

Mansel BW, Williams MA (2015) Internal stress drives slow glassy dynamics and quake-like behaviour in ionotropic pectin gels. Soft Matter 11(35):7016–7023

Mansel BW, Keen S, Patty PJ, Hemar Y, Williams MA (2013) A practical review of microrheological techniques. In: Rheology-new concepts, applications and methods. IntechOpen, London

Mansel BW, Chu CY, Leis A, Hemar Y, Chen HL, Lundin L, Williams MA (2015) Structural investigations of rheologically characterized hydrogen-bonded low-methoxyl pectin networks. Biomacromolecules 16(10):3209–3216

Mansel BW, Irani AH, Ryan TM, McGillivray DJ, Chen HL, Williams MA (2019a) Resolving solution conformations of the model semi-flexible polyelectrolyte homogalacturonan using molecular dynamics simulations and small-angle x-ray scattering. Eur Phys J E 42(2):19

Mansel BW, Ryan TM, Chen HL, Williams MA (2019b) Polysaccharide conformations measured by solution state x-ray scattering. Chem Phys Lett 739:136951

Marszalek PE, Li H, Oberhauser AF, Fernandez JM (2002) Chair-boat transitions in single polysaccharide molecules observed with force-ramp AFM. Proc Natl Acad Sci U S A 99(7):4278–4283

Mason TG, Ganesan K, Van Zanten JH, Wirtz D, Kuo SC (1997) Particle tracking microrheology of complex fluids. Phys Rev Lett 79(17):3282

McKee JR, Huokuna J, Martikainen L, Karesoja M, Nykänen A, Kontturi E, Ikkala O (2014) Molecular engineering of fracture energy dissipating sacrificial bonds into cellulose nanocrystal nanocomposites. Angew Chem Int Ed 53(20):5049–5053

Mercadante D, Melton LD, Jameson GB, Williams MA (2014) Processive pectin methylesterases: the role of electrostatic potential, breathing motions and bond cleavage in the rectification of Brownian motions. PLoS One 9(2):e87581

Mikshina PV, Petrova AA, Faizullin DA, Zuev YF, Gorshkova TA (2015) Tissue-specific rhamnogalacturonan I forms the gel with hyperelastic properties. Biochemistry 80(7):915–924

Mikshina PV, Makshakova ON, Petrova AA, Gaifullina IZ, Idiyatullin BZ, Gorshkova TA, Zuev YF (2017) Gelation of rhamnogalacturonan I is based on galactan side chain interaction and does not involve chemical modifications. Carbohydr Polym 171:143–151

Milani P, Braybrook SA, Boudaoud A (2013) Shrinking the hammer: micromechanical approaches to morphogenesis. J Exp Bot 64(15):4651–4662

Mohammadinejad R, Maleki H, Larrañeta E, Fajardo AR, Nik AB, Shavandi A, Cabane E (2019) Status and future scope of plant-based green hydrogels in biomedical engineering. Appl Mater Today 16:213–246

Mohnen D (2008) Pectin structure and biosynthesis. Curr Opin Plant Biol 11(3):266–277

Morris VJ, Chilvers GR (1984) Cold setting alginate-pectin mixed gels. J Sci Food Agric 35(12):1370–1376

Morris ER, Powell DA, Gidley MJ, Rees DA (1982) Conformations and interactions of pectins: I. Polymorphism between gel and solid states of calcium polygalacturonate. J Mol Biol 155(4):507–516

Munarin F, Guerreiro SG, Grellier MA, Tanzi MC, Barbosa MA, Petrini P, Granja PL (2011) Pectin-based injectable biomaterials for bone tissue engineering. Biomacromolecules 12(3):568–577

Munarin F, Tanzi MC, Petrini P (2012a) Advances in biomedical applications of pectin gels. Int J Biol Macromol 51(4):681–689

Munarin F, Petrini P, Tanzi MC, Barbosa MA, Granja PL (2012b) Biofunctional chemically modified pectin for cell delivery. Soft Matter 8(17):4731–4739

Neal JA, Mozhdehi D, Guan Z (2015) Enhancing mechanical performance of a covalent self-healing material by sacrificial noncovalent bonds. J Am Chem Soc 137(14):4846–4850

Nishi K, Kilfoil ML, Schmidt CF, MacKintosh FC (2018) A symmetrical method to obtain shear moduli from microrheology. Soft Matter 14(19):3716–3723

Norsker M, Jensen M, Adler-Nissen J (2000) Enzymatic gelation of sugar beet pectin in food products. Food Hydrocoll 14(3):237–243

O'Brien AB, Philp K, Morris ER (2009) Gelation of high-methoxy pectin by enzymic de-esterification in the presence of calcium ions: a preliminary evaluation. Carbohydr Res 344(14):1818–1823
Oakenfull D, Scott A (1984) Hydrophobic interaction in the gelation of high methoxyl pectins. J Food Sci 49(4):1093–1098
Owen JL, Kent LM, Ralet MC, Cameron RG, Williams MAK (2017) A tale of two pectins: Diverse fine structures can result from identical processive PME treatments on similar high DM substrates. Carbohydr Polym 168:365–373
Owens HS, Swenson HA, Schultz TH (1954) Factors influencing gelation with pectin. Nat Plant Hydrocolloids 3:10–15
Parker A, Vigouroux F, Reed WF (2000) Dissolution kinetics of polymer powders. AIChE J 46(7):1290–1299
Peaucelle A (2014) AFM-based mapping of the elastic properties of cell walls: at tissue, cellular, and subcellular resolutions. JoVE 89:e51317
Peaucelle A, Braybrook SA, Le Guillou L, Bron E, Kuhlemeier C, Höfte H (2011) Pectin-induced changes in cell wall mechanics underlie organ initiation in Arabidopsis. Curr Biol 21(20):1720–1726
Pelloux J, Rusterucci C, Mellerowicz EJ (2007) New insights into pectin methylesterase structure and function. Trends Plant Sci 12(6):267–277
Pereira RF, Barrias CC, Bártolo PJ, Granja PL (2018) Cell-instructive pectin hydrogels cross-linked via thiol-norbornene photo-click chemistry for skin tissue engineering. Acta Biomater 66:282–293
Pérez S, Mazeau K, du Penhoat CH (2000) The three-dimensional structures of the pectic polysaccharides. Plant Physiol Biochem 38(1-2):37–55
Picout DR, Ross-Murphy SB, Etherington N, Harding SE (2001) Pressure cell assisted solution characterization of polysaccharides. 1. Guar gum. Biomacromolecules 2(4):1301–1309
Picout DR, Ross-Murphy SB, Jumel K, Harding SE (2002) Pressure cell assisted solution characterization of polysaccharides. 2. Locust bean gum and tara gum. Biomacromolecules 3(4):761–767
Pilnik W, Voragen AG (1993) Pectic enzymes in fruit and vegetable juice manufacture. In: Enzymes in food processing, 3rd edn. Academic, New York, pp 363–399
Plazinski W (2011) Molecular basis of calcium binding by polyguluronate chains. Revising the egg-box model. J Comput Chem 32(14):2988–2995
Poon WC, Andelman D (2006) Soft condensed matter physics in molecular and cell biology. CRC Press, Boca Raton
Powell DA, Morris ER, Gidley MJ, Rees DA (1982) Conformations and interactions of pectins: II. Influence of residue sequence on chain association in calcium pectate gels. J Mol Biol 155(4):517–531
Renard CMG, Thibault JF (1996) Pectins in mild alkaline conditions: β-elimination and kinetics of demethylation. In: Progress in biotechnology, vol 14. Elsevier, London, pp 603–608
Rexová-Benková L, Markovič O (1976) Pectic enzymes. In: Advances in carbohydrate chemistry and biochemistry, vol 33. Academic, New York, pp 323–385
Rombouts FM, Thibault JF (1986) Sugar beet pectins: chemical structure and gelation through oxidative coupling. In: Chemistry and function of pectins, vol 310. American Chemical Society, Washington, DC, pp 49–60
Round AN, Rigby NM, MacDougall AJ, Morris VJ (2010) A new view of pectin structure revealed by acid hydrolysis and atomic force microscopy. Carbohydr Res 345(4):487–497
Schols HA, Voragen AGJ (1996) Complex pectins: structure elucidation using enzymes. In: Progress in biotechnology, vol 14. Elsevier, London, pp 3–19
Schuster E, Cucheval A, Lundin L, Williams MA (2011) Using SAXS to reveal the degree of bundling in the polysaccharide junction zones of microrheologically distinct pectin gels. Biomacromolecules 12(7):2583–2590

Schuster E, Lundin L, Williams MA (2012) Investigating the relationship between network mechanics and single-chain extension using biomimetic polysaccharide gels. Macromolecules 45(11):4863–4869

Secchi E, Roversi T, Buzzaccaro S, Piazza L, Piazza R (2013) Biopolymer gels with "physical" cross-links: gelation kinetics, aging, heterogeneous dynamics, and macroscopic mechanical properties. Soft Matter 9(15):3931–3944

Secchi E, Munarin F, Alaimo MD, Bosisio S, Buzzaccaro S, Ciccarella G, Piazza R (2014) External and internal gelation of pectin solutions: microscopic dynamics versus macroscopic rheology. J Phys Condens Matter 26(46):464106

Sikorski P, Mo F, Skjåk-Bræk G, Stokke BT (2007) Evidence for egg-box-compatible interactions in calcium– alginate gels from fiber X-ray diffraction. Biomacromolecules 8(7):2098–2103

Sila DN, Van Buggenhout S, Duvetter T, Fraeye I, De Roeck A, Van Loey A, Hendrickx M (2009) Pectins in processed fruits and vegetables: part II—structure–function relationships. Compr Rev Food Sci Food Saf 8(2):86–104

Slavov A, Garnier C, Crépeau MJ, Durand S, Thibault JF, Bonnin E (2009) Gelation of high methoxy pectin in the presence of pectin methylesterases and calcium. Carbohydr Polym 77(4):876–884

Sletmoen M, Skjåk-Bræk G, Stokke BT (2004) Single-molecular pair unbinding studies of Mannuronan C-5 epimerase AlgE4 and its polymer substrate. Biomacromolecules 5(4):1288–1295

Sletmoen M, Skjåk-Bræk G, Stokke BT (2005) Mapping enzymatic functionalities of mannuronan C-5 epimerases and their modular units by dynamic force spectroscopy. Carbohydr Res 340(18):2782–2795

Stokke BT, Draget KI, Smidsrød O, Yuguchi Y, Urakawa H, Kajiwara K (2000) Small-angle X-ray scattering and rheological characterization of alginate gels. 1. Ca– alginate gels. Macromolecules 33(5):1853–1863

Ström A (2006) Characterisation of pectin fine-structure and its effect on supramolecular properties, PhD Thesis, University of Cork, Ireland

Ström A, Williams MA (2003) Controlled calcium release in the absence and presence of an ion-binding polymer. J Phys Chem B 107(40):10995–10999

Ström A, Ribelles P, Lundin L, Norton I, Morris ER, Williams MA (2007) Influence of pectin fine structure on the mechanical properties of calcium– pectin and acid– pectin gels. Biomacromolecules 8(9):2668–2674

Ström A, Schuster E, Goh SM (2014) Rheological characterization of acid pectin samples in the absence and presence of monovalent ions. Carbohydr Polym 113:336–343

Sun B, Zhang X, Zhou G, Li P, Zhang Y, Wang H, Zhao Y (2017) An organic nonvolatile resistive switching memory device fabricated with natural pectin from fruit peel. Org Electron 42:181–186

Sun B, Zhu S, Mao S, Zheng P, Xia Y, Yang F, Zhao Y (2018) From dead leaves to sustainable organic resistive switching memory. J Colloid Interface Sci 513:774–778

Thakur BR, Singh RK, Handa AK, Rao MA (1997) Chemistry and uses of pectin—a review. Crit Rev Food Sci Nutr 37(1):47–73

Thakur S, Chaudhary J, Kumar V, Thakur VK (2019) Progress in pectin based hydrogels for water purification: trends and challenges. J Environ Manag 238:210–223

Thibault JF, Ralet MC (2008) Pectins, their origin, structure and functions. Adv Diet Fibre Technol 2008:369–378

Tho I, Kjøniksen AL, Knudsen KD, Nyström B (2006) Effect of solvent composition on the association behavior of pectin in methanol–water mixtures. Eur Polym J 42(5):1164–1172

Toft K, Grasdalen H, Smidsrod O (1986) Synergistic gelation of alginates and pectins. ACS Symp Ser 310:117–132

Tsoga A, Richardson RK, Morris ER (2004) Role of co-solutes in gelation of high-methoxy pectin. Part 1. Comparison of sugars and polyols. Food Hydrocoll 18(6):907–919

Valentine MT, Kaplan PD, Thota D, Crocker JC, Gisler T, Prud'homme RK, Weitz DA (2001) Investigating the microenvironments of inhomogeneous soft materials with multiple particle tracking. Phys Rev E 64(6):061506

van der Gucht J (2018) Grand challenges in soft matter physics. Front Phys 6:87

Vanitha T, Khan M (2019) Role of pectin in food processing and food packaging. In: Pectins-extraction, purification, characterization and applications. IntechOpen, London

Vincent RR, Williams MA (2009) Microrheological investigations give insights into the microstructure and functionality of pectin gels. Carbohydr Res 344(14):1863–1871

Vincent RR, Pinder DN, Hemar Y, Williams MAK (2007) Microrheological studies reveal semiflexible networks in gels of a ubiquitous cell wall polysaccharide. Phys Rev E 76(3):031909

Vincent RR, Cucheval A, Hemar Y, Williams MAK (2009) Bio-inspired network optimization in soft materials—Insights from the plant cell wall. Eur Phys J E 28(1):79–87

Vincent RRR, Mansel BW, Kramer A, Kroy K, Williams MAK (2013) Micro-rheological behaviour and nonlinear rheology of networks assembled from polysaccharides from the plant cell wall. New J Phys 15(3):035002

Vliet T (1995) Large deformation and fracture behaviour of gels. Faraday Discuss 101:359–370

Voragen F, Schols H, Visser RG (eds) (2013) Advances in pectin and pectinase research. Springer, Cham

Walkinshaw MD, Arnott S (1981a) Conformations and interactions of pectins: I. X-ray diffraction analyses of sodium pectate in neutral and acidified forms. J Mol Biol 153(4):1055–1073

Walkinshaw MD, Arnott S (1981b) Conformations and interactions of pectins: II. Models for junction zones in pectinic acid and calcium pectate gels. J Mol Biol 153(4):1075–1085

Wang Q, Wood PJ, Cui W, Ross-Murphy SB (2001) The effect of autoclaving on the dispersibility and stability of three neutral polysaccharides in dilute aqueous solutions. Carbohydr Polym 45(4):355–362

Wattier C, Turbant A, Sargos-Vallade L, Pelloux J, Rustérucci C, Cherqui A (2019) New insights into diet breadth of polyphagous and oligophagous aphids on two Arabidopsis ecotypes. Insect Sci 26(4):753–769

Willats WG, Orfila C, Limberg G, Buchholt HC, van Alebeek GJW, Voragen AG, Knox JP (2001) Modulation of the degree and pattern of methyl-esterification of pectic homogalacturonan in plant cell walls implications for pectin methyl esterase action, matrix properties, and cell adhesion. J Biol Chem 276(22):19404–19413

Willats WG, Knox JP, Mikkelsen JD (2006) Pectin: new insights into an old polymer are starting to gel. Trends Food Sci Technol 17(3):97–104

Williams MA, Marshall AT, Anjukandi P, Haverkamp RG (2007) Investigation of the effects of fine structure on the nanomechanical properties of pectin. Phys Rev E 76(2):021927

Williams MAK, Vincent RR, Pinder DN, Hemar Y (2008) Microrheological studies offer insights into polysaccharide gels. J Non-Newt Fluid Mech 149(1-3):63–70

Wilson LG, Poon WC (2011) Small-world rheology: an introduction to probe-based active microrheology. Phys Chem Chem Phys 13(22):10617–10630

Witten TA (1999) Insights from soft condensed matter. In: Bederson B (ed) More things in heaven and earth. Springer, New York

Xiao C, Anderson CT (2013) Roles of pectin in biomass yield and processing for biofuels. Front Plant Sci 4:67

Yoo SH, Fishman ML, Savary BJ, Hotchkiss AT (2003) Monovalent salt-induced gelation of enzymatically deesterified pectin. J Agric Food Chem 51(25):7410–7417

Yoo SH, Lee BH, Savary BJ, Lee S, Lee HG, Hotchkiss AT (2009) Characteristics of enzymatically-deesterified pectin gels produced in the presence of monovalent ionic salts. Food Hydrocoll 23(7):1926–1929

Zerzour R, Kroeger J, Geitmann A (2009) Polar growth in pollen tubes is associated with spatially confined dynamic changes in cell mechanical properties. Dev Biol 334(2):437–446

Zhang H, Zhang F, Wu J (2013) Physically crosslinked hydrogels from polysaccharides prepared by freeze–thaw technique. React Funct Polym 73(7):923–928

Chapter 8
Digestion and Metabolism of Pectin

Kathrin Haider and Peter Wilde

8.1 Introduction

8.1.1 Pectin as a Source of Soluble Fibre and Influence on Digestion

Pectin is considered as a dietary fibre which are generally defined as carbohydrate polymers with three or more monomeric units, which are not hydrolysed by the endogenous enzymes in the small intestine of humans and are either naturally occurring in the food as consumed or have been shown to have a beneficial physiological effect (Codex Alimentarius Commission 2009). When isolated and added to food during processing, pectin is approved as dietary fibre by the FDA due to its ability to maintain normal blood cholesterol levels and additionally by EFSA due to its proven ability to reduce post-prandial rise in blood glucose (EFSA Panel on Dietetic Products, U.S. Food and Drug Administration 2018). Soluble dietary fibres are regarded as beneficial for health for a number of reasons, firstly their ability to retain water and to resist digestive enzymes in the upper gut, which is thought to slow down the digestion and absorption of glucose and lipids. In the lower gut, they are fermented by the gut microbiota to produce beneficial metabolites such as short chain fatty acids (SCFA). This fermentation can lead to the proliferation of beneficial microbiota, and the metabolites provide nutrition for the gut epithelium, are thought to mitigate the development of colon cancer and act as immunosuppressors. They also act as signalling molecules to improve a range of physiological functions including insulin sensitivity, lipid metabolism and satiety. The molecular structure of pectin is thought to be important for its ability to interact with cell surface proteins such as lectins, which are important for cell signalling and anti-cancer activity.

K. Haider · P. Wilde (✉)
Quadram Institute Bioscience, Norwich Research Park, Norwich, UK
e-mail: peter.wilde@quadram.ac.uk

V. Kontogiorgos (ed.), *Pectin: Technological and Physiological Properties*,
https://doi.org/10.1007/978-3-030-53421-9_8

Low molecular weight, unbranched glycan chains with the appropriate sugar composition are more able to interact with surface bound proteins and trigger responses (Maxwell 2015). Even though pectins are not metabolised by humans per se, they interfere with physiological processes at many stages of the digestion like gastric emptying, entero-hepatic circulation, absorption of nutrients and intestinal permeability and fermentation products (Popov 2016). Since intrinsic digestive enzymes will not hydrolyse pectin, pectin gels find use in the delivery of probiotics (Cabrera et al. 2011; Naqash et al. 2017).

8.1.2 Sources of Pectin

Pectins are very common in a normal western diet with consumption reported as around 4–5 g each day (Willats et al. 2006) or 2.5 ± 1.2 g/day in postmenopausal women aged 50–79 years in the US (Navarro et al. 2016). This is not only due to a high abundance in fruits and vegetables, but also because it is used in food technology as a gelling agent, stabiliser and emulsifier. Pectin's physicochemical and functional properties depend on its structural and molecular make-up as well as on extrinsic factors like ionic strength, pH, and temperature. The degree of polymerisation, i.e., the molecular weight, and the number and distribution of methoxyl groups (degree of methylation, DM) define its solubility, whereas the latter in combination with pH and presence of a cosolute also defines its gelling properties. There is huge structural diversity of pectins, however, pectins used as food additives most commonly derive from citrus peel and apple pomace (Willats et al. 2006) which are usually rather simple molecules lacking side branched sugar chains (Popov 2016) as opposed to a diet rich in foods with different native pectin structures. The molecular weight, degree of branching and methylation all influence the gelling and water holding capacity of the pectin. This in turn will influence the behaviour in the upper gut in terms of slowing down gastric emptying and digestion. The pectin source will affect these properties as described in the previous chapters. In addition, when consumed in the whole food, pectin has to be liberated and solubilised from the cell walls for it to be effective in promoting gel formation and influencing digestion. Also, the molecular changes occurring during ripening, result in changes to the molecular structure of pectin, and this was found to reduce molecular weight and branching, which affects, among others, the putative anti-cancer properties of papaya (Prado et al. 2017).

The most complex polysaccharide in our diet is rhamnogalacturonan-II (RG-II) which contains 13 different sugars and 21 distinct glycosidic linkages arranged as a backbone with six different side chains (Ndeh et al. 2017) and can be found in fruits like apple (Doco et al. 1997), grape berry (Vidal et al. 2001) and vegetables like bamboo (Phyllostachys edulis) shoot (Kaneko et al. 1997), carrot (Doco et al. 1997), red beet (*Beta vulgaris L. var conditiva*) tubers (Strasser and Amadò 2002), radish root (Matoh et al. 1993), and tomato (Doco et al. 1997). Another common source of RG-II intake is red wine which contains between 100 and 150 mg/L (Patrice et al.

1997; O'Neill et al. 2004). One of the pectins found in sugar beet is RG-II (Ishii and Matsunaga 1996) but its special property to act, unlike other pectins, as an emulsifier is thought to be due to its high protein fraction and hydrophobicity (Gromer et al. 2009; Einhorn-Stoll 2018). Other plant sources such as pumpkin or various tropical fruits as well as pulp waste from different origins are currently investigated alongside innovate extraction procedures to either maintain the natural structure as intact as possible or to modify the extracted pectins according to the desired physicochemical properties by e.g., demethoxylations using acidic, alkaline-ammonia or enzymatic treatments, resulting in different low-methoxylated or amidated pectins (Einhorn-Stoll 2018). We are still in the process of elucidating the linkages and conformations of pectins as shown by the recent work on identifying the enzymes used by *Bacteroides thetaiotaomicron* to break down RG-II (Ndeh et al. 2017). Likewise research is under way to understand the macrostructure within plant cell walls, suggesting e.g., unique interactions between rhamnogalacturonan-I and cellulose in carrot cell walls which were not found in tomato or strawberry (Canteri et al. 2019).

8.1.3 Processed vs. Whole Food Sources

Purified, extracted pectins have been found to have health benefits, and are classed as a form of dietary fibre by FDA and EFSA (EFSA Panel on Dietetic Products, U.S. Food and Drug Administration 2018). The availability of the pectin is dependent on the ability of processing, preparation or consumption/digestion to release the pectin from the cell wall matrix from a particular source. Therefore so called natural or whole food sources of pectin may not be made available if they are too strongly embedded in the cell wall matrix. Pectin is conventionally extracted with acids at elevated temperature (see Chap. 4) so it would be expected that some pectins are released from intact cell wall structures during gastric digestion (Ulmius et al. 2011).

Pectin can also influence the release and bioavailability of other bioactive compounds like polyphenols. In a study where apple purée was compared to an extract, the latter showed a higher maximum plasma concentration and time to peak for the flavanol epicatechin. The authors suggested three potential reasons: (1) binding of epicatechin to apple fibre or entrapment within cells which may reduce bioaccessibility, (2) increased viscosity of the bolus due to the apple pectin, (3) reduced rates of enterocyte epicatechin uptake due to competition with other polyphenols (Hollands et al. 2013). However, co-ingestion of apple pectin and a polyphenol-rich apple concentrate was more effective in colon fermentation and on beneficial effects in lipid metabolism than separate ingestion (Aprikian et al. 2003). Pectin is relatively process-sensitive and cooking food causes structural changes (Lovegrove et al. 2017) like tissue softening and cell separation during heating (Brett and Waldron 1996). This may affect the molecular weight and structure of the pectin released from the matrix, affecting their functionality.

8.2 Digestion

8.2.1 Introduction to Digestion

Digestion of food starts in the mouth where food is masticated, if necessary, and mixed with salivary fluid which contains the starch degrading enzyme α-amylase (Whelan 1960), minerals such as calcium, and a range of proteins. Mastication is a relatively short step in the digestive process, however it is very important, as it will start off starch digestion, which will continue in the stomach before the pH becomes too low. Also, oro-sensory exposure and duration influences gastric emptying (Morey et al. 2016), satiation (Cecil et al. 1998) and appetite (Wijlens et al. 2012), which seems to be determined by the food texture and not taste intensity (Lasschuijt et al. 2017). The feeling of satiation defines the point when consuming a meal ceases whereas satiety is the feeling of fullness after eating till new feeling of hunger is perceived (Benelam 2009). Swallowed food then passes into the stomach where it is mixed with gastric secretions containing hydrochloric acid, which gradually reduces the pH, protease (pepsin) and gastric lipase. The stomach stores the food and gradually empties it at a rate that is dependent on the nutrient composition, as influenced by the oro-sensory processes, and nutrient sensing cells in the duodenum. Once the stomach fills and begins to expand, stretch receptors in the stomach wall signal to the brain to initiate feelings of satiation, and reduction in levels of the hormone ghrelin, reduces feelings of hunger. The structure and viscosity of the food in the stomach can also reduce gastric emptying, as the exit of the stomach, the pylorus, will sieve out large, solid particles. Thus, viscoelastic structures in the stomach can slow down gastric emptying and prolong feelings of satiation and reduced appetite. This is a crucial step in digestion which will control the amount and rate of food delivered to the small intestine and is strongly influenced by the structure and gelling behaviour of the food consumed (Guerra et al. 2012) (Fig. 8.1).

Food is gradually emptied into the duodenum, where enzymes secreted by the pancreas, with the aid of bile from the liver, hydrolyse starch, glycogen, sugar, protein and fat ready for absorption. Humans do not possess enzymes capable of digesting other carbohydrates such as dietary fibre, and so they will pass largely undigested through the small intestine. However, dietary fibre may increase viscosity, form structures or interact with the digestion process to slow down digestion or absorption steps. Slowing digestion will prolong the presence of nutrients further down the small intestine, where increasing numbers of nutrient sensing cells will increase secretion of satiety promoting hormones (Wanders et al. 2011). Finally, food that is not digested by endogenous enzymes will pass into the colon, where the majority of the gut microbiota are located. The diverse range of bacteria secrete a whole range of enzymes, which in combination, are capable of fermenting and metabolising the fermentable compounds remaining, these include resistant protein and starch, dietary fibres and other compounds such as phytochemicals. Fermentation of dietary fibre produces metabolites, the most common being the short chain fatty acids such as acetate, propionate and butyrate which have a range of health benefits.

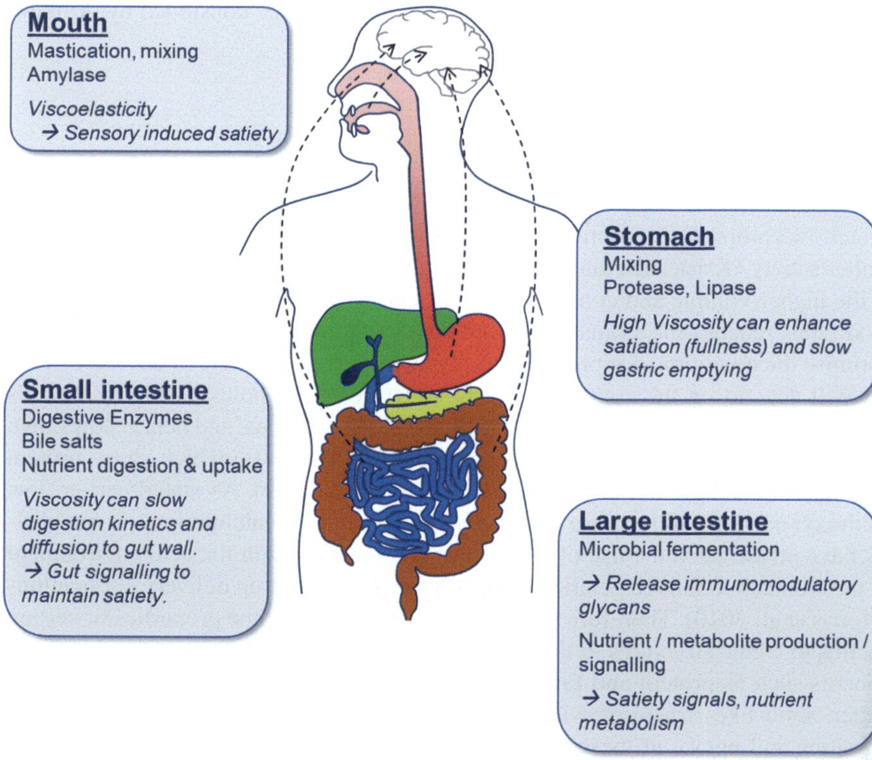

Fig. 8.1 The key stages of digestion in humans summarising the main functions of each. Possible role of pectin in italics. Dotted arrows indicate gut signalling pathways to the brain

8.2.2 Oral, Gastric and Small Intestinal Conditions

8.2.2.1 Mouth

In the mouth, the presence of calcium in the saliva may induce gelation of low methoxy pectins (Voragen et al. 2009), although there are no papers specifically studying this, some researchers have utilised this property of pectin to optimise structuring for controlled drug release and mucoadhesion. Increasing texture in the mouth, coupled with other oro-sensory stimuli, are known to influence expectations of satiation and satiety (Fiszman and Tarrega 2017). Slowing down consumption by forcing mastication will slow the rate of energy intake, which in turn contribute to the reduction in appetite (Hawton et al. 2018), but the mechanisms involved and the underlying physiological basis are not fully understood. Studies showed that the more solid meals are, the earlier the meal was terminated, which could be due to the energy density per g food, i.e., the dietary fibre content. In a study comparing a pectin-rich food, also the way of consumption was important, meaning that whole apples and apple soup eaten with a spoon have a comparable effect on hunger and satiation, but were both higher than the effect of apple juice consumed by drinking (Mattes 2005; de Graaf 2012).

8.2.2.2 Stomach

In the stomach, the main satiating factor is gastric volume via the stimulation of stretch receptors. Consumption of viscous or gelling fibres like pectin leads to a higher satiety (Kristensen and Jensen 2011; Wanders et al. 2011), presumably due to the higher volume and contact time with gastric sensors (de Graaf 2012). When food bolus reaches the stomach, it is gradually acidified with hydrochloric acid to optimise the conditions for protein hydrolysis by pepsin. The rate of decrease in the pH will depend on the composition and structure of the bolus. A highly viscous bolus can reduce the extent of mixing with the acid secretions, and hence slow down the pH change within the bolus (Marciani et al. 2000, 2001) which slows protein hydrolysis and digestion (Mulet-Cabero et al. 2017, 2019). As stated earlier, low methoxyl pectins will increase gelation in the presence of calcium ions. Some studies have investigated the use of calcium in combination with the acid environment of the stomach to induce gelation of pectin for sustained drug delivery applications (Morris et al. 2010). Therefore it may be possible to utilise the properties of certain pectins to create structures in the stomach that may reduce digestion of other components such as protein and fat, and slow down their release from the stomach as larger, solid-like structures will not pass through the pylorus. This will prolong digestion and uptake of these nutrients and thus prolong satiety. One study looked at using this approach to prolong gastric emptying in mice, but with limited success (Khramova et al. 2019). Delayed gastric emptying will prolong nutrient absorption in the small intestine stimulating the release of appetite-regulating peptides (Wanders et al. 2011). This mechanism has been demonstrated for pectin and alginate, both of which have high water holding capacities and are viscous, soluble fibres (Rigaud et al. 1998; Guillon and Champ 2000, 2002). Their physico-chemical behaviour has been shown to lead to a reduction in both appetite and energy intake (Wanders et al. 2011, 2014; Jensen et al. 2012). Wanders et al. showed that pectins with different properties do not have the same effect on gastric emptying and appetite. The gelled pectin reduced appetite ratings significantly compared to a viscous pectin, a bulking pectin, and the control without dietary fibre. However, the way of administration also affects the outcome. The gelled pectin consumed as a gel reduced the healthy volunteers' appetite to a bigger extent than the same pectin ingested in the form of capsules or in a liquid drink (Wanders et al. 2014).

8.2.2.3 Small Intestine

Acidic chyme is emptied from the stomach into the small intestine, the main site of nutrient digestion and absorption. Pectin is not digested by the amylolytic enzymes secreted from the pancreas, and so effectively passes through the small intestine undigested. However, the solution conditions such as pH and ionic strength/composition will influence its solubility and viscosity, which may play a role in pectin's functionality during digestion. Increased viscosity can slow down, via reduced mixing, delayed diffusion and viscous entrapment, the digestion kinetics and transport of digestion products to the gut wall for their uptake. There are some small intestinal regions which have distinct microbial populations which are thought to be important for human health and disease, especially during the digestion of dietary fibre, however these small intestinal microbiota require further research to fully understand their importance in the breakdown of pectin and the resultant effects on health (Guerra et al. 2012). Between 90% of pectin in healthy human subjects (Saito et al. 2005), and 85% (Sandberg et al. 1983) to 68% (Holloway et al. 1983) in ileostomy patients reach the large intestine. These findings are estimated by measuring galacturonic acid, a hydrolytic cleavage product of the homogalacturonan component of pectic polysaccharides, so there may be more conformational changes other than the full break down via polysaccharide fermentation (Saito et al. 2005).

In the small intestine, bile salts and mixed micelles play an important role in the digestion, transport and uptake of lipids. Pectin as a viscous, soluble fibre is thought to entrap bile salts or mixed micelles in a viscous network at different stages of the digestive pathway and can therefore influence lipid absorption (Gunness and Gidley 2010; Grundy et al. 2018) and thus reducing hyperlipaemia, a risk factor in cardiovascular disease. Furthermore dietary fibre like pectin can form a complex or association with bile salts during digestion and reduce their reabsorption and enterohepatic circulation, the loss of which has to be compensated for by de novo synthesis from cholesterol in the liver (Mertens et al. 2017). These interactions with bile salts are thought to lead to lower plasma cholesterol levels as well as a reduction in lipolysis, however, different extents of interaction were observed depending on the botanical origin and structural properties of the pectins (Cervantes-Paz et al. 2017). A meta-analysis including 67 controlled trials with fibres derived from oat, psyllium, guar gum or pectin, showed that intake of these fibres resulted in a reduction of 0.045 mmol/l total cholesterol per gram soluble fibre when fibre supplementation was in the practical range of 2–10 g/day (Brown et al. 1999).

The cholesterol lowering effect of pectin could also be due to lower postprandial glucose levels via reducing the rate of intestinal absorption of glucose and its decrease in insulin production. Insulin is an activator of 3-hydroxy 3-methylglutaryl coenzyme A reductase (HMG-Co AR), which regulates cholesterol synthesis and has increased activity at high insulin levels (Popov 2016). Soluble fibres have also been shown to affect the diffusion of molecules through the mucus layer (Mackie et al. 2016a, b). Mucin forms a protective macromolecular network on the surface of the gut wall, but the network contains pores which allow diffusion of digesta to the gut wall. The presence of soluble fibre has been shown in vitro to fill the pores and thus reduce the diffusion of particles and micelles into the mucous, thus further reducing the rate of nutrient absorption. However, four weeks pectin supplementation did not alter intestinal barrier function and mucosal defense capacity in healthy subjects (Wilms et al. 2019). Another mechanism of soluble fibres influencing insulinaemia and postprandial glycaemia could be a delay in starch degradation in the small intestine by either formation of a fibre-starch complex or fibre-enzyme complex or by alteration of enzyme and substrate diffusion due to increased viscosity of the reaction mixture (Ou et al. 2001; Slaughter et al. 2001; Shelat et al. 2010, 2011; Repin et al. 2018). A decelerated sugar absorption from the small intestinal lumen due to increased viscosity as discussed before could not be confirmed by the in vitro study of Repin et al. nor by Dhital et al. (2015) and Repin et al. (2017). Participants in a study by Wanders et al. had higher fasting glucose concentrations after repeated intake of 10 g pectin over a period of 15 days (Wanders et al. 2013). A reason could be added sugars in commercial pectins (Popov 2016). Studies are contradictory and exact mechanisms are still to be elucidated. Nevertheless, EFSA Panel on Dietetic Products (2006) granted a health claim when at least 10 g of pectin are consumed with the meal it can be stated that it "contributes to the reduction of the blood glucose rise" (EFSA Panel on Dietetic Products, Brouns et al. 2012). Once advances with regards to pectin source and type (DE and M_W) are made, the EFSA claim would ideally require a minimum level of characterization, including DE and $\mathrm{M_W}$ (Brouns et al. 2012).

In addition to its interactions with bile salts and the viscosity related effects, pectins can also interfere in other ways with the fat digestion in the small intestine. Depending on its molecular structure and physicochemical properties, pectins can stabilise or destabilise emulsions, and thus change the available surface area for the lipase to act on which leads to a higher or lower lipolysis rate, respectively. Further, pectins can bind calcium that is necessary for lipid digestion. Steric factors inhibiting the interfacial anchoring of lipase and colipase as well as resistance of interfacial films at the oil/water interface can hinder efficient break down of dietary fat. With pectins this is thought to be due to proteins found in the pectin structure. Some emulsifier can bind lipolysis products and induce accumulation at the interface which leads to a reduced lipase action. (Golding and Wooster 2010; Pilosof 2017; Cervantes-Paz et al. 2017). It is likely that more than one of these mechanisms are exerted by certain pectin structures, hence more research is needed to understand the complexity of the physiological processes occurring in the human digestive tract.

8.2.2.4 Large Intestine

Material which avoids hydrolysis by the human digestive enzymes in the small intestine, which is mainly polysaccharides and protein, becomes available for fermentation by the gut microbiota in the colon, where also water and electrolytes are re-absorbed (Guerra et al. 2012). Pectin is slightly degraded in the small intestine, but is mainly broken down in the large intestine (Holloway et al. 1983). Trillions of microbes reside in our large intestine in a complex community with a whole range of different glycoside hydrolase, polysaccharide lyase and carbohydrate esterase enzyme families. This gives flexibility in the fermentation substrate availability and hence all fibre consumption will lead to modifications in the microbiota composition, however, the implications on health or disease of these abundance shifts are still not entirely clear (Makki et al. 2018). Breaking down previously undigested material involves different bacteria and their enzymes, as well as cross-feeding to produce the fermentation products hydrogen, methane, carbon dioxide, lactate, succinate but most importantly short chain fatty acids (SCFAs) like acetate, propionate and butyrate (Slavin 2013). The fermentation products are very important determinants of the health benefits (Williams et al. 2017). Locally, they induce changes in the colon like lowering of the pH and protection of the mucus layer, avoiding penetrability and subsequent increased susceptibility to infections (Makki et al. 2018). Whether fibre consumption increases microbial diversity is inconsistent (Williams et al. 2017) is inconsistent. A recent systematic review and meta-analysis showed that dietary fibre interventions in randomised controlled trials had no real effect on the diversity of the gut microbiota but did increase abundance of *Bifidobacterium* and *Lactobacillus* spp. as well as faecal butyrate concentration (Whelan 1960). *Bifidobacterium* increased almost 25% in the presence of pectin compared with the control in an in vitro fermentation study (Yang et al. 2013). Increase in faecal mass, which has a beneficial effect on the stool bulking, can be either due to not all fibres being fermentable or an increase in bacterial mass due to fermentation (Slavin 2013). However, it is still under debate whether the latter has a substantial effect on stool output or stool softening (McRorie Jr. and McKeown 2017; Makki et al. 2018). The effect of viscosity on laxation is not well investigated (Dikeman and Fahey 2006). Most viscous fibres, like pectins, are fermented and do not alter the rheological properties in the large intestine unlike psyllium, which is regarded as a non-fermenatable fibre (McRorie and Chey 2016). Butyrate is mainly used locally as an energy source by colonocytes, but SCFAs are also found in the peripheral circulation to be transported to the liver and peripheral tissue where they act as signalling molecules and regulate host metabolism. In this way they also exert their immune and cell proliferation modulating effect (Makki et al. 2018). Butyrate with its inflammation-modulating effect is currently under study as a colon cancer-preventive agent (Chen and Vitetta 2018), for example, via starch esterified with butyrate (Clarke et al. 2011). Both dietary fibres and SCFAs, mainly acetate and butyrate, are important for mucus production stimulation and secretion (Makki et al. 2018).

The fermentation of dietary fibres by the microbiota has various systemic effects on the human body. Propionate is cleared by the liver (Slavin 2013) and depletes plasma cholesterol by inhibiting hepatic cholesterol synthesis (Gunness and Gidley 2010; Davison and Temple 2018). High fibre intake is associated with a lower risk for type 2 diabetes which is thought to be due to several mechanisms. One is via an effect on the immune system via lymphocytes and toll-like receptor 5 in the intestinal mucosa, which are associated with inflammation, insulin sensitivity and glucose homeostasis. The fermentation metabolites can also influence the energy metabolism pathways when inducing lipolysis leading to increased fat storage, influencing hepatic de novo synthesis of triglycerides and modification of bile acid metabolites (Gunness and Gidley 2010; Davison and Temple 2018). Acetate and propionate produced from guar gum fermentation could at least partially explain the increase in peripheral glucose disposal by an increase in the GLP-1 secretion in a mouse model (den Besten et al. 2015). In contrast to this, another mouse model showed that SCFA production and inulin but not guar gum could prevent diet-induced obesity and insulin resistance (Weitkunat et al. 2017) and studies in humans show that guar gum reduces the blood glucose rise directly after the meal (Gatenby et al. 1996).

8.3 Metabolism

Pectin is metabolised by the gut microbiota to yield various metabolites as described above and is therefore in faeces not present or to a very low extent (Holloway et al. 1983). The extent of fermentation is not always the same and depends in in vitro models on the fermentation conditions, DE of pectin (Dongowski and Lorenz 1998), composition of neutral sugars, distribution of homogalacturonan and rhamnogalacturonan fractions, degree of branching, and the presence of amide groups, whereas the SCFA and propionate production was highest with high methoxyl pectins (Larsen et al. 2019). Studies have detected galacturonans (Nyman and Asp 1982; Hill 1995), which may be involved in resorption of lead ions from the colon and in its transport (Dongowski and Lorenz 1998). An in vitro study of nine structurally diverse pectins from various sources found changes microbiota abundances which were structure specific. Depending on the molecular make-up of the substrates, bacterial taxa which are associated with certain health outcomes were either increased or decreased (Larsen et al. 2019). In the future, this may help us to stimulate targeted production of certain fermentation products, like SCFAs or short glycan chains which have specific bioactivities such as modulating the immune response (Rösch et al. 2017). The immunomodulatory effect is suggested to be not only caused by fermentation but also by blocking innate immune receptors directly. It was shown recently that pectin binds and inhibits, toll-like receptor (TLR) 2 and specifically inhibits the proinflammatory TLR2–TLR1 pathway while the tolerogenic TLR2–TLR6 pathway remains unaltered depending the concentration and level of methyl esterification of pectin. Pectins with low methyl esterification have more unesterified regions and are more charged, which led to more interaction with TLR2 through electrostatic forces exhibiting the anti-inflammatory effect (Sahasrabudhe et al. 2018). Anti-cancer activities of pectin have been known for decades (Willats et al. 2006; Leclere et al. 2013) through mechanisms involving inhibition of tumor growth, inducing apoptosis, suppressing metastasis, and modulating immunological responses (Zhang et al. 2015), however, structure-activity relationship studies are needed in the future since many studies have used pectin preparations with different structural domains, which does not allow mechanistic links with defined molecular structures to be made (Willats et al. 2006). Methylation may also have an effect on overall gene activity. When 6.0 g citrus pectin was ingested, the plasma methanol level doubled in healthy subjects after 2 h. With these increased levels of methanol half the genes selected for analysis which encoded glycoproteins had an altered expression (almost 50:50 whether up- or down-regulated). 9 of the 32 proteins participated in cell-cell communication processes which may indicate that methanol concentration alters the expression of genes responsible for intercellular communication in white blood cells (Shindyapina et al.

2014). Furthermore, pectin has been shown to stimulate protein metabolism in the gut (Pirman et al. 2007).

8.4 Conclusions

There is increasing evidence that pectins are not simply an inert thickening agent, but possess a range of health benefits. These effects are mediated by their influence on digestive processes, their resistance to digestion by the intrinsic digestive enzymes in humans and their subsequent fermentation by our microbiota. The water holding capacity and viscosity enhancing properties can influence appetite, and also the diffusion processes controlling the digestion and absorption of nutrients which in turn may positively influence cholesterol levels and glycaemic responses. As a soluble fibre, pectin is by definition resistant to digestion, but fermentation by the microbiota yielding metabolites implicated with a wide range of health benefits. In addition, the specific glycan side chains have been linked with immunomodulatory effects, but these will be discussed in more detail in the next chapter.

References

Aprikian O, Duclos V, Guyot S, Besson C, Manach C, Bernalier A, Morand C, Remesy C, Demigne C (2003) Apple pectin and a polyphenol-rich apple concentrate are more effective together than separately on cecal fermentations and plasma lipids in rats. J Nutr 133(6):1860–1865

Benelam B (2009) Satiation, satiety and their effects on eating behaviour. Nutr Bull 34(2):126–173

Brett CT, Waldron KW (1996) Physiology and biochemistry of plant cell walls. Chapman and Hall, London

Brouns F, Theuwissen E, Adam A, Bell M, Berger A, Mensink RP (2012) Cholesterol-lowering properties of different pectin types in mildly hyper-cholesterolemic men and women. Eur J Clin Nutr 66(5):591–599

Brown L, Rosner B, Willett WW, Sacks FM (1999) Cholesterol-lowering effects of dietary fiber: a meta-analysis. Am J Clin Nutr 69(1):30–42

Cabrera JC, Cambier P, van Cutsem P (2011) Drug encapsulation in pectin hydrogel beads- a systematic study of simulated digestion media. Int J Pharm Pharm Sci 3(suppl. 5):292–299

Canteri MHG, Renard C, Le Bourvellec C, Bureau S (2019) ATR-FTIR spectroscopy to determine cell wall composition: application on a large diversity of fruits and vegetables. Carbohydr Polym 212:186–196

Cecil JE, Francis J, Read NW (1998) Relative contributions of intestinal, gastric, oro-sensory influences and information to changes in appetite induced by the same liquid meal. Appetite 31(3):377–390

Cervantes-Paz B, Ornelas-Paz J, Ruiz-Cruz S, Rios-Velasco C, Ibarra-Junquera V, Yahia EM, Gardea-Béjar AA (2017) Effects of pectin on lipid digestion and possible implications for carotenoid bioavailability during pre-absorptive stages: A review. Food Res Int 99:917–927

Chen JZ, Vitetta L (2018) Inflammation-modulating effect of butyrate in the prevention of colon cancer by dietary fiber. Clin Colorectal Cancer 17(3):E541–E544

Clarke JM, Topping DL, Christophersen CT, Bird AR, Lange K, Saunders I, Cobiac L (2011) Butyrate esterified to starch is released in the human gastrointestinal tract. Am J Clin Nutr 94(5):1276–1283
Codex Alimentarius Commission (2009) Report of the 30th session of the Codex Commitee on Nutrition and Foods for Special Dietary Uses. J. F. W. F. S. Programme
Davison KM, Temple NJ (2018) Cereal fiber, fruit fiber, and type 2 diabetes: explaining the paradox. J Diabetes Complicat 32(2):240–245
de Graaf C (2012) Texture and satiation: the role of oro-sensory exposure time. Physiol Behav 107(4):496–501
den Besten G, Gerding A, van Dijk TH, Ciapaite J, Bleeker A, van Eunen K, Havinga R, Groen AK, Reijngoud DJ, Bakker BM (2015) Protection against the metabolic syndrome by guar gum-derived short-chain fatty acids depends on peroxisome proliferator-activated receptor gamma and glucagon-like peptide-1. PLoS One 10(8):e0136364
Dhital S, Gidley MJ, Warren FJ (2015) Inhibition of alpha-amylase activity by cellulose: kinetic analysis and nutritional implications. Carbohydr Polym 123:305–312
Dikeman CL, Fahey GC (2006) Viscosity as related to dietary fiber: a review. Crit Rev Food Sci Nutr 46(8):649–663
Doco T, Williams P, Vidal S, Pellerin P (1997) Rhamnogalacturonan II, a dominant polysaccharide in juices produced by enzymic liquefaction of fruits and vegetables. Carbohydr Res 297(2):181–186
Dongowski G, Lorenz A (1998) Unsaturated oligogalacturonic acids are generated by in vitro treatment of pectin with human faecal flora. Carbohydr Res 314(3):237–244
EFSA Panel on Dietetic Products (2006) N. a. A. N. Scientific Opinion on the substantiation of health claims related to pectins and reduction of post-prandial glycaemic responses (ID 786), maintenance of normal blood cholesterol concentrations (ID 818) and increase in satiety leading to a reduction in energy intake (ID 4692) pursuant to Article 13(1) of Regulation (EC) No 1924/2006. EFSA J 8(10):1747
Einhorn-Stoll U (2018) Pectin-water interactions in foods – from powder to gel. Food Hydrocoll 78:109–119
Fiszman S, Tarrega A (2017) Expectations of food satiation and satiety reviewed with special focus on food properties. Food Funct 8(8):2686–2697
Gatenby SJ, Ellis PR, Morgan LM, Judd PA (1996) Effect of partially depolymerized guar gum on acute metabolic variables in patients with non-insulin-dependent diabetes. Diabet Med 13(4):358–364
Golding M, Wooster TJ (2010) The influence of emulsion structure and stability on lipid digestion. Curr Opin Colloid Interface Sci 15(1-2):90–101
Gromer A, Kirby AR, Gunning AP, Morris VJ (2009) Interfacial structure of sugar beet pectin studied by atomic force microscopy. Langmuir 25(14):8012–8018
Grundy MM, Fardet A, Tosh SM, Rich GT, Wilde PJ (2018) Processing of oat: the impact on oat's cholesterol lowering effect. Food Funct 9(3):1328–1343
Guerra A, Etienne-Mesmin L, Livrelli V, Denis S, Blanquet-Diot S, Alric M (2012) Relevance and challenges in modeling human gastric and small intestinal digestion. Trends Biotechnol 30(11):591–600
Guillon F, Champ M (2000) Structural and physical properties of dietary fibres, and consequences of processing on human physiology. Food Res Int 33(3-4):233–245
Guillon F, Champ MM (2002) Carbohydrate fractions of legumes: uses in human nutrition and potential for health. Br J Nutr 88(Suppl 3):S293–S306
Gunness P, Gidley MJ (2010) Mechanisms underlying the cholesterol-lowering properties of soluble dietary fibre polysaccharides. Food Funct 1(2):149–155
Hawton K, Ferriday D, Rogers P, Toner P, Brooks J, Holly J, Biernacka K, Hamilton-Shield J, Hinton E (2018) Slow down: behavioural and physiological effects of reducing eating rate. Nutrients 11(1):50

Hill MJ (1995) Bacterial fermentation of complex carbohydrate in the human colon. Eur J Cancer Prev 4(5):353–358

Hollands WJ, Hart DJ, Dainty JR, Hasselwander O, Tiihonen K, Wood R, Kroon PA (2013) Bioavailability of epicatechin and effects on nitric oxide metabolites of an apple flavanol-rich extract supplemented beverage compared to a whole apple puree: a randomized, placebo-controlled, crossover trial. Mol Nutr Food Res 57(7):1209–1217

Holloway WD, Tasman-Jones C, Maher K (1983) Pectin digestion in humans. Am J Clin Nutr 37(2):253–255

Ishii T, Matsunaga T (1996) Isolation and characterization of a boron-rhamnogalacturonan-II complex from cell walls of sugar beet pulp. Carbohydr Res 284(1):1–9

Jensen MG, Knudsen JC, Viereck N, Kristensen M, Astrup A (2012) Functionality of alginate based supplements for application in human appetite regulation. Food Chem 132(2):823–829

Kaneko S, Ishii T, Matsunaga T (1997) A boron-rhamnogalacturonan-II complex from bamboo shoot cell walls. Phytochemistry 44(2):243–248

Khramova DS, Vityazev FV, Saveliev NY, Burkov AA, Beloserov VS, Martinson EA, Litvinets SG, Popov SV (2019) Pectin gelling in acidic gastric condition increases rheological properties of gastric digesta and reduces glycaemic response in mice. Carbohydr Polym 205:456–464

Kristensen M, Jensen MG (2011) Dietary fibres in the regulation of appetite and food intake. Importance of viscosity. Appetite 56(1):65–70

Larsen N, Bussolo de Souza C, Krych L, Barbosa Cahú T, Wiese M, Kot W, Hansen KM, Blennow A, Venema K, Jespersen L (2019) Potential of pectins to beneficially modulate the gut microbiota depends on their structural properties. Front Microbiol 10:223

Lasschuijt MP, Mars M, Stieger M, Miquel-Kergoat S, de Graaf C, Smeets PAM (2017) Comparison of oro-sensory exposure duration and intensity manipulations on satiation. Physiol Behav 176:76–83

Leclere L, Van Cutsem P, Michiels C (2013) Anti-cancer activities of pH- or heat-modified pectin. Front Pharmacol 4:128

Lovegrove A, Edwards CH, De Noni I, Patel H, El SN, Grassby T, Zielke C, Ulmius M, Nilsson L, Butterworth PJ, Ellis PR, Shewry PR (2017) Role of polysaccharides in food, digestion, and health. Crit Rev Food Sci Nutr 57(2):237–253

Mackie A, Rigby N, Harvey P, Bajka B (2016a) Increasing dietary oat fibre decreases the permeability of intestinal mucus. J Funct Foods 26:418–427

Mackie AR, Macierzanka A, Aarak K, Rigby NM, Parker R, Channell GA, Harding SE, Bajka BH (2016b) Sodium alginate decreases the permeability of intestinal mucus. Food Hydrocoll 52:749–755

Makki K, Deehan EC, Walter J, Backhed F (2018) The impact of dietary fiber on gut microbiota in host health and disease. Cell Host Microbe 23(6):705–715

Marciani L, Gowland PA, Spiller RC, Manoj P, Moore RJ, Young P, Al-Sahab S, Bush D, Wright J, Fillery-Travis AJ (2000) Gastric response to increased meal viscosity assessed by echo-planar magnetic resonance imaging in humans. J Nutr 130(1):122–127

Marciani L, Gowland PA, Spiller RC, Manoj P, Moore RJ, Young P, Fillery-Travis AJ (2001) Effect of meal viscosity and nutrients on satiety, intragastric dilution, and emptying assessed by MRI. Am J Physiol Gastrointest Liver Physiol 280(6):G1227–G1233

Matoh T, Ishigaki K-i, Ohno K, Azuma J-i (1993) Isolation and characterization of a boron-polysaccharide complex from radish roots1. Plant Cell Physiol 34(4):639–642

Mattes R (2005) Soup and satiety. Physiol Behav 83(5):739–747

Maxwell EG, Colquhoun IJ, Chau HK, Hotchkiss AT, Waldron KW, Morris VJ, Belshaw NJ (2015) Rhamnogalacturonan I containing homogalacturonan inhibits colon cancer cell plif-eration by decreasing ICAM1 expression. Carbohydrate Polymers 132:546–553. https://doi.org/10.1016/j.carbpol.2015.06.082

McRorie JW, Chey WD (2016) Fermented fiber supplements are no better than placebo for a laxative effect. Dig Dis Sci 61(11):3140–3146

McRorie JW Jr, McKeown NM (2017) Understanding the Physics of functional fibers in the gastrointestinal tract: an evidence-based approach to resolving enduring misconceptions about insoluble and soluble fiber. J Acad Nutr Diet 117(2):251–264
Mertens KL, Kalsbeek A, Soeters MR, Eggink HM (2017) Bile acid signaling pathways from the enterohepatic circulation to the central nervous system. Front Neurosci 11:617
Morey S, Shafat A, Clegg ME (2016) Oral versus intubated feeding and the effect on glycaemic and insulinaemic responses, gastric emptying and satiety. Appetite 96:598–603
Morris GA, Kök SM, Harding SE, Adams GG (2010) Polysaccharide drug delivery systems based on pectin and chitosan. Biotechnol Genet Eng Rev 27(1):257–284
Mulet-Cabero A-I, Rigby NM, Brodkorb A, Mackie AR (2017) Dairy food structures influence the rates of nutrient digestion through different in vitro gastric behaviour. Food Hydrocoll 67:63–73
Mulet-Cabero A-I, Mackie AR, Wilde PJ, Fenelon MA, Brodkorb A (2019) Structural mechanism and kinetics of in vitro gastric digestion are affected by process-induced changes in bovine milk. Food Hydrocoll 86:172–183
Naqash F, Masoodi FA, Rather SA, Wani SM, Gani A (2017) Emerging concepts in the nutraceutical and functional properties of pectin-a review. Carbohydr Polym 168:227–239
Navarro SL, Neuhouser ML, Cheng T-YD, Tinker LF, Shikany JM, Snetselaar L, Martinez JA, Kato I, Beresford SAA, Chapkin RS, Lampe JW (2016) The Interaction between dietary fiber and fat and risk of colorectal cancer in the women's health initiative. Nutrients 8(12):779
Ndeh D, Rogowski A, Cartmell A, Luis AS, Baslé A, Gray J, Venditto I, Briggs J, Zhang X, Labourel A, Terrapon N, Buffetto F, Nepogodiev S, Xiao Y, Field RA, Zhu Y, O'Neill MA, Urbanowicz BR, York WS, Davies GJ, Abbott DW, Ralet M-C, Martens EC, Henrissat B, Gilbert HJ (2017) Complex pectin metabolism by gut bacteria reveals novel catalytic functions. Nature 544:65
Nyman M, Asp N-G (1982) Fermentation of dietary fibre components in the rat intestinal tract. Br J Nutr 47(3):357–366
O'Neill MA, Ishii T, Albersheim P, Darvill AG (2004) Rhamnogalacturonan II: structure and function of a borate cross-linked cell wall pectic polysaccharide. Annu Rev Plant Biol 55(1):109–139
Ou SY, Kwok KC, Li Y, Fu L (2001) In vitro study of possible role of dietary fiber in lowering postprandial serum glucose. J Agric Food Chem 49(2):1026–1029
Patrice P, Malcolm AON, Cécile P, Marie-Thérèse C, Alan GD, Peter A, Michel M (1997) Lead complexation in wines with the dimers of the grape pectic polysaccharide rhamnogalacturonan II. OENO One 31(1):1093
Pilosof AMR (2017) Potential impact of interfacial composition of proteins and polysaccharides stabilized emulsions on the modulation of lipolysis. The role of bile salts. Food Hydrocoll 68:178–185
Pirman T, Ribeyre MC, Mosoni L, Rémond D, Vrecl M, Salobir J, Mirand PP (2007) Dietary pectin stimulates protein metabolism in the digestive tract. Nutrition 23(1):69–75
Popov SV (2016) Pectins as biological modulators of human physiological reactions. In: Wiederschain G (ed) Glycobiology and human diseases. CRC Press, Boca Raton
Prado SBRD, Ferreira GF, Harazono Y, Shiga TM, Raz A, Carpita NC, Fabi JP (2017) Ripening-induced chemical modifications of papaya pectin inhibit cancer cell proliferation. Sci Rep 7(1):16564
Repin N, Kay BA, Cui SW, Wright AJ, Duncan AM, Douglas Goff H (2017) Investigation of mechanisms involved in postprandial glycemia and insulinemia attenuation with dietary fibre consumption. Food Funct 8(6):2142–2154
Repin N, Cui SW, Goff HD (2018) Rheological behavior of dietary fibre in simulated small intestinal conditions. Food Hydrocoll 76:216–225
Rigaud D, Paycha F, Meulemans A, Merrouche M, Mignon M (1998) Effect of psyllium on gastric emptying, hunger feeling and food intake in normal volunteers: a double blind study. Eur J Clin Nutr 52(4):239–245

Rösch C, Taverne N, Venema K, Gruppen H, Wells JM, Schols HA (2017) Effects of in vitro fermentation of barley β-glucan and sugar beet pectin using human fecal inocula on cytokine expression by dendritic cells. Mol Nutr Food Res 61(1):1600243

Sahasrabudhe NM, Beukema M, Tian L, Troost B, Scholte J, Bruininx E, Bruggeman G, van den Berg M, Scheurink A, Schols HA, Faas MM, de Vos P (2018) Dietary fiber pectin directly blocks toll-like receptor 2–1 and prevents doxorubicin-induced ileitis. Front Immunol 9:383

Saito D, Nakaji S, Fukuda S, Shimoyama T, Sakamoto J, Sugawara K (2005) Comparison of the amount of pectin in the human terminal ileum with the amount of orally administered pectin. Nutrition 21(9):914–919

Sandberg AS, Ahderinne R, Andersson H, Hallgren B, Hultén L (1983) The effect of citrus pectin on the absorption of nutrients in the small intestine. Human nutrition. Clin Nutr 37(3):171–183

Shelat KJ, Vilaplana F, Nicholson TM, Wong KH, Gidley MJ, Gilbert RG (2010) Diffusion and viscosity in arabinoxylan solutions: implications for nutrition. Carbohydr Polym 82(1):46–53

Shelat KJ, Vilaplana F, Nicholson TM, Gidley MJ, Gilbert RG (2011) Diffusion and rheology characteristics of barley mixed linkage β-glucan and possible implications for digestion. Carbohydr Polym 86(4):1732–1738

Shindyapina AV, Petrunia IV, Komarova TV, Sheshukova EV, Kosorukov VS, Kiryanov GI, Dorokhov YL (2014) Dietary methanol regulates human gene activity. PLoS One 9(7):e102837

Slaughter SL, Ellis PR, Butterworth PJ (2001) An investigation of the action of porcine pancreatic α-amylase on native and gelatinised starches. Biochim Biophys Acta 1525(1):29–36

Slavin J (2013) Fiber and prebiotics: mechanisms and health benefits. Nutrients 5(4):1417–1435

So D, Whelan K, Rossi M, Morrison M, Holtmann G, Kelly JT, Shanahan ER, Staudacher HM, Campbell KL (2018) Dietary fiber intervention on gut microbiota composition in healthy adults: a systematic review and meta-analysis. Am J Clin Nutr 107(6):965–983

Strasser GR, Amadò R (2002) Pectic substances from red beet (Beta vulgaris L. var. conditiva). Part II. Structural characterisation of rhamnogalacturonan II. Carbohydr Polym 48(3):263–269

U.S. Food and Drug Administration (2018) Review of the scientific evidence on the physiological effects of certain non-digestible carbohydrates. U.S. Department of Health and Human Services, Center for Food Safety and Applied Nutrition, Silver Spring

Ulmius M, Johansson-Persson A, Norden TI, Bergenstahl B, Onning G (2011) Gastrointestinal release of beta-glucan and pectin using an in vitro method. Cereal Chem 88(4):385–390

Vidal S, Williams P, O'Neill MA, Pellerin P (2001) Polysaccharides from grape berry cell walls. Part I: tissue distribution and structural characterization of the pectic polysaccharides. Carbohydr Polym 45(4):315–323

Voragen AGJ, Coenen GJ, Verhoef RP, Schols HA (2009) Pectin, a versatile polysaccharide present in plant cell walls. Struct Chem 20(2):263–275

Wanders AJ, van den Borne JJ, de Graaf C, Hulshof T, Jonathan MC, Kristensen M, Mars M, Schols HA, Feskens EJ (2011) Effects of dietary fibre on subjective appetite, energy intake and body weight: a systematic review of randomized controlled trials. Obes Rev 12(9):724–739

Wanders AJ, Mars M, Borgonjen-van den Berg KJ, de Graaf C, Feskens EJM (2013) Satiety and energy intake after single and repeated exposure to gel-forming dietary fiber: post-ingestive effects. Int J Obes 38:794

Wanders AJ, Feskens EJ, Jonathan MC, Schols HA, de Graaf C, Mars M (2014) Pectin is not pectin: a randomized trial on the effect of different physicochemical properties of dietary fiber on appetite and energy intake. Physiol Behav 128:212–219

Weitkunat K, Stuhlmann C, Postel A, Rumberger S, Fankhanel M, Woting A, Petzke KJ, Gohlke S, Schulz TJ, Blaut M, Klaus S, Schumann S (2017) Short-chain fatty acids and inulin, but not guar gum, prevent diet-induced obesity and insulin resistance through differential mechanisms in mice. Sci Rep 7(1):6109

Whelan WJ (1960) The action patterns of α-amylases. Starch 12(12):358–364

Wijlens AGM, Erkner A, Alexander E, Mars M, Smeets PAM, de Graaf C (2012) Effects of oral and gastric stimulation on appetite and energy intake. Obesity 20(11):2226–2232

Willats WGT, Knox JP, Mikkelsen JD (2006) Pectin: new insights into an old polymer are starting to gel. Trends Food Sci Technol 17(3):97–104

Williams BA, Grant LJ, Gidley MJ, Mikkelsen D (2017) Gut fermentation of dietary fibres: physico-chemistry of plant cell walls and implications for health. Int J Mol Sci 18:10

Wilms E, Jonkers DMAE, Savelkoul HFJ, Elizalde M, Tischmann L, de Vos P, Masclee AAM, Troost FJ (2019) The impact of pectin supplementation on intestinal barrier function in healthy young adults and healthy elderly. Nutrients 11(7):1554

Yang J, Martínez I, Walter J, Keshavarzian A, Rose DJ (2013) In vitro characterization of the impact of selected dietary fibers on fecal microbiota composition and short chain fatty acid production. Anaerobe 23:74–81

Zhang W, Xu P, Zhang H (2015) Pectin in cancer therapy: a review. Trends Food Sci Technol 44(2):258–271

Chapter 9
Pectin Bioactivity

Huihuang Ding and Steve W. Cui

9.1 Introduction

Pectin is widely distributed in plants, commonly find in fruits, legumes, oil seeds and tubers. Pectins from different plant origins are important ingredients in processed foods, pharmaceutical, cosmetic, or feed products as binders or texturizing agents. US Food and Drug Administration (FDA) has approved the generally recognised as safe (GRAS) status for pectin; European Food Safety Authority (EFSA) also stated "there is no safety concern for the use of pectin and amidated pectin as food additives for the general population and that there is no need for a numerical acceptable daily intake" (Akhmedov et al. 2017).

As a major type of prebiotics and bioactive polysaccharides, pectins have gained increased interests in the past three decades: numerous studies have been reported for its health promoting functions. Pectins have a wide range of bioactivities, such as anti-cancer, immune enhancement, gastroprotection, antibacterial and anticoagulant. Recent studies also demonstrated that pectin is a promising material as the carrier or encapsulation matrix for targeted delivering of functional components or exert synergistic bioactivities. In this chapter, the bioactivity of pectins from various sources are classified, and the structure-function relationship is elucidated. In addition, the potential applications of bioactive pectins are summarised.

H. Ding (✉) · S. W. Cui
Guelph Research and Development Centre, Agriculture and Agri-Food Canada, Guelph, ON, Canada
e-mail: dingh@uoguelph.ca; Steve.Cui@canada.ca

V. Kontogiorgos (ed.), *Pectin: Technological and Physiological Properties*,
https://doi.org/10.1007/978-3-030-53421-9_9

9.2 Bioactivity of Pectins and Hydrolysates/Derivatives

The utilisation of pectins as prebiotics are discussed in another chapter. In this section, bioactivities, such as anticancer, immunomodulatory, and antibacterial/antivirus activities of pectins are emphasised.

9.2.1 *Anticancer Activity*

The anticancer properties of pectins and the application in cancer therapy have been widely reported by inhibiting the cancer cells growth and suppressing the development of precancerous conditions (Khotimchenko 2010; Zhang et al. 2015; Lara-Espinoza 2018). Recent studies provided more details on the anticancer activity and correlated with its structural characteristics. For example, rhamnogalacturonan-I (RG-I) rich pectins from ginseng flower (Cui et al. 2019) and ginseng (Shi et al. 2017) were reported to exert strong binding activities to galectin-3. RG-I from *Opuntia ficus-indica* after heat modification could selectively inhibit LANS cancer cells without affecting normal cells (Lefsih et al. 2018). Citrus pectins (both of HG and RG-I rich extracts) inhibited the growth of HT-29 colon cancer cells (Ma et al. 2018), and suppressed hemagglutination mediated by galectin-3 (Zhang et al. 2016). Pectic oligosaccharides from citrus could reduce the colon cancer risk and prevent progression of cancer cells (Tan et al. 2018). RG-I from the flowers of *Lonicera japonica* inhibited the growth of BxPC-3 and PANC-1 pancreatic cancer cells with inhibitory rates of 66.7% and 52.1%, respectively, at the concentration of 1 mg/mL (Lin et al. 2016). Potato RG-I could inhibit the proliferation of HT-29 cells (Cheng et al. 2013). More reports of pectin sources and their anticancer activities are summarised in Table 9.1.

The proposed anticancer activity mechanisms mainly involve the following: (1) binding of pectin to proteins responsible for the adhesion of tumor cells; (2) apoptosis induction in tumor cells; (3) inhibition of galectin-3, which protects cancer cells from apoptosis; (4) inhibition on the migration of cancer cell; and (5) slowdown in cancer progression and/or aggregation (Ovodov 2009; Khotimchenko 2010; Zhang et al. 2015). Although there is an increasing usage of pectin in cancer therapy, the underlining mechanism of the apoptosis induction and indirect inhibitory effects on cancer cells by pectin is still unclear. Some studies proposed a different or even contradictory mechanisms on the anticancer activity. For example, Stegmayr et al. reported that galactan and free galactose cannot inhibit galectins, and the inhibitory effect by galactose-containing side chains was not demonstrated (Stegmayr et al. 2016); pectins with methoxylation degrees of 37% and 70.4% were even reported to stimulated tumor growth in the tested mouse model (Jacobasch et al. 2008).

The anticancer mechanisms could be directly or indirectly correlated with their antibacterial, anti-inflammatory, prebiotic and immunostimulating activities (Chen et al. 2006; Flint et al. 2008; Georgiev et al. 2012). Pectic oligosaccharides could

Table 9.1 Anticancer activity of pectin

Pectin source	Anticancer activity	Reference
Viscum album L.	Anticancer activity: targeted cytotoxicity against NK-insensitive tumor cells	Zhu et al. (1994)
–	Anticancer activity: cytotoxicity against HCMV-infected cells	Steinmassl and Anderer (1996)
Citrus	Anticancer activity: induced apoptosis in human prostate cancer cells	Jackson et al. (2007)
Potato, citrus	Anticancer activity: specific binding to the human galectin-3	Gunning et al. (2009)
Potato	Anticancer activity: inhibited the proliferation of HT-29 cells and induced significant G2/M cell cycle arrest	Cheng et al. (2013)
Ginseng	Anticancer activity: inhibited galectin-3-mediated hemagglutination, cancer cell adhesion and homotypic aggregation	Gao et al. (2013)
Different sources	Anticancer activity: reduced proliferation of colon cancer cells	Maxwell et al. (2015a, b)
Flowers of *Lonicera japonica*	Anti-cancer activity: inhibited BxPC-3 and PANC-1 pancreatic cancer cells	Lin et al. (2016)
Citrus	Anti-cancer activity: inhibited hemagglutination mediated by galectin-3	Zhang et al. (2016)
Pumpkin	Anticancer activity	Zhao et al. (2017)
Ginseng	Anticancer activity: inhibitory activity on galectin-3	Shi et al. (2017)
Papaya	Anticancer activity: decreased cell viability and induced necroptosis	do Prado et al. (2017)
Citrus	Anticancer activity: inhibited Caco-2 cells	Ai et al. (2018)
Opuntia ficus-indica	Anti-cancer activity: selectively cytotoxic activity against LANS cancer cells	Lefsih et al. (2018)
Citrus	Anticancer activity: inhibited HT-29 colon cancer cells	Ma et al. (2018)
Ginseng	Anticancer activity: exerted stronger effects on cell migration	(Fan et al. 2018)
Ginseng flower	Anticancer activity: exhibited strong binding activities to galectin-3	(Cui et al. 2019)
Saussurea laniceps	Antivirus and anticancer activity: inhibited the secretion of HBsAg and HBeAg in HepG2.2.15 cells	Chen et al. (2019a)
Ficus carica L. skin	Antioxidant and cytotoxic inhibitory activities: inhibited HepG2 and A549 cells	Gharibzahedi et al. (2019a, b)

affect galectin network and apoptotic proteins in the colon by regulating gut microbiota (Zhang et al. 2015). Tan et al. (2018) reviewed the indirect inhibitory effects of pectic oligosaccharides against colon cancer cells: the antioxidant and anti-inflammatory activities could regulate potential oxidative and inflammation-activated pathways of colon cancer cells; the antibacterial activities could inhibit some aggressive bacterial infection, hence to control colon cancer risk.

9.2.2 Immunomodulatory Activity

As presented in Table 9.2, ginseng pectin exerted both anticancer and immunomodulatory activities, showing therapeutic effects on diabetes mellitus hyperglycaemia, oxidative stress, and immunological dysfunction (Chen et al. 2017). RG-I from bee pollen of *Nelumbo nucifera* could increase lymphocyte proliferation and macrophage phagocytosis (Li et al. 2018). Wang et al. (2018) reported the immunostimulating

Table 9.2 Immunostimulating activity of pectin

Pectin source	Immunostimulating activity	Reference
Asteraceae	Anti-complementary and immunostimulating activity: induced chemotaxis of human macrophages, T cells and NK cells	Nergard et al. (2005)
Centella asiatica (L.) Urban	Immunostimulating activity	Wang et al. (2003, 2005)
White cabbage	Immunostimulating and complement-fixing activities	Westereng et al. (2006)
Biophytum petersianum Klotzsch	Immunostimulating activity: against Peyer's patch immunocompetent cells	Grønhaug et al. (2011)
Cola cordifolia bark	Immunostimulating and complement-fixing activities	Austarheim et al. (2012)
Panax ginseng	Immunostimulating activity	Zhang et al. (2012)
Ilex paraguariensis	Anti-infective activity: affected neutrophil influx, and inhibited tissue expression of iNOS and COX-2	Dartora et al. (2013)
Lycium ruthenicum Murr. fruit	Immunostimulating activity: promoted macrophage proliferation in vitro	Peng et al. (2014)
Elderberries	Complement fixing activity and macrophage stimulating activity	Ho et al. (2015)
Hallabong peel	Immunostimulating activity and anti-inflammation	Lee et al. (2016)
Barley leaf	Immunostimulating activity	Kim et al. (2017)
Sweet pepper	Immunostimulating activity: increased TNF-alpha, IL-1 beta and IL-10 cytokine secretion by THP-1 macrophages	do Nascimento et al. (2017)
Ginseng	Immunostimulating activity: showed therapeutic effects on diabetes mellitus hyperglycemia, oxidative stress, immunological dysfunction, and neoplasms	Chen et al. (2017)
Hovenia dulcis peduncles	Immunostimulating activity: stimulated the proliferation of RAW264.7 cells	Wang et al. (2017)
Bee pollen of *Nelumbo nucifera*	Immunostimulating activity: increased lymphocyte proliferation and macrophage phagocytosis, and induced production of NO	Li et al. (2018)
Dendrobium nobile Lindl	Immunostimulating activity	Wang et al. (2018)

activity of highly branched RG-I (50% substitution) from *Dendrobium nobile* Lindl. The immunostimulating activity of partially hydrolysed pectins from sweet pepper could increase TNF-alpha, IL-1 beta and IL-10 cytokine secretion by THP-1 macrophages (do Nascimento et al. 2017). RG-I rich extract from barley leaf could facilitate the production of IgA via induction of various IgA-producing cytokines in small intestinal immune system (Kim et al. 2017). Hallabong peel pectins exhibited immunostimulating and anti-inflammatory activities, which enhanced NK cell function and attenuated pro-inflammatory cytokine levels by the supplement in yogurt (Lee et al. 2016). RG-I from *Ilex paraguariensis* reduced the accumulation of neutrophil influx in lungs, and decreased tissue expression of iNOS and COX-2 (Dartora et al. 2013). Pectins from the fruits of *Lycium ruthenicum* Murr promoted macrophage proliferation and enhanced the secretion of nitrogen monoxide *in vitro* (Peng et al. 2014). The reported studies of immunostimulating activities of pectins and sources are listed in Table 9.2.

9.2.3 Anti-Inflammatory, Antioxidant, and Antimicrobial Activities

As mentioned previously, the anti-inflammatory, antibacterial, and antioxidant activities of pectin could be essential for its anticancer and/or immunomodulatory activities. Pectin rich extracts (RG-I and RG-II with Mw of 19 kDa) from *Saussurea laniceps* were reported to have antivirus and anti-cancer activity, which inhibited the secretion of HBsAg and HBeAg in HepG2.2.15 cells (Chen et al. 2019a). Pectin from *Suaeda fruticosa* leaves was reported to exert antioxidant, anti-inflammatory, antinociceptive activities (Mzoughi et al. 2018). Gharibzahedi et al. (2019a, b) reported the antioxidant and cytotoxic inhibitory activities of pectin from fig skin, which inhibited the growth of HepG2 and A549 cells. Apple pectin derivatives had anti-radical activities against DPPH (Rha et al. 2011), and apple pectin also exerted anti-inflammation activity, which alleviated the inflammation in colitis (Lim et al. 2003). Ogutu and Mu (2017) reported the antioxidant capacity of sweet potato pectin after sonication treatment; the ORAC and FRAP values increased fivefolds and threefolds, respectively. Mikhaleva et al. (2011) concluded that the antioxidant activity of extracted or hydrolysed pectin could be contributed by large amounts of phenolic compounds. This is in agreement with the observation that carbohydrates generally do not have anti-oxidant activities (Hu et al. 2016). The antibacterial/antivirus activities could also be correlated to the functional substituted groups in the pectin. Akhmedov et al. (2017) reported the antimicrobial activity of sulphamic pectin derivatives on *S. aureus*, *S. epidermidis*, *E. coli*, *P. vulgaris*, *S. faecalis*, *S. pyogenes*, *S. faecalis*, and *P. aeruginosa*. Fan et al. (2012a) reported the antimicrobial activity of pectin derivatives, which inhibited *S. aureus*, *E. coli*, *and B. subtilis*. The earlier research confirmed the anti-bacterial properties of low methoxylated HG pectin (i.e., zosterin), which inhibited the growth of gram-negative and gram-positive

Table 9.3 Anti-inflammatory, antioxidant, antibacterial, and antivirus activities of pectin

Pectin source	Bioactivity	Reference
Marshmallow root	Anti-inflammatory and antitussive activities (on cat)	Nosalova et al. (1993)
Apple	Anti-inflammation activity: alleviated the inflammation in colitis	Lim et al. (2003)
Different pectins (cowberry, bergenan, lemnan, silenan, zosteran, etc.)	Anti-inflammatory activity: inhibited the development of large intestinal inflammation	Markov et al. (2011)
Commercial pectins (comaruman, bergenan, potamogetonan)	Antioxidant activity	Mikhaleva et al. (2011)
Derivative of apple pectin	Antioxidant activity: anti-radical activities against DPPH	Rha et al. (2011)
Derivative of commercial pectin	Antimicrobial activity: inhibited three bacteria (*S. aureus*, *E. coli*, and *B. subtilis*)	Fan et al. (2012a)
Passion fruit peel, orange pomace, and soy hull	Antioxidant activity	de Moura et al. (2017)
Sulphamic pectin derivatives	Antimicrobial activity: against *S. aureus*, *S. epidermidis*, *E. coli*, *P. vulgaris*, *S. faecalis*, *S. pyogenes*, *S. faecalis*, and *P. aeruginosa*	Akhmedov et al. (2017)
Sweet potato	Antioxidant capacity: The ORAC value increased five hold; FRAP value increased threefolds	Ogutu and Mu (2017)
Suaeda fruticosa leaves	Antioxidant, anti-inflammatory, antinociceptive activities	Mzoughi et al. (2018)
Saussurea laniceps	Antivirus and anticancer activity: inhibited the secretion of HBsAg and HBeAg in HepG2.2.15 cells	Chen et al. (2019a)
Hawthorn	Antioxidant activity	Chen et al. (2019b)
Ficus carica L. skin	Antioxidant and cytotoxic inhibitory activities: inhibited HepG2 and A549 cells	Gharibzahedi et al. (2019a, b)

bacteria, such as *S. aureus*, *E. coli*, *Y. pseudotuberculosis*, *S. typhimurium*, and *P. Aeruginosa* (Zaporozhets et al. 1991). The anti-inflammatory, antioxidant and antimicrobial activities of pectins are summarised in Table 9.3.

9.2.4 Gastroprotection and Other Bioactivities

In addition to the prebiotic activities, pectins also demonstrate gastroprotection activities, including (but not limited to) antidiarrheal, antiulcer, and antiplatelet. Maria-Ferreira et al. (2014) studied the antiulcerogenic and gastroprotective activities of pectins from *Acmella oleracea* (L.), which accelerated the healing of gastric ulcer, and reduced inflammation and oxidative stress. RG-I from *Acmella oleracea* with type II arabinogalactan side chains could inhibit ethanol-induced gastric ulcers in rats (Nascimento et al. 2013). Pectin oligosaccharides from Actinidia arguta exerted prebiotic and antiglycation activities in vitro (Zhu et al. 2019).

Klosterhoff et al. (2018) reported the in vivo anti-fatigue and antioxidant activities of HG rich pectin (52.1% galacturonic acid, and molecular weight of 75 kDa) from acerola. The detoxification and metal removal activities of citrus, sugar beet, and seaweed pectins have been summarised by Lara-Espinoza et al. (2018). Fan et al. (2012b) reported the anticoagulant activity of pectin derivative. Zou et al. (2015) reported the anti-complementary activity of pectins (RG-I and HG) from root bark, stem bark, and leaves of *Terminalia macroptera*. RG-I from white cabbage (Westereng et al. 2009) and Asteraceae (Nergard et al. 2005) were also reported to exert anti-complementary activity. Pectin as the coating agent improved fibroblast proliferation, bone and connective tissue regeneration (Folkert et al. 2019). RG-I from apple and lupin pectins induced bone matrix formation and mineralization (Gurzawska et al. 2012). Citric pectin sulphamic derivatives were reported to have antithrombotic activity (Cipriani et al. 2009). Okra pectin exerted the antiadhesive activity against *H. pylori* (Lengsfeld et al. 2004). The selective antiadhesive activity of ginseng pectins against *H. pylori* and *P. gingivalis* was also reported (Lee et al. 2006). The gastroprotective, anticomplementary, antiadhesive and other bioactivities of pectins are summarised in Table 9.4.

Table 9.4 Gastroprotection and other bioactivities of pectin

Pectin source	Gastroprotection and other bioactivities	Reference
Plantago major L.	Anticomplementary activity	Samuelsen et al. (1996)
Okra	Antiadhesive activity: against *H. pylori*	Lengsfeld et al. (2004)
Panax ginseng	Antiadhesive activity: against *H. pylori* and *P. gingivalis*	Lee et al. (2006)
White cabbage, kale, and red kale	Anticomplementary activity	Samuelsen et al. (2007)
White cabbage	Anticomplementary activity	Westereng et al. (2009)
Derivative of citrus pectin	Anticoagulant and antithrombotic activities	Cipriani et al. (2009)
Derivative of citrus pectin	Anticoagulant and antimicrobial activities	Bae et al. (2009)
Althaea officinalis L.	Antitussive activity: cough suppressive activity (on guinea pig)	Šutovská et al. (2009)
Sugar beet	Prebiotics	Holck et al. (2011)
Derivative of commercial pectin	Anticoagulant activity	Fan et al. (2012b)
Apple and lupin	Induced bone matrix formation and mineralization	Gurzawska et al. (2012)

(continued)

Table 9.4 (continued)

Pectin source	Gastroprotection and other bioactivities	Reference
Asteraceae	Gastroprotective effects: inhibited ethanol-induced gastric ulcers (in rats)	Nascimento et al. (2013)
Cola cordifolia leaf	Gastroprotective and healing effects (wound and gastric ulcer healing), and anticomplementary activity	Austarheim et al. (2014)
Acmella oleracea (L.)	Antiulcerogenic and gastroprotective activities: accelerated the gastric ulcer healing, accompanied by increasing of cellular proliferation and gastric mucus content, reducing inflammatory parameters and oxidative stress	Maria-Ferreira et al. (2012, 2014)
–	Improved osseointegration by RG-I coating of implants	Svava et al. (2014)
Flaxseed	Prebiotics: improved SCFA production in vitro	Ding et al. (2015)
Potato	Prebiotics: gut health for weaning piglets	Strube et al. (2015)
Elderberries	Anticomplementary and macrophage stimulating activities	Ho et al. (2015)
Citrus	Anticoagulant activity	Hu et al. (2015)
Root bark, stem bark, and leaves of *Terminalia macroptera*	Anti-complementary activity	Zou et al. (2015)
Potato	Prebiotics: stimulated the growth of *Bifidobacterium* spp. and *Lactobacillus* spp.	Khodaei et al. (2016)
Acmella oleracea	Gut health: against colitis, increased transepithelial electrical resistance, accelerate the tight junction reassembly	Baggio et al. (2017)
Wine, apple	Prebiotics: increased microbial diversity in the human gut microbiota	Ndeh et al. (2017)
Acerola	Antifatigue and antioxidant activities	Klosterhoff et al. (2018)
Sugar beet	Prebiotics: stimulated lactic acid bacteria in a species-dependent manner	Prandi et al. (2018)
Potato	Coating: improved bone and connective tissue regeneration	Folkert et al. (2019)
Different sources	Prebiotics	Larsen et al. (2019)
Actinidia arguta	Antiglycation and prebiotic activities	Zhu et al. (2019)

9.3 Structure-Function Relationship

9.3.1 Rhamnogalacturonan-I (RG-I) Pectin

RG-I pectins are composed of rhamnogalacturonan backbone with 20–80% of rhamnose residues branched at *O*-4 position, and side chains may be composed of arabinan, galactan, and arabinogalactan depending on pectin source (Cui et al. 2019). It was reported that highly branched structure of ginseng pectins (incl. arabinan, galactan, arabinogalactan I, and arabinogalactan II) contributed to the anti-cancer and immunomodulatory activities (Shi et al. 2017; Cui et al. 2019). The anti-cancer activities, such as inhibition of galectin-3-mediated hemagglutination, cancer cell adhesion and homotypic aggregation, and binding of galectin-3 to T-cells, was associated to galactan side chains; these activities were proportional to the length of arabinan side chains up to 4 units, and affected by substitution location (Gao et al. 2013). Pumpkin RG-I with long beta-1,4-D-galactan side chains was found to be associated quickly while dissociated slowly with galectin-3; the mechanism was mainly due to the interaction between RG-I and galectin-3 (Zhao et al. 2017). Compared with RG-I rich pectin from ginseng flower, HG linked with rhamnogalacturonan II (RG-II) showed weaker or no interactions with galectin-3 (Cui et al. 2019). Cheng et al. (2013) studied four structurally distinct pectins from potato, and concluded that the anti-cancer activity could be correlated to RG-I domain. RG-I with beta-1,4-D-galactan side chains were more active than the other RG-I-rich and HG-rich citrus pectins (Zhang et al. 2016). Grønhaug et al. (2011) also confirmed that galactan side chains played an important role for expression of immunomodulating activity against both Peyer's patch immunocompetent cells and macrophages. Galactose in the side chains of bee pollen RG-I was correlated to the increase of macrophage phagocytosis, while arabinose was unimportant for NO production (Li et al. 2018). The arabinogalactan side chains of ginseng RG-I could be the essential structures for stimulating NO secretion and lymphocyte proliferation (Zhang et al. 2012). RG-I coating of implants showed the improvement of osseointegration, which was positively correlated to high content of 1,4-linked galactose and a low content of arabinose (Svava et al. 2014). High molecular weight (Mw) pectic arabinogalactan (Mw 1150 kDa) showed anticomplementary activity and an induction of B-cell proliferation, while extracted low-Mw pectin (Mw 20 kDa) showed low anticomplementary activity and no influence on B- or T-cell proliferation (Nergard et al. 2005). The anticomplementary activity of HG and RG-I from *Terminalia macroptera* were mainly attributed to RG-I; and higher molecular weight fractions exerted higher activity (Zou et al. 2015). The correlation of bioactivity to RG-I structural characteristics are listed in Table 9.5.

9.3.2 Homogalacturonan (HG) and Rhamnogalacturonan-II (RG-II) Pectins

Both of HG and RG-II pectins have homogalacturonan backbone: HG is recognised as the smooth regions of pectins without side chains, while RG-II has complex side chains containing rhamnose and some unique residues, such as apiose, aceric acid (3-C-carboxy-5-deoxy-L-xylose), and KDO (3-deoxy-D-manno-octu-

Table 9.5 Correlation of bioactivity to RG-I structure

Bioactivity	Structure-function relationship	Reference
Anticancer activity	Interaction of RG-I with effector-cell or tumor-cell	Zhu et al. (1994)
Anticomplementary activity	Removal of arabinose branch decreased the activity	Samuelsen et al. (1996)
Anticancer activity	Interaction of RG-I with effector-cell or virus-infected cell with chemical specificity	Steinmassl and Anderer (1996)
Anti-complementary and immunostimulating activity	Pectic arabinogalactan showed higher complement fixation activity and immunostimulating activities than extracted pectin	Nergard et al. (2005)
Anticomplementary activity	The activity was correlated to galactan side chains, but not on molecular weight distribution	Samuelsen et al. (2007)
Immunostimulating activity	Galactan side chains mainly contributed to the activities	Grønhaug et al. (2011)
Immunostimulating and anticomplementary activities	Two RG-I fractions had different activities; complement-fixing activity and macrophage activation involved different mechanisms	Austarheim et al. (2012)
Immunostimulating activity	Arabinogalactan side chains were essential structures to stimulating NO secretion and lymphocyte proliferation	Zhang et al. (2012)
Induced bone matrix formation and mineralization	RG-I contributed to bone matrix formation and mineralization	Gurzawska et al. (2012)
Anticancer activity	RG-I domain contributed to the anticancer activity	Cheng et al. (2013)
Anticancer activity	The structural characteristics of galactan side chains (i.e., length and location) were essential for the bioactivity	Gao et al. (2013)
Improved osseointegration	Higher content of 1,4-linked galactose and lower content of arabinose in RG-I coating contributed to the activity	Svava et al. (2014)
Prebiotics	Uronic acid content and degree of branch affected SCFA generation; less branched RG-I with higher uronic acid content had higher total SCFA production during 72 h in vitro fermentation	Ding et al. (2015)
Immunostimulating activity and anti-inflammation	RG-I enhanced NK cell function and reduced pro-inflammatory cytokine levels	Lee et al. (2016)
Anti-cancer activity	RG-I pectins with galactan side chains were more active than the other RG-I rich and HG rich pectins	Zhang et al. (2016)
Gut health	RG-I contributed to improved intestinal epithelial barrier function	Baggio et al. (2017)
Antimicrobial activity	Sulphamic groups contributed to the activity	Akhmedov et al. (2017)
Anticancer activity	Combination of HG and RG-I rich pectins exerted stronger activity on cell migration	Fan et al. (2018)

(continued)

Table 9.5 (continued)

Bioactivity	Structure-function relationship	Reference
Immunostimulating activity	Galactose side chains residues might contribute to macrophage phagocytosis, whereas arabinose was unimportant for NO production	Li et al. (2018)
Prebiotics	Fractions rich in arabinans and a low degree of polymerization had higher activity to stimulate lactic acid bacteria	Prandi et al. (2018)
Anticancer activity	The structural characteristics contributed to the inhibitory activities: RG-I showed higher activity (interactions with galectin-3) than HG and RG-II	Cui et al. (2019)

losic acid) (Cui et al. 2019). In the complex food matrix or nutritional supplements, RG-I, HG or RG-II could exert synergistic bioactivities. Fan et al. (2018) reported the combination of HG and RG-I-rich pectins exerted stronger anti-cancer activity than either HG- or RG-I-rich pectin alone. The anti-cancer activity was attributed to the inhibition of cancer cell migration, which was not correlated with the binding of galectin-3 (Fan et al. 2018). Ai et al. (2018) studied the anti-cancer activity of RG-I, RG-II, and galacturonic acid oligomers from citrus, and found that RG-II-domain had higher inhibition effects on Caco-2 cells than RG-I domain. The anti-cancer activity of RG-I and RG-II rich extracts (Mw of 19 kDa) from *Saussurea laniceps* was due to the interaction between pectin and HBsAg or HBeAg in HepG2.2.15 cells, which did not involve cellular pathways (Chen et al. 2019a). Okra pectin extracts contained rhamnogalacturonans and HG components, and the antiadhesive activity could be associated with the three-dimensional structure (combination of glycoproteins and high uronic acid region) (Lengsfeld et al. 2004). The HG rich pectin (47.5% of uronic acid) from *Suaeda fruticosa* leaves was reported to have antioxidant, anti-inflammatory, antinociceptive activities (Mzoughi et al. 2018). Maxwell et al. (2015a, b) tested several pectin extracts with different structural characteristics for the anti-cancer activity, and found that HG component might be essential for the anti-proliferation of DLD1 and HCT116 colon cancer cells. Wang et al. (2014) found that the purified HG from green tea exhibited *in vitro* phagocytosis-enhancing activity in human leukemia cells (HL-60). Popov and Ovodov (2013) have summarised previous studies on pectin immunomodulatory effects: the immunosuppressive activity could be associated with homogalacturonan (HG) backbone, and HG with 80% galacturonic acid residues was found to decrease macrophage activity; branched galacturonan had biphasic immunomodulatory activities, which could increase phagocytosis and antibody production. The correlation of bioactivity to HG or RG-II structure, and possible correlation between bioactivity and pectin conformation are listed in Table 9.6.

Table 9.6 Correlation of bioactivity to HG and RG-II structure and pectin conformation

Bioactivity	Structure-function relationship	Reference
Antiadhesion activity	Three-dimensional structure (combination of glycoproteins and high uronic acid region) contributed to the activity	Lengsfeld et al. (2004)
Immunostimulating and anticomplementary activities	Arabinose and galactose rich segments contributed to complement-fixing activity	Westereng et al. (2006)
Anticancer activity	Neutral side chain (with terminal galactose at the nonreducing end) contributed to the activity; modified pectin could provide optimise bioactivity	Gunning et al. (2009)
Anticomplementary activity	Multiple side chains contributed to the efficient complement-fixing activity; molecular weight distribution had little effect	Westereng et al. (2009)
Antitussive activity	Rhamnogalacturonan contributed to the activities	Šutovská et al. (2009)
Anti-inflammatory activity	The activity was determined by galacturonan backbone	Markov et al. (2011)
Gastroprotective effects	The possible mechanisms involved: binding of pectin to the mucosa surface, diminishing acid and pepsin secretion, increasing mucus synthesis/radical scavenging	Nascimento et al. (2013)
Anti-infective activity	The structural characteristics (e.g., molecular weight, acidic monosaccharide, and degrees of methyl-esterification) might contribute to the activity	Dartora et al. (2013)
Anticomplementary activity and gastroprotective and healing effects (wound and gastric ulcer healing)	The structural characteristics (Mw, side chains) and viscosity might contribute to the activities; no macrophage activating effects after removal of lipopolysaccharide; the pectic oligosaccharides did not show complement-fixing activity	Austarheim et al. (2014)
Antiulcerogenic and gastroprotective activities	Rhamnogalacturonan might contribute to cellular proliferation, and reduction of inflammation	Maria-Ferreira et al. (2012, 2014)
Anticancer activity	HG might be essential for the anti-proliferative activity	Maxwell et al. (2015a, b)
Complement fixing activity	High Mw fractions had higher complement fixation activity; the ramified region contributed to the activities	Zou et al. (2015)
Immunostimulating activity	The structural characteristics contributed to the activities by facilitating the production of IgA in small intestinal immune system	Kim et al. (2017)
Anticancer activity	The structural characteristics contributed to the activities: quick association and slow dissociation between pectin and galectin-3	Zhao et al. (2017)
Immunostimulating activity	Degree of methyl esterification, molecular weight and presence of side chains stimulated THP-1 macrophages	do Nascimento et al. (2017)

(continued)

Table 9.6 (continued)

Bioactivity	Structure-function relationship	Reference
Immunostimulating activity	The structural characteristics (monosaccharide, peptide content, degrees of esterification and methylation, etc.) contributed to the activities	Chen et al. (2017)
Prebiotics	RG-II hydrolysis in the gut increased microbial diversity	Ndeh et al. (2017)
Immunostimulating activity	The branches and molecular weight contributed to the activity	Wang et al. (2017)
Anticancer activity	RG-II had higher inhibition effects than RG-I domain	Ai et al. (2018)
Antivirus and anticancer activity	The structural characteristics contributed to the inhibitory activities: interaction between SLP-4 and HBsAg or HBeAg	Chen et al. (2019a)
Prebiotics	High methoxyl pectins generated highest SCFA	Larsen et al. (2019)
Antiglycation and prebiotic activities	The in vitro antiglycation activity was proportional to the galacturonic acid contents, and inversely proportional to the degree of branch	Zhu et al. (2019)

9.3.3 Partially Hydrolysed Pectins, Pectic Oligosaccharides, and Pectin Derivatives

The bioactivities of modified or partially hydrolysed pectins, and related mechanisms and structure-function relationship have been extensively studied by Morris and his team (Gunning et al. 2009; Maxwell et al. 2012; Morris et al. 2011, 2013). In these studies, the selective binding of pectin fragments to galectin-3 was confirmed by fluorescence microscopy, flow cytometry, and atomic force spectroscopy. The neutral side chains (with terminal galactose at the nonreducing end) were the main contributing structures for the demonstrated bioactivity; the hydrolysed pectins were found to be more effective. Ma et al. (2018) found that ultrasound and pectinase treatment accumulated higher content of galactose in pectin hydrolysates without changing primary structures, which contributed to an improved inhibitory activity against HT-29 colon cancer cells. Pectinase treatment significantly increased tea polysaccharide induced IL-6 production in macrophages, which was closely related to the changes of pectin structure and/or synergistic effects with released functional components (Ndeh et al. 2017). Partially hydrolysed pectins from sweet pepper increased TNF-alpha, IL-1 beta and IL-10 cytokine secretion by THP-1 macrophages, and the study also found that degree of methyl esterification, molecular weight and presence of side chains involved in the modulation of cytokine secretion by THP-1 macrophage (do Nascimento et al. 2017). Chen et al. (2019b) applied the ultrasonic treatment on extracted hawthorn pectin, and results showed that the antioxidant activity of the partially hydrolysed pectin was significantly improved. Ogutu and Mu (2017) found the increase of antioxidant capacity of sweet potato pectin after sonication treatment, which might be correlated to the debranched structure and exposure of functional groups.

The in vitro antiglycation activity of pectic oligosaccharides from *Actinidia arguta* was proportional to the galacturonic acid content, and inversely proportional to its branching degree. The degree of esterification was less important for both antiglycation and prebiotic activity of pectic oligosaccharides (Zhu et al. 2019). Another study indicated that limited-hydrolysis produced pectic oligosaccharides of ginseng pectin had comparable antiadhesive activity, while the activity was lost after complete hydrolysis (Lee et al. 2006). Holck et al. (2011) reported that probiotics were sensitive to pectic oligosaccharides, and significantly different microbiota profiles in the gut were observed. The prebiotic activity of this pectin was positively correlated with arabinose/rhamnose and galactose/rhamnose molar ratios and the neutral sugar content, especially galactose and arabinose (Zhu et al. 2019). The correlation of bioactivity to pectic oligosaccharides, pectin derivatives or substituted groups are listed in Table 9.7.

Based on the reported studies, the structure-function relationship could be summarised as follows:

1. Pectin bioactivities are widely correlated to galactan and arabinogalactan side chains, rhamnogalacturonan and homogalacturonan backbones, substituted groups, and their synergistic effects.
2. RG-I galactan side chains contributed to the anticancer and immunostimulating activities (e.g., increase of macrophage phagocytosis, enhanced binding of galectin-3 to T-cells, and inhibition of galectin-3-mediated hemagglutination, cancer cell adhesion and homotypic aggregation).
3. RG-I arabinogalactan side chains contributed to anticomplementary activity, and higher molecular weight fractions exerted higher activity; and stimulated NO secretion and lymphocyte proliferation, and induction of B-cell proliferation.
4. Pectin conformation (e.g., combination of glycoproteins and high uronic acid region) contributed to antiadhesive and anticomplementary activities.
5. HG backbone might contribute to the anti-proliferation of DLD1 and HCT116 colon cancer cells, and phagocytosis-enhancing activity in human leukemia cells (HL-60).
6. HG contributed to anti-inflammatory activity by inhibiting the development of large intestinal inflammation.
7. RG-II contributed to the inhibition effects on Caco-2 cells, and RG-I and RG-II rich extracts showed anti-cancer activity by interacting with HBsAg or HBeAg in HepG2.2.15 cells.
8. RG-I coating with high content of 1,4-linked galactose and a low content of arabinose could help improve bone and connective tissue regeneration.
9. The enrichment of pectin neutral side chains, especially with terminal galactose at the nonreducing end, and controlled or limited hydrolysis of pectin may increase anticancer and immunostimulating activities (e.g., binding with galec-

Table 9.7 Correlation of bioactivity to hydrolysed or modified-pectin structural characteristics

Bioactivity	Structure-function relationship	Reference
Immunostimulating activity	Degraded pectin derivatives had higher immunostimulating activities than extracted pectin; carboxyl and acetyl groups contributed to the activities	Wang et al. (2003, 2005)
Antiadhesive activity	Structure contributed to the activity; pectic oligosaccharide could mimic core carbohydrate moiety in host–pathogen interactions	Lee et al. (2006)
Anticancer activity	Heat treatment enhanced apoptotic activity	Jackson et al. (2007)
Anticancer activity	Neutral side chain (with terminal galactose at the nonreducing end) contributed to the activity; modified pectin could provide optimise bioactivity	Gunning et al. (2009)
Anticoagulant and antithrombotic activities	High-molecular-weight sulfated pectins showed higher antithrombotic activity than low-molecular-weight fractions	Cipriani et al. (2009)
Anticoagulant and antimicrobial activities	Sulfate groups played an important role in improving the activities	Bae et al. (2009)
Prebiotics	Slight structural difference of pectic oligosaccharides had significantly different gut bacterial profiles	Holck et al. (2011)
Antioxidant activity	Large amounts of common phenols mainly contributed to the activities	Mikhaleva et al. (2011)
Antioxidant activity	Hydroxamic acid substituted group contributed to the activities	Rha et al. (2011)
Antimicrobial activity	Polycationic substitutions contributed to the inhibition of *S. aureus*, *E. coli*, and *B. subtilis*	Fan et al. (2012a)
Anticoagulant activity	Pectin sulfate groups contributed to the activities	Fan et al. (2012b)
Anticoagulant activity	Sulfated pectin contributed to the anticoagulant activity	Hu et al. (2015)
Anticancer activity	Partial hydrolysis of RG-I contributed to the activity	Shi et al. (2017)
Anticancer activity	Partial depolymerisation (decreased branching) enhanced anti-cancer properties; the bioactivities were lost after complete hydrolysis.	do Prado et al. (2017)
Antioxidant activity	Phenolic compounds with antioxidant capacity provide pectins with additional properties	de Moura et al. (2017)
Antimicrobial activity	Sulphamic groups contributed to the activity	Akhmedov et al. (2017)
Antioxidant capacity	Partial depolymerisation by sonication contributed to enhanced activity	Ogutu and Mu (2017)
Immunostimulating activity	Pectin and its derivatives exhibited different immunological activities in vitro; side chains and acetyl groups mainly contributed to the activity	Wang et al. (2018)
Antioxidant activity	Ultrasonic treatment significantly improved the antioxidant activity	Chen et al. (2019b)

tin-3, against HT-29 colon cancer cells, increase TNF-alpha, IL-1 beta and IL-10 cytokine secretion by THP-1 macrophages, etc.).

10. The substituted groups (e.g., carboxyl, acetyl, sulfate) contributed to immunostimulating, anticoagulant, antioxidant, and antimicrobial activities.
11. Probiotics could be more sensitive to pectic oligosaccharides; and uronic acid content, degree of branch, and substitution of pectin may have higher impact on short-chain fatty acid (SCFA) generation.

9.4 Application of Bioactive Pectins

Pectins have long been utilised in various food products, such as beverage, meat, dairy, bakery, edible films, and food preservatives. Bioactive pectins not only provide excellent stabilizing, texturing, emulsifying, or film-forming capabilities in aforementioned food systems, but also exert specific bioactivities for health-oriented consumers. Bioactive pectins have been widely utilised as prebiotics, blood cholesterol lowering supplements. Substantial amount of evidences indicated some pectins are the effective ingredients with anticancer, immunomodulatory, hypoglycemic, antiallergic, anti-inflammatory, anticoagulant, and antiulcer activities. The general applications of pectins in the food, nutrition, and pharmaceutical industries have been widely discussed in previous studies (Yamada 1996; Thakur et al. 1997; Sriamornsak 2003; Sande 2005; Ovodov 2009; Khotimchenko 2010; Maxwell et al. 2012; Mishra et al. 2012; Popov and Ovodov 2013; Espitia et al. 2014; Zhang et al. 2015; Lara-Espinoza et al. 2018).

Pectins could also be utilised as carrier or encapsulation material for the delivery or controlled release of drugs or other bioactive compounds. Sande (2005) evaluated the pectin structural characteristics for successful formulations, and pectin recipes for matrix tablets, hydrogel beads, microparticles, and coatings were summarised. Svava et al. (2014) reported the improved osseointegration by RG-I coating of implants. Munarin et al. (2011) studied modified pectin as an injectable cell vehicle for bone tissue regeneration. Pectin coating improved the stability of nanoparticles under gastrointestinal conditions (Chang et al. 2017). Pectin could also be utilised as biomedical particles for wound healing and lung regeneration for diabetic patients (Jayakumar et al. 2014).

Pectin from orange peel as the carrier could enhance anti-inflammatory activity (Reena et al. 2017). Krivorotova et al. (2017) used commercial pectin as the carrier for nisin to exert better antimicrobial activity, and found that low degree of esterification exhibit higher antimicrobial activity than high methoxyl pectin particles. Gence et al. (2018) studied commercial pectin as the carrier for carotenoid, and found that the degree of methoxylation increased carotenoid bioaccessibility. Pectin from pomelo fruit peel as the carrier displayed broad spectrum antimicrobial activity, antioxidant potential, and non-cytotoxic nature (Patil et al. 2016). Ye et al. (2018) reported the prolonged release ability of pectin carrier with good biological security. The benefits of pectin as the carrier also include good hydrophilicity, sustained antimicrobial nature, promote cell growth and proliferation, and lead to rapid healing (Tummalapalli et al. 2016), showing higher heat resistance and greater electrolyte tolerance (Yao et al. 2015). The potential applications of bioactive pectins and sources are combined in Table 9.8.

9.5 Conclusions

Pectins have been widely applied in food, feed or pharmaceutical industries for their functionality. Many pectins and their hydrolysates/derivatives have been observed anticancer, immunostimulation, anti-inflammatory, gastroprotection, antibacterial, antiadhesive, and complement fixing activities. The immunostimulating, antibacterial, anti-inflammatory, and prebiotic activities of pectin synergistically contribute to the anticancer activity. Moreover, some pectins could improve regeneration of bone and connective tissue, wound healing, and bioaccessibility of encapsulated ingredients with good biological compatibility, and greater heat resistance and electrolyte tolerance.

Pectin bioactivities are closely correlated to the structural characteristics, such as RG-I galactan and arabinogalactan side chains, rhamnogalacturonan and homogalacturonan backbones, substituted groups, and their synergistic effects. RG-I galactan side chains mainly contributed to the anticancer and immunostimulating activities, and pectin conformation contributed to antiadhesive and anticomplementary activities. HG contributed to anti-inflammatory activity, both of HG and RG-II could be correlated to anti-proliferation of colon cancer cells. The substituted groups (e.g., carboxyl, acetyl, sulfate) contributed to immunostimulating, anticoagulant, antioxidant, and antimicrobial activities. Moreover, the enrichment of pectin side chains may increase anticancer and immunostimulating activities. It is worth noting that the information on clinical trial and dose-function correlation, and long-term effects are still very limited. Several point-to-point structure function correlations are unclear or even contradictory in some studies, which requires further investigation.

Table 9.8 Potential applications and sources of bioactive pectins

Category	Application	Pectin Sources
Food supplement or medicine	Anticancer	*Viscum album*, potato, ginseng, ginseng flower, *Lonicera japonica* flower, citrus, pumpkin, papaya, *Opuntia ficus-indica*, *Saussurea laniceps*, fig (*Ficus carica* L.)
	Immunoregulation	*Centella asiatica* (L.) Urban, white cabbage, *Biophytum petersianum* Klotzsch, *Cola cordifolia* bark, Panax ginseng, mate (*Ilex paraguariensis*), *Lycium ruthenicum* Murr, elderberries, hallabong (*Citrus sphaerocarpa*) peel, barley leaf, sweet pepper, *Hovenia dulcis* peduncles, bee pollen of *Nelumbo nucifera*, *Dendrobium nobile* Lindl
	Prebiotics	Sugar beet, flaxseed, potato, *Actinidia arguta*
	Gastroprotection	*Spilanthes oleracea*, *Acmella oleracea* L.
	Antioxidant	Apple, passion fruit peel, orange pomace, soy hull, sweet potato, acerola, *Suaeda fruticosa* leaves, hawthorn, *Ficus carica* L.
	Anti-inflammatory	Apple, hallabong (*Citrus sphaerocarpa*) peel, *Suaeda fruticosa* leaves
	Anti-complementary	*Plantago major* L., Asteraceae, White cabbage, Kale, *Cola cordifolia* leaves, elderberries, *Terminalia macroptera*
	Detoxification/metal removal	Seaweed, citrus, sugar beet, cabbage, carrot, apple
	Antibacterial	Zosterin, citrus
	Antivirus	*Saussurea laniceps*
	Anticoagulant	Citrus
	Blood cholesterol lowering	Prickly pear (*Opuntia* spp.), grapefruit
	Antiulcer	*Acmella oleracea* (L.)
	Antituberculous	Citrus pectin
	Antithrombotic activity	Citric pectin
	Antinociceptive	*Suaeda fruticosa* leaves
	Antiadhesive	Okra, panax ginseng
	Anti-infectious	*Ilex paraguariensis*
	Antifatigue	Acerola
	Bone matrix formation	Apple, lupin
	Hypoglycemic	*Actinidia arguta*
Drug carrier	Colon-deliver Medicine	Citrus, pomelo fruit peel, orange peel
	Tissue Repair	Citrus, potato
Feed supplements	Anti-inflammatory and antitussive supplement	Marshmallow root
	Cough suppressive supplement	*Althaea officinalis* L.

References

Ai L, Chung Y-C, Lin S-Y, Lee K-C, Lai PF-H, Xia Y, Wang G, Cui SW (2018) Active pectin fragments of high in vitro antiproliferation activities toward human colon adenocarcinoma cells: rhamnogalacturonan II. Food Hydrocolloids 83:239–245

Akhmedov O, Shomuratov SA, Turaev A, Vaili A (2017) Synthesis and study of antimicrobial activity of sulphamic pectin derivatives. Chem Sustain Dev 25(2):139–143

Austarheim I, Christensen BE, Hegna IK, Petersen BO, Duus JO, Bye R, Michaelsen TE, Diallo D, Inngjerdingen M, Paulsen BS (2012) Chemical and biological characterization of pectin-like polysaccharides from the bark of the Malian medicinal tree Cola cordifolia. Carbohydr Polym 89(1):259–268

Austarheim I, Christensen BE, Aas HTN, Thöle C, Diallo D, Paulsen BS (2014) Chemical characterization and complement fixation of pectins from Cola cordifolia leaves. Carbohydr Polym 102:472–480

Bae IY, Joe YN, Rha HJ, Lee S, Yoo S-H, Lee HG (2009) Effect of sulfation on the physicochemical and biological properties of citrus pectins. Food Hydrocoll 23(7):1980–1983

Baggio CH, Lopes F, Nascimento A, Cipriani T, McKay D, MacNaughton W (2017) Modulation of Intestinal Epithelial Barrier Function by Rhamnogalacturonan. FASEB J 31(1):995

Chang C, Wang T, Hu Q, Zhou M, Xue J, Luo Y (2017) Pectin coating improves physicochemical properties of caseinate/zein nanoparticles as oral delivery vehicles for curcumin. Food Hydrocoll 70:143–151

Chen C-H, Sheu M-T, Chen T-F, Wang Y-C, Hou W-C, Liu D-Z, Chung T-C, Liang Y-C (2006) Suppression of endotoxin-induced proinflammatory responses by citrus pectin through blocking LPS signaling pathways. Biochem Pharmacol 72(8):1001–1009

Chen Q, Zhu L, Tang Y, Zhao Z, Yi T, Chen H (2017) Preparation-related structural diversity and medical potential in the treatment of diabetes mellitus with ginseng pectins. Ann N Y Acad Sci 1401:75

Chen W, Zhu X, Ma J, Zhang M, Wu H (2019a) Structural elucidation of a novel pectin-polysaccharide from the petal of saussurea laniceps and the mechanism of its anti-HBV activity. Carbohydr Polym 223:115077

Chen X, Qi Y, Zhu C, Wang Q (2019b) Effect of ultrasound on the properties and antioxidant activity of hawthorn pectin. Int J Biol Macromol 131:273–281

Cheng H, Zhang Z, Leng J, Liu D, Hao M, Gao X, Tai G, Zhou Y (2013) The inhibitory effects and mechanisms of rhamnogalacturonan I pectin from potato on HT-29 colon cancer cell proliferation and cell cycle progression. Int J Food Sci Nutr 64(1):36–43

Cipriani TR, Gracher AHP, de Souza LM, Fonseca RJ, Belmiro CL, Gorin PA, Sassaki GL, Iacomini M (2009) Influence of molecular weight of chemically sulfated citrus pectin fractions on their antithrombotic and bleeding effects. Thromb Haemost 101(05):860–866

Cui L, Wang J, Huang R, Tan Y, Zhang F, Zhou Y, Sun L (2019) Analysis of pectin from Panax ginseng flower buds and their binding activities to galectin-3. Int J Biol Macromol 128:459–467

Dartora N, de Souza LM, Paiva SM, Scoparo CT, Iacomini M, Gorin PA, Rattmann YD, Sassaki GL (2013) Rhamnogalacturonan from Ilex paraguariensis: a potential adjuvant in sepsis treatment. Carbohydr Polym 92(2):1776–1782

de Moura FA, Macagnan FT, Dos Santos LR, Bizzani M, de Oliveira Petkowicz CL, Da Silva LP (2017) Characterization and physicochemical properties of pectins extracted from agroindustrial by-products. J Food Sci Technol 54(10):3111–3117

Ding HH, Cui SW, Goff HD, Gong J (2015) Short-chain fatty acid profiles from flaxseed dietary fibres after in vitro fermentation of pig colonic digesta: Structure–function relationship. Bioact Carbohydr Diet Fibre 6(2):62–68

do Nascimento GE, SMB W, Ramirez MI, Iacomini M, Cordeiro LMC (2017) The influence of sweet pepper pectin structural characteristics on cytokine secretion by THP-1 macrophages. Food Res Int 102:588–594 do Prado SB, Ferreira GF, Harazono Y, Shiga TM, Raz A, Carpita NC, Fabi JP (2017) Ripening-induced chemical modifications of papaya pectin inhibit cancer cell proliferation. Sci Rep 7(1):16564

Espitia PJP, Du WX, de Jesús Avena-Bustillos R, Soares NDFF, McHugh TH (2014) Edible films from pectin: Physical-mechanical and antimicrobial properties-A review. Food hydrocolloids 35:287–296

Fan L, Cao M, Gao S, Wang W, Peng K, Tan C, Wen F, Tao S, Xie W (2012a) Preparation and characterization of a quaternary ammonium derivative of pectin. Carbohydr Polym 88(2):707–712

Fan L, Gao S, Wang L, Wu P, Cao M, Zheng H, Xie W, Zhou J (2012b) Synthesis and anticoagulant activity of pectin sulfates. J Appl Polym Sci 124(3):2171–2178

Fan Y, Sun L, Yang S, He C, Tai G, Zhou Y (2018) The roles and mechanisms of homogalacturonan and rhamnogalacturonan I pectins on the inhibition of cell migration. Int J Biol Macromol 106:207–217

Flint HJ, Bayer EA, Rincon MT, Lamed R, White BA (2008) Polysaccharide utilization by gut bacteria: potential for new insights from genomic analysis. Nat Rev Microbiol 6(2):121

Folkert J, Mieszkowska A, Gaber T, Miksch K, Dirscherl K, Gurzawska K (2019) Surface nanocoating with plant-derived pectins improves fibroblast response in vitro. Starch 71(3-4):1800162

Gao X, Zhi Y, Sun L, Peng X, Zhang T, Xue H, Tai G, Zhou Y (2013) The inhibitory effects of a rhamnogalacturonan I (RG-I) domain from ginseng pectin on galectin-3 and its structure-activity relationship. J Biol Chem 288(47):33953–33965

Gence L, Servent A, Poucheret P, Hiol A, Dhuique-Mayer C (2018) Pectin structure and particle size modify carotenoid bioaccessibility and uptake by Caco-2 cells in citrus juices vs. concentrates. Food Funct 9(6):3523–3531

Georgiev Y, Ognyanov M, Yanakieva I, Kussovski V, Kratchanova M (2012) Isolation, characterization and modification of citrus pectins. J BioSci Biotechnol 1(3):223–233

Gharibzahedi SMT, Smith B, Guo Y (2019a) Pectin extraction from common fig skin by different methods: the physicochemical, rheological, functional, and structural evaluations. Int J Biol Macromol 136:275–283

Gharibzahedi SMT, Smith B, Guo Y (2019b) Ultrasound-microwave assisted extraction of pectin from fig (Ficus carica L.) skin: optimization, characterization and bioactivity. Carbohydr Polym 222:114992

Grønhaug TE, Kiyohara H, Sveaass A, Diallo D, Yamada H, Paulsen BS (2011) Beta-d-(1→4)-galactan-containing side chains in RG-I regions of pectic polysaccharides from Biophytum petersianum Klotzsch. contribute to expression of immunomodulating activity against intestinal Peyer's patch cells and macrophages. Phytochemistry 72(17):2139–2147

Gunning AP, Bongaerts RJ, Morris VJ (2009) Recognition of galactan components of pectin by galectin-3. FASEB J 23(2):415–424

Gurzawska K, Svava R, Syberg S, Yihua Y, Haugshøj KB, Damager I, Ulvskov P, Christensen LH, Gotfredsen K, Jørgensen NR (2012) Effect of nanocoating with rhamnogalacturonan-I on surface properties and osteoblasts response. J Biomed Mater Res Pt A 100(3):654–664

Ho GTT, Ahmed A, Zou Y-F, Aslaksen T, Wangensteen H, Barsett H (2015) Structure–activity relationship of immunomodulating pectins from elderberries. Carbohydr Polym 125:314–322

Holck J, Hjernø K, Lorentzen A, Vigsnæs LK, Hemmingsen L, Licht TR, Mikkelsen JD, Meyer AS (2011) Tailored enzymatic production of oligosaccharides from sugar beet pectin and evidence of differential effects of a single DP chain length difference on human faecal microbiota composition after in vitro fermentation. Process Biochem 46(5):1039–1049

Hu Y, Ye X, Yin X, Chen S (2015) Sulfation of citrus pectin by pyridine-sulfurtrioxide complex and its anticoagulant activity. LWT- Food Sci Technol 60(2):1162–1167

Hu S, Yin J, Nie S, Wang J, Phillips GO, Xie M, Cui SW (2016) *In vitro* evaluation of the antioxidant activities of carbohydrates. Bioact Carbohydr Diet Fibre 7(2):19–27

Jackson CL, Dreaden TM, Theobald LK, Tran NM, Beal TL, Eid M, Gao MY, Shirley RB, Stoffel MT, Kumar MV (2007) Pectin induces apoptosis in human prostate cancer cells: correlation of apoptotic function with pectin structure. Glycobiology 17(8):805–819

Jacobasch G, Dongowski G, Florian S, Müller-Schmehl K, Raab B, Schmiedl D (2008) Pectin does not inhibit intestinal carcinogenesis in APC-deficient Min/+ mice. J Agric Food Chem 56(4):1501–1510

Jayakumar GC, Usharani N, Kawakami K, Rao JR, Nair BU (2014) Preparation of antibacterial collagen–pectin particles for biotherapeutics. RSC Adv 4(81):42846–42854

Khodaei N, Fernandez B, Fliss I, Karboune S (2016) Digestibility and prebiotic properties of potato rhamnogalacturonan I polysaccharide and its galactose-rich oligosaccharides/oligomers. Carbohydr Polym 136:1074–1084

Khotimchenko YS (2010) Antitumor properties of nonstarch polysaccharides: fucoidans and chitosans. Russ J Mar Biol 36(5):321–330

Kim H, Yu K-W, Hong H-D, Shin K-S (2017) Effect of arabinoxylan-and rhamnogalacturonan I-rich polysaccharides isolated from young barley leaf on intestinal immunostimulatory activity. J Funct Foods 35:384–390

Klosterhoff RR, Kanazawa LK, Furlanetto AL, Peixoto JV, Corso CR, Adami ER, Iacomini M, Fogaça RT, Acco A, Cadena SM (2018) Anti-fatigue activity of an arabinan-rich pectin from acerola (*Malpighia emarginata*). Int J Biol Macromol 109:1147–1153

Krivorotova T, Staneviciene R, Luksa J, Serviene E, Sereikaite J (2017) Impact of pectin esterification on the antimicrobial activity of nisin-loaded pectin particles. Biotechnol Prog 33(1):245–251

Lara-Espinoza C, Carvajal-Millán E, Balandrán-Quintana R, López-Franco Y, & Rascón-Chu A (2018) Pectin and pectin-based composite materials: Beyond food texture. Molecules 23(4): 942

Larsen N, Bussolo de Souza C, Krych L, Barbosa Cahu T, Wiese M, Kot W, Hansen KM, Blennow A, Venema K, Jespersen L (2019) Potential of pectins to beneficially modulate the gut microbiota depends on their structural properties. Front Microbiol 10:223

Lee J-H, Shim JS, Lee JS, Kim M-K, Chung M-S, Kim KH (2006) Pectin-like acidic polysaccharide from Panax ginseng with selective antiadhesive activity against pathogenic bacteria. Carbohydr Res 341(9):1154–1163

Lee M-H, Kim M, Kim M, Kwak JH, Chang DH, Yu WK, Lee S-H, Lee JH (2016) Consumption of dairy yogurt with the polysaccharide rhamnogalacturonan from the peel of the Korean citrus hallabong enhances immune function and attenuates the inflammatory response. Food Funct 7(6):2833–2839

Lefsih K, Giacomazza D, Passantino R, Costa MA, Bulone D, Mangione MR, Guarrasi V, Mingoia F, San Biagio PL, Madani K (2018) Biochemical and biophysical characterization of water-soluble pectin from Opuntia ficus-indica and its potential cytotoxic activity. Phytochemistry 154:47–55

Lengsfeld C, Titgemeyer F, Faller G, Hensel A (2004) Glycosylated compounds from okra inhibit adhesion of Helicobacter pylori to human gastric mucosa. J Agric Food Chem 52(6):1495–1503

Li S, Yang G, Yan J, Wu D, Hou Y, Diao Q, Zhou Y (2018) Polysaccharide structure and immunological relationships of RG-I pectin from the bee pollen of Nelumbo nucifera. Int J Biol Macromol 111:660–666

Lim BO, Lee SH, Park DK, CHOUE RW (2003) Effect of dietary pectin on the production of immunoglobulins and cytokines by mesenteric lymph node lymphocytes in mouse colitis induced with dextran sulfate sodium. Biosci Biotechnol Biochem 67(8):1706–1712

Lin L, Wang P, Du Z, Wang W, Cong Q, Zheng C, Jin C, Ding K, Shao C (2016) Structural elucidation of a pectin from flowers of *Lonicera japonica* and its antipancreatic cancer activity. Int J Biol Macromol 88:130–137

Ma X, Wang D, Chen W, Ismail BB, Wang W, Lv R, Ding T, Ye X, Liu D (2018) Effects of ultrasound pretreatment on the enzymolysis of pectin: Kinetic study, structural characteristics and anti-cancer activity of the hydrolysates. Food Hydrocoll 79:90–99

Maria-Ferreira D, da Silva LM, Nascimento A, Iacomini M, Cipriani T, de Paula Werner MF, Baggio CH (2012) Isolation of a rhamnogalacturonan polysaccharide from Spilanthes oleracea with gastroprotective and healing effects. Planta Med 78(11):58

Maria-Ferreira D, Da Silva LM, Mendes DAGB, de Almeida CD, Nascimento AM, Iacomini M, Cipriani TR, Santos ARS, de Paula Werner MF, Baggio CH (2014) Rhamnogalacturonan from Acmella oleracea (L.) RK Jansen: gastroprotective and ulcer healing properties in rats. PLoS One 9(1):e84762

Markov P, Popov S, Nikitina I, Ovodova R, Ovodov YS (2011) Anti-inflammatory activity of pectins and their galacturonan backbone. Russ J Bioorg Chem 37(7):817–821

Maxwell EG, Belshaw NJ, Waldron KW, Morris VJ (2012) Pectin–an emerging new bioactive food polysaccharide. Trends Food Sci Technol 24(2):64–73

Maxwell EG, Morris VJ, Waldron KW, Belshaw NJ (2015a) Potato rhamnogalacturonan I reduces colon cancer cell proliferation via a galectin-3 independent mechanism. Proc Nutr Soc 74:1

Maxwell EG, Colquhoun IJ, Chau HK, Hotchkiss AT, Waldron KW, Morris VJ, Belshaw NJ (2015b) Rhamnogalacturonan I containing homogalacturonan inhibits colon cancer cell proliferation by decreasing ICAM1 expression. Carbohydr Polym 132:546–553

Mikhaleva NY, Borisenkov M, Gyunter E, Popeiko O, Ovodov YS (2011) Effect of successive acid and enzymatic hydrolysis on the structure and antioxidant activity of pectins. Russ J Bioorg Chem 37(7):822–828

Mishra RK, Banthia AK, Majeed ABA (2012) Pectin based formulations for biomedical applications: a review. Asian Journal of Pharmaceutical and Clinical Research 5(4):1–7

Morris VJ, Gromer A, Kirby AR, Bongaerts RJ, Gunning AP (2011) Using AFM and force spectroscopy to determine pectin structure and (bio) functionality. Food Hydrocoll 25(2):230–237

Morris VJ, Belshaw NJ, Waldron KW, Maxwell EG (2013) The bioactivity of modified pectin fragments. Bioact Carbohydr Diet Fibre 1(1):21–37

Munarin F, Guerreiro S, Grellier M, Tanzi MC, Barbosa M, Petrini P, Granja P (2011) Pectin-based injectable biomaterials for bone tissue engineering. Biomacromolecules 12(3):568–577

Mzoughi Z, Abdelhamid A, Rihouey C, Le Cerf D, Bouraoui A, Majdoub H (2018) Optimized extraction of pectin-like polysaccharide from Suaeda fruticosa leaves: characterization, antioxidant, anti-inflammatory and analgesic activities. Carbohydr Polym 185:127–137

Nascimento AM, de Souza LM, Baggio CH, Werner MF, Maria-Ferreira D, da Silva LM, Sassaki GL, Gorin PA, Iacomini M, Cipriani TR (2013) Gastroprotective effect and structure of a rhamnogalacturonan from Acmella oleracea. Phytochemistry 85:137–142

Ndeh D, Rogowski A, Cartmell A, Luis AS, Baslé A, Gray J, Venditto I, Briggs J, Zhang X, Labourel A (2017) Complex pectin metabolism by gut bacteria reveals novel catalytic functions. Nature 544(7648):65

Nergard CS, Matsumoto T, Inngjerdingen M, Inngjerdingen K, Hokputsa S, Harding SE, Michaelsen TE, Diallo D, Kiyohara H, Paulsen BS (2005) Structural and immunological studies of a pectin and a pectic arabinogalactan from Vernonia kotschyana Sch. Bip. ex Walp.(Asteraceae). Carbohydr Res 340(1):115–130

Nosalova G, Strapkova A, Kardosova A, Capek P (1993) Antitussive activity of a rhamnogalacturonan isolated from the roots of *Althaea officinalis L var Robusta*. J Carbohydr Chem 12(4-5):589–596

Ogutu FO, Mu T-H (2017) Ultrasonic degradation of sweet potato pectin and its antioxidant activity. Ultrason Sonochem 38:726–734

Ovodov YS (2009) Current views on pectin substances. Russ J Bioorg Chem 35(3):269–284

Patil SN, Paradeshi JS, Chaudhari PB, Mishra SJ, Chaudhari BL (2016) Bio-therapeutic potential and cytotoxicity assessment of pectin-mediated synthesized nanostructured cerium oxide. Appl Biochem Biotechnol 180(4):638–654

Peng Q, Xu Q, Yin H, Huang L, Du Y (2014) Characterization of an immunologically active pectin from the fruits of Lycium ruthenicum. Int J Biol Macromol 64:69–75

Popov SV, Ovodov YS (2013) Polypotency of the immunomodulatory effect of pectins. Biochemist 78(7):823–835

Prandi B, Baldassarre S, Babbar N, Bancalari E, Vandezande P, Hermans D, Bruggeman G, Gatti M, Elst K, Sforza S (2018) Pectin oligosaccharides from sugar beet pulp: molecular characterization and potential prebiotic activity. Food Funct 9(3):1557–1569

Reena K, Prabakaran M, Leeba B, Gajendiran M, Antony SA (2017) Green synthesis of pectin-gold-PLA-PEG-PLA nanoconjugates: In vitro cytotoxicity and anti-inflammatory activity. J Nanosci Nanotechnol 17(7):4549–4557
Rha HJ, Bae IY, Lee S, Yoo S-H, Chang P-S, Lee HG (2011) Enhancement of anti-radical activity of pectin from apple pomace by hydroxamation. Food Hydrocoll 25(3):545–548
Samuelsen AB, Paulsen BS, Wold JK, Otsuka H, Kiyohara H, Yamada H, Knutsen SH (1996) Characterization of a biologically active pectin from Plantago major L. Carbohydr Polym 30(1):37–44
Samuelsen AB, Westereng B, Yousif O, Holtekjølen AK, Michaelsen TE, Knutsen SH (2007) Structural features and complement-fixing activity of pectin from three Brassica oleracea varieties: white cabbage, kale, and red kale. Biomacromolecules 8(2):644–649
Sande SA (2005) Pectin-based oral drug delivery to the colon. Expert Opin Drug Deliv 2(3):441–450
Sriamornsak P (2003) Chemistry of pectin and its pharmaceutical uses: A review. Silpakorn University International Journal 3(1–2):206–228
Shi H, Yu L, Shi Y, Lu J, Teng H, Zhou Y, Sun L (2017) Structural characterization of a rhamnogalacturonan I domain from ginseng and its inhibitory effect on galectin-3. Molecules 22(6):1016
Stegmayr J, Lepur A, Kahl-Knutson B, Aguilar-Moncayo M, Klyosov AA, Field RA, Oredsson S, Nilsson UJ, Leffler H (2016) Low or no inhibitory potency of the canonical galectin carbohydrate-binding site by pectins and galactomannans. J Biol Chem 291(25):13318–13334
Steinmassl M, Anderer FA (1996) Enhancement of human NK and LAK cytotoxicity against HCMV-infected cells by rhamnogalacturonan: specificity of reaction. Viral Immunol 9(1):27–34
Strube ML, Jensen TK, Meyer AS, Boye M (2015) In situ prebiotics: enzymatic release of galacto-rhamnogalacturonan from potato pulp in vivo in the gastrointestinal tract of the weaning piglet. AMB Express 5(1):66
Šutovská M, Nosáľová G, Šutovský J, Fraňová S, Prisenžňáková L, Capek P (2009) Possible mechanisms of dose-dependent cough suppressive effect of Althaea officinalis rhamnogalacturonan in guinea pigs test system. Int J Biol Macromol 45(1):27–32
Svava R, Gurzawska K, Yihau Y, Haugshøj KB, Dirscherl K, Levery SB, Jørgensen NR, Gotfredsen K, Damager I, Ulvskov P (2014) The structurally effect of surface coated rhamnogalacturonan I on response of the osteoblast-like cell line SaOS-2. J Biomed Mater Res Pt A 102(6):1961–1971
Thakur BR, Singh RK, Handa AK, Rao MA (1997) Chemistry and uses of pectin—a review. Critical Reviews in Food Science & Nutrition 37(1):47–73
Tan HD, Chen W, Liu QS, Yang GJ, Li KK (2018) Pectin oligosaccharides ameliorate colon cancer by regulating oxidative stress-and inflammation-activated signaling pathways. Front Immunol 9:1504
Tummalapalli M, Berthet M, Verrier B, Deopura B, Alam M, Gupta B (2016) Drug loaded composite oxidized pectin and gelatin networks for accelerated wound healing. Int J Pharm 505(1-2):234–245
Wang X-S, Dong Q, Zuo J-P, Fang J-N (2003) Structure and potential immunological activity of a pectin from Centellaasiatica (L.) Urban. Carbohydr Res 338(22):2393–2402
Wang XS, Liu L, Fang JN (2005) Immunological activities and structure of pectin from Centella asiatica. Carbohydr Polym 60(1):95–101
Wang H, Wei G, Liu F, Banerjee G, Joshi M, Bligh S, Shi S, Lian H, Fan H, Gu X (2014) Characterization of two homogalacturonan pectins with immunomodulatory activity from green tea. Int J Mol Sci 15(6):9963–9978
Wang M, Liu Y, Qiang M, Wang J (2017) Structural elucidation of a pectin–type polysaccharide from Hovenia dulcis peduncles and its proliferative activity on RAW264. 7 cells. Int J Biol Macromol 104:1246–1253
Wang JH, Zhang YD, Luo JP (2018) Structure elucidation of a pectin from Dendrobium nobile Lindl. and its immunological activity. Biotechnol Biotechnol Equip 32(3):744–750
Westereng B, Yousif O, Michaelsen TE, Knutsen SH, Samuelsen AB (2006) Pectin isolated from white cabbage–structure and complement-fixing activity. Mol Nutr Food Res 50(8):746–755

Westereng B, Coenen GJ, Michaelsen TE, Voragen AG, Samuelsen AB, Schols HA, Knutsen SH (2009) Release and characterization of single side chains of white cabbage pectin and their complement-fixing activity. Mol Nutr Food Res 53(6):780–789
Yamada H (1996) Pectin and Peetinase. Prog Biotech 4:173–190
Yo X, Bunt C, Cornish J, Quek SY, Wen J (2015) Oral delivery of bovine lactoferrin using pectin- and chitosan-modified liposomes and solid lipid particles: improvement of stability of lactoferrin. Chem Biol Drug Des 86(4):466–475
Ye PJ, Huang C, Yang S, Gao P, Li ZP, Tang SY, Xiang Y, Liu YF, Chen YP, He DX (2018) Facile fabrication of a novel hybrid nanoparticles by self-assembling based on pectin-doxorubicin conjugates for hepatocellular carcinoma therapy. Artif Cells Nanomed Biotechnol 46(3):S661–S670
Zaporozhets T, Besednova N, Liamkin G, Loenko I, Popov A (1991) Antibacterial and therapeutic effectiveness of a pectin from sea grass Zostera. Antibiot Chemoter 36(4):24–26
Zhang X, Li S, Sun L, Ji L, Zhu J, Fan Y, Tai G, Zhou Y (2012) Further analysis of the structure and immunological activity of an RG-I type pectin from Panax ginseng. Carbohydr Polym 89(2):519–525
Zhang T, Zhou P, Zhan Y, Shi X, Lin J, Du Y, Li X, Deng H (2015) Pectin/lysozyme bilayers layer-by-layer deposited cellulose nanofibrous mats for antibacterial application. Carbohydr Polym 117:687–693
Zhang T, Lan Y, Zheng Y, Liu F, Zhao D, Mayo KH, Zhou Y, Tai G (2016) Identification of the bioactive components from pH-modified citrus pectin and their inhibitory effects on galectin-3 function. Food Hydrocoll 58:113–119
Zhao J, Zhang F, Liu X, Ange KS, Zhang A, Li Q, Linhardt RJ (2017) Isolation of a lectin binding rhamnogalacturonan-I containing pectic polysaccharide from pumpkin. Carbohydr Polym 163:330–336
Zhu HG, Zollner TM, Kleinfranke A, Anderer FA (1994) Enhancement of MHC-unrestricted cytotoxic activity of human CD56+ CD3- natural-killer (NK) cells and CD3+ T-cells by rhamnogalacturonan - target-cell specificity and activity against NK-insensitive targets. J Cancer Res Clin Oncol 120(7):383–388
Zhu R, Wang C, Zhang L, Wang Y, Chen G, Fan J, Jia Y, Yan F, Ning C (2019) Pectin oligosaccharides from fruit of Actinidia arguta: structure-activity relationship of prebiotic and antiglycation potentials. Carbohydr Polym 217:90–97
Zou YF, Barsett H, Ho GTT, Inngjerdingen KT, Diallo D, Michaelsen TE, Paulsen BS (2015) Immunomodulating pectins from root bark, stem bark, and leaves of the Malian medicinal tree Terminalia macroptera, structure activity relations. Carbohydr Res 403:167–173

Chapter 10
Pectin as Drug-Release Vehicle

Ali Assifaoui and Odile Chambin

10.1 Introduction

The use of pectin as a vehicle for drug release has increased significantly in the last years for many reasons: (1) their high availability from renewable sources, biodegradability and biocompatibility (Ashford et al. 1994; Liu et al. 2003; Sriamornsak 1998), (2) the positive effect of pectin supplementation on the health benefits (reduction of cholesterol, prebiotic effect, cancer prevention, toxic metal removal) (Zhang et al. 2015; Lara-Espinoza et al. 2018; Chen et al. 2018; Tian et al. 2016), or (3) their various physical-chemical properties (emulsifying and gelling abilities). Moreover, other properties (muco-adhesiveness, ability to dissolve in alkaline environments, lower sensitivity to gastric pH, high resistance to degradation by enzymes (proteases and amylases) of the upper gastro-intestinal (GI) tract, and degradation by colon enzymes (pectinases)) make pectin a good candidate to formulate different drug delivery systems for various routes of drug administration (nasal, ocular, gastric and more specifically large intestine) (Chambin et al. 2006b; Das and Ng 2010; Maestrelli et al. 2008). In this chapter, we will present an overview of pectin drug delivery systems classified by their administration routes.

A. Assifaoui (✉) · O. Chambin
School of Pharmacy, Université de Bourgogne Franche Comté, UMR PAM, PCAV Team, Dijon, France
e-mail: ali.assifaoui@u-bourgogne.fr; odile.chambin@u-bourgogne.fr

V. Kontogiorgos (ed.), *Pectin: Technological and Physiological Properties*,
https://doi.org/10.1007/978-3-030-53421-9_10

10.2 Different Drug Delivery Systems by Various Administration Routes

Routes of administration can be classified according to the target action of the dosage form, which can be systemic or local delivery (Sultatos 2007). Therefore, administration may be enteral (oral, sublingual, buccal and rectal), topical (skin, eye, ear or nose) or parenteral (Fig. 10.1). In addition to active pharmaceutical ingredient (API), the carrier or excipient is very important: if it plays a technological role, it can also improve drug delivery as it could provide controlled and sustained release of drug and it influences the stability of the dosage form. Pectin as a natural polysaccharide exhibits interesting properties for drug delivery. It can be added as a diluent to form tablets or to fill capsules but also it can be used to control and to target the API release in various dosage forms. Its ability to form gel enhances the contact time of API with the physiological part of the body, particularly in the gastro-intestinal tract by targeting the colon, and in ocular, nasal, and vaginal mucosal tissues.

10.2.1 Oral Drug Delivery System (ODDs)

Oral drug delivery (ODDs) is the most preferred and convenient route of drug administration due to high patient compliance, cost-effectiveness, flexibility in the design of dosage form and ease of production. Therefore, this administration route leaves pharmaceutical industry with a huge challenge to find new therapeutic entities that can be delivered by this route of administration. Successful oral delivery needs to select the active pharmaceutical ingredient (API), which hits a pharmaceutical target, as well as to identify a formulation for that drug that provides the required pharmacokinetic (PK) profile to result in a desired pharmacodynamic (PD) outcome (Mrsny 2012). When it is swallowed, the ODDs have to pass through the different parts of the GI tract and need to "survive" before being absorbed, despite the harsh environments of stomach, duodenum, and early jejunum with bile salts. Thus, ODDs will first meet the strongly acidic environment (pH 2–3) of the stomach that contains a variety of enzymes (pepsin, lipases, and proteases). This step could

Fig. 10.1 Routes of drug administration

last from 1 to 3 h depending on the content of the stomach (fasting or fed). Then ODDs will pass the first segment of the small intestine (duodenum) where pancreatic enzymes, bicarbonate and bile salts are also released. During transit, the pH of the GI tract lumen rises from the acidic levels of the stomach, to 5–6 in the duodenum, to neutrality (~7) in the mid-to-distal jejunum. Considering the complexity of GI tract, a variety of methods have been developed to allow passage of labile drugs through the stomach in a protective structure and then release them later in the GI tract to allow their absorption. However, to deliver an API to the colon remains a real challenge. Among these formulations, oral pectin drug-delivery systems offer a solution for some of the difficulties encountered in the GI tract.

10.2.1.1 Pectin in Pharmaceutical Tablets

Presently, oral tablets still account for over 80% of all dosage forms used in human medicines. Tablets are defined as a solid unit dosage form, which contain the API with suitable excipients, and are prepared by compression of a constant volume of particles, moulding, or extrusion. The excipients including diluents, binders, fillers, lubricants, disintegrants, sweeteners, and aroma compounds play an important role on tablet production from process aspects as well as from biopharmaceutical point of view with their ability to dissolve and release the API in the body (Swarbrick and Boylan 2002). Polysaccharides could be used as an excipient with different roles: as matrix tablet or as coating to control the drug release. For matrix tablets, pectin and API are intimately mixed. In 1994, Ashford et al. (1994) presented some findings with pectin matrix showing that the type of pectin, the presence of calcium, and the solubility of calcium salt influence the drug release. Both types of pectin (high methoxy (HM) and low methoxy (LM)) could maximise colonic specificity by providing optimal protection of drug during its transit to the colon and a high susceptibility to enzymic attack. Kim et al. (1998) studied the potential for direct compression of pectin. Pectin has been shown to be hard, rigid, and poorly compactible with a lack of plastic deformation. As a result, the authors suggested to use a mixture of 50/50 with microcrystalline cellulose. Ahrabi et al. (2000) investigated some formulation factors to decrease early release in upper part of gastrointestinal tract and improve compactability. Amidated pectin gives harder tablets and could be also mixed with ethylcellulose or coated to improve all performances. Other authors also recommended mixtures with dextrates (Mura et al. 2015), hydroxypropylmethylcellulose (HPMC) (O'Brien et al. 2009), or chitosan (Atyabi et al. 2005). Chomto and Nunthanid (2017) have recently studied the physicochemical and powder characteristics (particle size, moisture content, flowability) of various citrus pectins and their abilities to form pharmaceutical matrix tablets. They have shown that factors such as molecular weight (M_W), degree of esterification (DE), degree of amidation (DA), crystalline structure and salt or acid forms of different pectins could affect the tableting process and thus the drug release from these tablets. Indeed, pectin matrix tablets formed a continuous gel layer which controls the diffusion of the API and its subsequent release. They have shown that all the studied pectins possessed fair flowability except pomelo pectin. Pectins with lower DE and higher M_w produced tablets with higher tensile strengths and slower erosion rate, reducing theophylline release. Thus, it can be concluded that it is difficult to obtain a suitable tablet with pectin alone as matrix-former and that pectin needs to be associated with other compounds to improve mechanical properties of tablets and to better control drug release and target the colon.

In order to use pectin as a coating agent, it is important to introduce it into the coat mixture. This mixture is then applied as a film around a tablet or a core to bring different properties: swallowing, taste masking, drug protection against light and moisture, controlled release in order to increase resistance to the environment (increase shelf life) and target the API release (enteric, sustained or colonic release). For Ashford et al. (1994), LM pectin is able to protect more the tablet in the upper gastrointestinal tract than HM pectin. Wakerly et al. (1997) showed that a mixture with other polymers as ethylcellulose further improved the properties of the tablets. Macleod et al. (1999) have shown the interest of a coating made of pectin HM, HPMC and chitosan to target the colon but when the amount of pectin increased, the coating layer became more fragile. Therefore, the correct ratio should be checked carefully and this was also the conclusion for pectin/chitosan coatings (Fernández-Hervás and Fell 1998) and pectin/HPMC coatings for the delivery of 5-aminosalicylic acid in the colon (Turkoglu and Ugurlu 2002). The coating layer could be also obtained by a double compression process. Hodges et al. (2009) have evaluated colon-targeting tablets of pectin-HPMC contained a radio labelled molecule, in six healthy male volunteers. They have observed that tablets released in the colon in all subjects but with different kinetic rate. They have attributed this to "quality" of the hydrogel layer which may initiate radiolabel release when the formation of this layer was disrupted. Thus, inadequate prior hydration of the hydrogel layer prevented access of pectinolytic enzymes and reduced fluid availability in the transverse colon may have retarded tablet disintegration and radiolabel diffusion. It is also possible to develop a dry powder coating technique using a pan coater. Using this technique, Tung et al. (2016) have obtained coated tablets of metronidazole by using this technique and observed a lag time in API release controlled by the coating level of the mixture pectin/HPMC.

10.2.1.2 Beads or Microparticles

Different beads or microparticles could be used with LM pectin. Figure 10.2 summarises different strategies to obtain beads and improves their properties, particularly concerning drug release. All these strategies will be developed in this part.

Cation Pectinate Beads

Low methoxy (LM) pectin is mainly used to form encapsulating matrices where both water-soluble and water-insoluble, either small organic compounds or active protein drugs can be entrapped (Table 10.1) (Assifaoui et al. 2011; Atyabi et al. 2005; Chambin et al. 2006a; Das and Ng 2010; Dhalleine et al. 2011; El-Gibaly

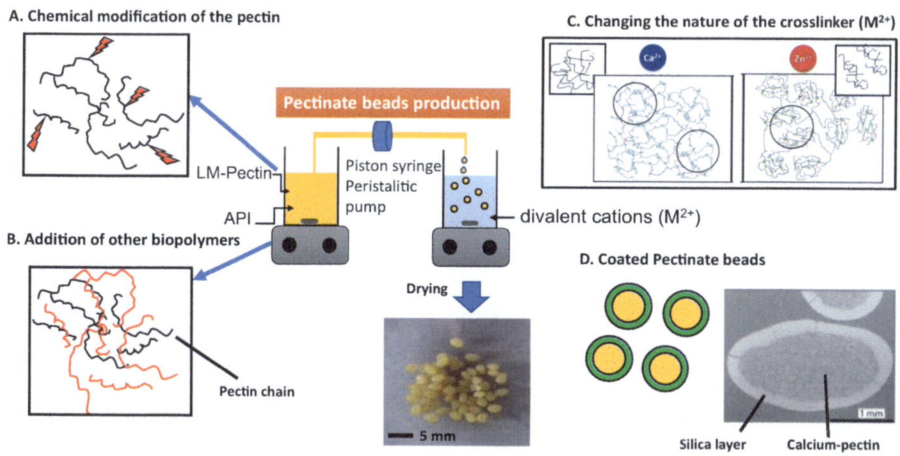

Fig. 10.2 Different strategies to modify pectinate beads in order to control the API release. (**a**) Chemical modification, (**b**) addition of other biopolymers (polysaccharide or proteins), (**c**) changing the nature of divalent cation, (**d**) applying a coated layer

2002; Hagesaether et al. 2008; Nguyen et al. 2014; Sriamornsak 1998, 1999; Sriamornsak and Nunthanid 1998). This feature is due to the LM pectin's ability to form instantly a gel when it is in contact with specific divalent cations (cross-linkers) such as Ca^{2+} and Zn^{2+} (ionotropic gelation) (Fig. 10.2c). This method, called "external gelation", induces the formation of anisotropic gel (beads) where a decrease of both the polymer concentration and the Young modulus from the periphery to the core of the bead is observed (Skjåk-Bræk et al. 1989; Maire du Poset et al. 2018). It is thus possible to produce calcium-pectinate (films, gels, beads, microspheres) where the yield and the encapsulation efficiency (EE) could be highly improved (Fig. 10.2).

The nature of divalent cation affects the gel structure at "molecular level" and thus the drug release at "macroscopic level". In vivo studies have shown that theophylline release from zinc-pectinate beads was delayed in comparison with calcium-pectinate microspheres (Dhalleine et al. 2011). This is due to the high swelling ratio of calcium pectinate beads compared to zinc pectinate beads (Assifaoui et al. 2011). Results from molecular dynamic (MD) simulations have shown that the association of carboxylate groups of the pectin (R-COO$^-$) and zinc ions is "egg-box" model; while this association does not follow the "egg-box" model for calcium-carboxylate (which is wrongly admitted in comparison to alginate model) (Fig. 10.2c) (Assifaoui et al. 2015). In the opposite of zinc-pectin interaction, the calcium-pectin interaction is less strong thus allowing high flexibility of the 3D network which can explain the high ability of such structure to swell (Huynh et al. 2016). Maire du Poset et al. (2018) have formulated an anisotropic gel made by the diffusion of iron (II) through a dialysis membrane to a polygalacturonate solution and have shown by X-ray absorption spectroscopy that iron inside the gel keeps its degree of oxidation. The presence of a high content of iron (II) in these

Table 10.1 Summary of pectin uses in different dosage forms

Administration route	Target	Dosage form	Pectin properties	References
Oral drug delivery systems	Insulin release	Nano- and microparticles of a chitosan-pectin complex	Coacervation	Maciel et al. (2017)
	Buccal drug delivery	Patches, tablets, films	Bioadhesive properties	Noreen et al. (2017), Kaur and Kaur (2012), Jug et al. (2012), Mura et al. (2015)
	Gastro-retentive dosage (stomach)	Floated systems (tablets, beads)	Pectin and calcium carbonate	Sriamornsak et al. (2007a)
	Colon drug delivery	Beads/ microparticles	Ionotropic gelation	Ashford et al. (1994), Sriamornsak et al. (2007a, b)
		Pectin beads coated by silica layer	Inorganic coating	Assifaoui et al. (2013)
		Pectin in matrix tablet	Matrix agent	Ahrabi et al. (2000), Mura et al. (2015), Chomto and Nunthanid (2017)
		Pectin in coated tablets	Coating agent	Wakerly et al. (1997), Turkoglu and Ugurlu (2002), Hodges et al. (2009), Tung et al. (2016)
Nasal delivery	Prolonged contact time with nasal mucosa	Nasal spray	In situ gel ability	Illum (2012), Nižić et al. (2019), de Vries and van der Baan (1989)
Ocular	Improve the drug bioavailability	Spray	Gel in contact with the electrolytes in the tear	Giunchedi et al. (1999)
Cutaneous drug delivery systems	Sustained drug release	Transdermal drug delivery systems	Polymer matrix	Mazzitelli et al. (2013), Shahzad et al. (2019)
	Absorb exudates	Wound dressing	Film-forming agent	Munarin et al. (2012)

(continued)

Table 10.1 (continued)

Administration route	Target	Dosage form	Pectin properties	References
Vaginal	Release of prebiotics and probiotics	Tablets made of microparticles	Microparticles obtained by crosslinking calcium to (pectin/hyaluronic acid)	Pliszczak et al. (2011)
	Antiretroviral drugs	Film	Film forming properties (mix with PVA)	Cautela et al. (2019)
	Moisturisers	Gel	Gelling properties	Caswell and Kane (2002)
Parenteral	Bio-engineering	Injectable gels	Internal gelation with calcium carbonate	Moreira et al. (2014)
	Cancer	Nanoparticles	Interaction with galectin 3	Glinsky and Raz (2009)

hydrogels make them perfectly suitable for iron fortification, since the iron requirement in terms of absorbed iron is around 1–2 mg/day. Sharma et al. (2012) have prepared thiolated pectin nanoparticles (NP) for ocular route, by ionotropic gelation of a very dilute solution of thiolate pectin (0.01% w/v) by using magnesium chloride as the cross-linker and timolol maleate as the model drug (see Sect. 10.2.2.3). Compared to calcium, Mg^{2+} allowed the formation of small NP (~240 nm) (Das and Ng 2010). The weak interaction between thiolated pectin and the Mg^{2+} probably induced the formation of a pre-gel phase, where the polymeric chains react with gelling agent to form small cluster. (Sharma et al. 2012). This could be explained by the gelling ability of pectins with divalent cations which follow the order: $Mg^{2+} << Zn^{2+} < Fe^{2+} < Ca^{2+} < Sr^{2+} < Ba^{2+}$ (Kohn 1975; Torre et al. 1995; Maire du Poset et al. 2018; Huynh et al. 2017).

Pectin Modification for Dosage Forms

The presence of numerous hydroxyl and carboxyl groups which are distributed along the backbone, as well as a certain amount of neutral sugars present as side chains make the pectin a suitable candidate for chemical and physical treatments (Fig. 10.2a). The functionalisation of the pectin will allow the production of "smart" dosage forms which can respond to the internal stimuli (pH, enzymes) or external stimuli (light, magnetic field) (Li and Mooney 2016). Different methods of chemical modification of pectin are described including substitution (amidation, alkylation, thiolation, sulfation), chain elongation (cross-linking and grafting) and depolymerization (acid or enzymatic hydrolysis) (Chen et al. 2015). Some examples will be developed below to illustrate the influence of these chemical modifications, but it will not be a detailed account of all the work on this area. Cationic derivatives of pectin could be prepared by pectin quaternisation by reacting with CHPTA (3-chloro-2-hydroxypropyltrimethylammonium chloride) in alkaline conditions.

After characterisation of the compound obtained (determination of substitution degree), the properties of these pectins were checked. A high degree of pectin substitution increased particularly the moisture absorption properties. This quaternisation brings also antibacterial properties against various strains of bacteria. Thus, the functionality of pectin could be improved for different applications in drug delivery, packaging, or cosmetic when a preservative effect is required (Fan et al. 2012). Another study modified pectin by immobilisation of a hydrophobic thiol-bearing ligand (4-aminothiophenol) to the backbone due to oxidation of thiol groups to disulphide bonds. This modification induced a lower swelling and disintegration behaviour (Perera et al. 2010). Then microparticles were obtained by spray-drying with metronidazole loaded as model drug. This pectin modification leads to an improved stability of microparticles and a colon-specific drug delivery with low cell toxicity. It could also improve local drug efficacy due to an increase of mucoadhesive properties with interaction of thiols and cysteine present in the mucus. Modification of pectin could be also achieved by acylation of hydroxyl groups of pectin by using several fatty acid anhydrides (Monfregola et al. 2011b). The properties of modified pectin were checked and indicated a reduced water sorption when palmitic anhydride was grafted in comparison with oleic and linoleic anhydride (Monfregola et al. 2011a). Octenyl succinic anhydride could be used with the same free solvent process to obtain modified pectin with various degrees of substitution. The affinity towards water was reduced as the degree of substitution increased but these modified pectins kept their ability to react with calcium during ionotropic gelation and allowed bead manufacturing (Mahe et al. 2018). Wang et al. (2019) have recently prepared chemical cross-linked pectin carriers by adding succinic anhydride and glutaric dialdehyde as the cross-linking reagents. They have shown that such formulation improved both encapsulation efficiency and loading efficiency of diclofenac sodium, a water-soluble drug, depending on the cross-linking conditions. Moreover, diclofenac was hardly released in the acidic media but rapidly released in alkaline media.

Mixed Biopolymer-Pectin Beads and Emulsion Gels

In order to optimise the encapsulation process (improve the encapsulation efficiency and the yield) or to delay the API release, many authors have used pectin in conjunction with other polysaccharides (e.g., chitosan, starch, HPMC) or proteins (e.g., whey proteins, caseins). These macromolecules can be used as coating agents (e.g., chitosan, HPMC) (see Sect. 10.2.1.1) or inside the bead allowing formation of an interpenetrating polymer network bringing additional features such as high affinity to bind the API or increase the gelling ability (Fig. 10.2b). When the API is hydrophobic, the approach used by many researchers is to formulate an oil-in-water emulsion where pectin is the continuous phase and the API is solubilised in the oil phase. The emulsion can be stabilised by using surfactants (Tween® or biopolymers such as proteins). Once the emulsion is stable, it can be transformed into emulsion gels by heat treatment or acidification with glucono-d-lactone (GDL) or addition of divalent salts ($CaCl_2$ or $ZnCl_2$), and even enzymatic cross-linking with glutaminase. Moreover, compared to emulsions alone, such formulations promote a better oxidative stability of lipids and a controlled release thanks to the presence of pectin network. Ruffin et al. (2014) have shown that the combination between pre-heated whey protein (as an emulsion stabiliser) and low methoxy pectin is a good strategy to enhance stability of vitamin-A during storage and also preserve it against low pH (gastric media) and to control the release of this lipophilic API.

Coated Beads (Organic/Inorganic)
In order to better control the drug release, calcium pectinate beads can be coated by applying organic (e.g., chitosan, HMPC) or inorganic (e.g., silica) layers. Kim et al. (2003) have produced chitosan-coated pectin beads and showed that the protein was released from these dosage forms only in simulated colonic fluid and not in simulated gastric fluid. The strong ionic interaction between NH_3^+ of chitosan and free carboxylate groups of pectin is responsible for the formation of a denser layer of chitosan, which enhanced the mechanical properties and mucoadhesiveness of pectin beads. It has been demonstrated that calcium-pectinate beads swell in the dissolution medium and subsequently disintegrate to quickly target colon (swelling-erosion-controlled process) (Das and Ng 2010; Assifaoui et al. 2011). To reduce swelling, the beads can be covered with silica, starting from tetraethoxysilane (TEOS), by a sol-gel process. Assifaoui et al. (2013) have shown that beyond 0.25M of TEOS, the silica layer (~100 μm) is thick enough to act as a barrier to water uptake and to reduce the swelling ratio of the beads (Fig. 10.2d).

Floated Calcium-Pectinate Beads
Pectin has been used in the development of gastro-retentive dosage forms which have the specificity to improve the delivery and performance of drugs that are locally active in the stomach (Auriemma et al. 2018; Bera et al. 2015; Elmowafy et al. 2009; Sriamornsak et al. 2007b, 2005, 2007a; Stops et al. 2008). Therefore, the formulation of floating oral dosage forms can achieve a site-specific therapeutic action, for example, antibiotic administration for *Helicobacter pylori* eradication in the treatment of peptic ulcer disease (Patil et al. 2016). Floating dosage forms can be made by a gelling process of hydrocolloid materials or by incorporating a floatation chamber, under vacuum or gas filled. Floating dosage forms require a sufficient high level of fluids in the stomach to allow the system to float and release the drug in the stomach. This limitation can be overcome by using a mucoadhesive polymer such as pectin (Patil et al. 2016; Bera et al. 2015; Prezotti et al. 2014; Thirawong et al. 2007). Floated pectinate beads were prepared by dispersing carbonate salt in pectin solution and then extruding into either neutral or acidified solution of calcium chloride. The reaction of carbonate salts with acid induced the formation of carbon dioxide initiating the formation of porous structured beads. Highly porous beads showed good floating ability (floating lag time and total floating time) with fast drug release.

10.2.2 Other Administration Routes

Even if oral route is the most important, some alternative routes are now investigated to restrain some disadvantages such as poor bioavailability due to pre-systemic metabolism by acidic pH and digestive enzymes, and first-past hepatic effects. They could be used for topic applications but also for systemic effects and some of them are less painful than parenteral administration.

10.2.2.1 Buccal Drug Delivery Systems

Buccal mucosa could be used for systemic drug delivery systems as it is well vascularised and thus could avoid degradation and first-pass metabolism by gastrointestinal tract. Many buccal devices are described such as tablets, patches and thin films (Noreen et al. 2017) and polysaccharides are very useful for such devices. Kaur and Kaur (2012) presented mucoadhesive patches made of pectin and chitosan by solvent casting method and loaded with carvedilol hydrochloride to treat hypertension and angina attack. Jug et al. (2012) developed mucoadhesive buccal patches with triclosan for antibacterial properties against *Streptococcus mutants*, cariogenic bacteria of the oral cavity. The patches were obtained with a mixture of amidated low methoxy pectin (ALMP), Carbopol® and a polymer of cyclodextrin. Various ratio of these three polymers were studied. ALMP influenced the cyclodextrin complexation of triclosan increasing the solubilization of the drug in β-cyclodextrin but decreasing its solubility in polymeric cyclodextrins. ALMP alone was unable to provide sustained drug release due to its high erosion rate but the mixture ALMP/Carbopol® (80:20 w/w) showed optimal swelling and erosion properties with good mucoadhesive properties. The release of triclosan was suitable for local delivery and the transbuccal permeation was negligible. The ALMP could also be used in combination with carboxylethylcellulose to obtain thin films called wafers for econazole buccal delivery in candidiasis (Mura et al. 2015). An experimental design was set to select the best formulation with predictable mucoadhesive and in situ residence properties suitable for therapeutic efficacy. Finally, Nguyen et al. (2013) studied the interactions between pectin coated liposomes and saliva and dental enamel to investigate their potential use to obtain a protective biofilm formed naturally on the teeth. Different pectins (LM, HM, ALMP) were used at different concentrations and the film structure were monitored by atomic force microscopy (AFM). Pectins are useful in dental therapy due to their ability to retain on the enamel surfaces.

10.2.2.2 Nasal Drug Delivery Systems

Nasal cavity is mainly used for treatment of local diseases such as nasal congestion, infection or allergic diseases. For example, nasal application of corticosteroids is widespread in the treatment of nasal inflammatory disorders such as allergic and non-allergic rhinitis, chronic rhinosinusitis and nasal polyposis (Alagusundaram et al. 2010). However, nasal route is developed nowadays for systemic delivery of small molecular weight drugs, particularly when a rapid action is suitable in migraine, pain, or smoking cessation (Illum 2012) thus avoiding the hepatic first-pass metabolism. Nasal administration presents another interest as it is suitable for patients who have difficulties to swallow tablets. Thanks to their gelling ability, pectins have the possibility to form an in situ gel system which can offer unique advantages in nasal drug delivery. Indeed, it can be easily sprayed as a liquid formulation in the nasal cavity and ensure prolonged contact time with nasal mucosa due to rapid gelation that reduces post-nasal drip, anterior leakage and mucociliary clearance of the administered API. Fentanyl pectin nasal spray (FPNS) is an approved nasal fentanyl formulation (PecFent®) that incorporates the gelation ability of pectins. This drug delivery system allows the API to be rapidly absorbed across the nasal membrane and directly into the blood stream. The pectin used in this formulation is a low methoxy (LM) pectin with a very low degree of esterification (DE) and the solution must be substantially free of gelling agents such metal ions to avoid the formation of a gel during storage. The controlled release of the drug could be linked to hydration/diffusion mechanisms of the mucosal tissue with interaction between pectin-free acid groups and mucin. Charlton et al. (2007) studied the mechanism of transport of two model drugs (mannitol and propranolol) through MDCK epithelial cells and they showed that pectin increased retention of propranolol, a lipophilic compound. Nižić et al. (2019) have also developed an in situ gelling system for effective nasal delivery of fluticasone by using a fine drug suspension which contains LM pectin, gellan gum and sodium hyaluronate as gelling/thickening agents and Tween 80 as a suspending agent. They have shown that parameters such formulation spraying ability and turbinate deposition, which are responsible of a well-designed formulation, can be tuned by understanding the interaction between the three polysaccharides. Nasal sprays containing nicotine have also been investigated in smoking cessation and with pectin, two different systems could be developed: (1) a rapid delivery of nicotine for rapid absorption and (2) a controlled release for sustained nicotine absorption (de Vries and van der Baan 1989).

10.2.2.3 Ocular Drug Delivery Systems

Topical application of drugs to eye is the most accessible and desired mode of treatment for management of ocular disorders. But it is limited by the shorter pre-corneal residence time of the drug inducing a low bioavailability, which requires frequent topical administration. To overcome this drawback, polymeric-microparticle delivery-systems have been developed (Giunchedi et al. 1999; Sanzgiri et al. 1993; Sharma et al. 2016). Guinchedi et al. have evaluated pectin microspheres prepared by spray-drying technique, as delivery system for piroxicam (Px) (Giunchedi et al. 1999). Px loaded in the pectin microspheres showed a faster in vitro dissolution rate with respect to solid micronised drug. The precorneal retention of fluorescein-loaded microspheres was evaluated in vivo in albino rabbits: an aqueous dispersion of fluorescent microspheres showed a significantly increased residence time in the eye (2.5 h) when compared with a fluorescein solution (0.5 h). In vivo tests in rabbits of dispersions of Px-loaded microspheres indicated a significant improvement of Px bioavailability in the aqueous humour (2.5-fold) when compared with commercial Px eyedrops. The potential advantages and limitations of this delivery system have been discussed. Their main interests were to increase bioavailability but also to reduce vision problems induced by other dosage forms. Another strategy is to use hydrogel to increase the residence time and pectin has been investigated. Pectin could also be used to ensure in situ gel formation where a liquid is applied and the gel forms immediately in contact with eyes due to the presence of electrolytes in the tear.

10.2.2.4 Cutaneous Drug Delivery Systems

Transdermal drug delivery systems have been developed since 1980 for systemic delivery of many drugs (e.g., estradiol, trinitrin, nicotin, fentanyl). They are sustained release dosage forms able to maintain stable drug plasmatic concentrations up to seven days so reducing the dosage for the patient, avoiding the gut and the liver whereas the treatment could be easily stopped by the removal of patch. Usually they are adhesive formulations and some skin irritation could appear. Therefore, Mazzitelli et al. (2013) proposed the development of a new transdermal drug delivery system based on natural polymers (LM-pectin and gelatin) using their gelation properties with different rheological properties and drug release profile for testosterone administration. The importance of their rheological characteristics to resist to many mechanical stresses during the manufacturing process and their use have been discussed. Pectin is also used for wound dressings production. Shahzad et al. (2019) developed scaffolds made of pectin and alginate due to their ability to form a film by cross-linking with calcium, their biocompatibility, and their biodegradability. The films were loaded with chitosan nanoparticles of cefazolin and provided a local effect in the wound with a sustained delivery for the treatment of infections. Long et al. (2019) have presented an interesting approach with a 3D-printed hydrogel of chitosan and pectin to deliver lidocaine. The wound dressings were produced by printing the crosslinking hydrogels using an extrusion-based printer followed by freeze-drying. 3D-printed hydrogel wound dressings exhibit many advantages: formulations can be customised with a specific dose of drug based on patient requirement, the dimension of the scaffold can be easily controlled, the process is simple, the bioadhesion of polysaccharides is suitable for its use, the porosity of the scaffold allows oxygen penetration in the wound, and the properties of hydrogel absorb wound exudates and maintain a moist environment necessary for wound healing. On the other hand, a wide variety of hydrocolloid pectin-based wound dressings has been patented (Munarin et al. 2012) and is nowadays commercially available for wound healing (CombiDERM®, DuoDERM®, Grranuflex®, Hydrocoll®, CitruGel®).

These systems are often without any other drugs but are also of interest when pectin applications are considered.

10.2.2.5 Vaginal Drug Delivery Systems

Mucoadhesive polymers could be used not only for vaginal administration for local effect but also for systemic drug delivery and vaccination. A solution containing ketokenazole, pectin, monyl phenol-9 and polyethylene glycol was investigated for a vaginal douche with fungicidal properties (Valenta 2005). Pliszczak et al. (2011) studied microparticles made of pectin and hyaluronic acid to deliver prebiotics and probiotics to restore vaginal ecosystem. Microparticles were produced by emulsification/gelation methods with calcium as crosslinking agent. Their size and bioadhesive properties have been evaluated by rheological measurements and the release of probiotics in a simulated vaginal fluid has been investigated. Finally, these microparticles could be used to obtain a vaginal tablet by compression. Composite films based on PVA (polyvinyl alcohol) and high methoxy pectin were developed by Cautela et al. (2019) as new drug delivery systems for the prevention of HIV transmission. The films were produced as single or double-layered films by solvent casting technique and they were loaded with spray-dried Eudragit® nanoparticles of antiretroviral drugs (tenofovir disoproxil fumarate and emtricitabine). Moreover, these films were evaluated for technological, physicochemical and biological properties and films seems to be promising for vaginal microbicide products. All films were shown to be safe and provided sustained protection from viral infection. Caswell and Kane (2002) have shown that the use of vaginal moisturiser based on pectin (Summer's Eve®) induced a significant less vaginal residue compared to Replens® vaginal moisturiser, based on polycarbophil, and in comparable relief of vaginal dryness.

10.2.2.6 Parenteral Drug Delivery Systems

Injectable gels could be used for specific site delivery and regenerative medicine. Moreira et al. (2014) proposed different injectable pectin hydrogels produced by internal gelation with calcium carbonate to control gelation kinetics, the homogeneity of gels and to reach suitable pH for cell viability. The rheological behaviour was tested in order to obtain the appropriate injectability. These devices seem to be promising to provide adequate microenvironment for cells and drug delivery. Pectin polyplexes were also investigated in gene delivery for the treatment of genetic disorders. Among non-viral vectors, polycationic polymers as polyethyleneimine and chitosan have shown high transfection efficacy but they also display high cytotoxicity. Pectin was shown to be suitable to coat polyethyleneimine polyplexes with lower toxicity and higher stability (Asnaghi et al. 2011). Pectin could be also modified with amine groups to form complexes with plasmid DNA for DNA delivery (Katav et al. 2008). Finally, pectin appears to be useful in cancer therapy by inhibiting metastasis and tumour growth thanks to the galectin-3 whereas modified pectin could be loaded with cytotoxic drugs to increase the efficacy of conventional chemotherapy (Glinsky and Raz 2009). Finally all the uses of pectin presented in this chapter are summarized in Table 10.1 according to the route of administration.

10.3 Conclusions

Pharmaceutical industries are attracted by the fact that pectin is a natural polymer, biocompatible and biodegradable. Thus, pectin is used to produce many drug delivery systems for different actions (systemic/local) by different administration routes (e.g., oral, nasal, cutaneous, vaginal, or ocular). Moreover, their ability to form a gel (ex situ or in situ) and mucoadhesive properties present opportunities for applications in the formulation of various dosage forms. Microparticles made by the interaction between LM pectin and divalent cations such as calcium is one of the most promising techniques and offer a great potential of use. Indeed, ionotropic gelation can be used to encapsulate various active molecules from peptide to probiotics and to control and target their release based on the physiological properties of the route of administration. This can be achieved by the modification of the 3D-structure of the pectin network in order to control the mesh size (porosity), the flexibility, and the interaction between the pectin network with the API as well as with the physiological liquid corresponding to dissolution media. Different strategies exist: chemical modification of the pectin, introduction of others biopolymers (proteins, polysaccharides), changing the nature of the crosslinker, application of a coating (organic or inorganic) to modulate the properties and the drug release, and to obtain a solution to adapt the dosage form to the therapeutic requirements. For pectin-based drug delivery systems, it has been demonstrated that such systems are unique since they can deliver API in the colon (in the cases of inflammatory bowel diseases, cancer of the colon, Crohn's diseases) due to the presence of pectinolytic enzymes. However, the efficacy and the specificity release in the colon must be proved by achieving a reliable in vitro and in vivo model and also clinical trials. A systematic comparison study among selective pectin is required to overcome the difficulties associated with the choice of pectin, dosage formulation and intestinal absorption barrier. Considerable efforts in the optimisation of pectin-based drug delivery systems are required. Relatively little progress on the increase of hydrophobicity of such dosage forms has been reported. However, finding efficient methods to produce pectin derivates, which are more water resistant but still enzymatically labile, remained a big challenge. The main commercial dosage forms using pectin as drug delivery are limited to topical applications (e.g., creams or nasal sprays). A lack of commercial oral dosage forms based on pectin (e.g., tablets, beads, microparticles) is also noticeable. This may be due to the large use of alginates by industry in comparison to pectins. For the pectinate beads/microparticles, technical problems related to the industrial production of calcium-pectinate beads (low rate of production) must be overcome. Indeed, such formulations remained at laboratory scale.

Engineering work must be developed to improve the scale-up of such drug delivery systems but also with studies on the drying process (less expensive and more efficient) which is crucial for stability during storage.

References

Ahrabi SF, Madsen G, Dyrstad K, Sande SA, Graffner C (2000) Development of pectin matrix tablets for colonic delivery of model drug ropivacaine. Eur J Pharm Sci 10(1):43–52

Alagusundaram M, Chengaiah B, Gnanaprakash K, Ramkanth DS, Madhusudhana C, Dhachinamoorthi D (2010) Nasal drug delivery system - an overview. Int J Res Pharm Sci 1:454–465

Ashford M, Fell J, Attwood D, Sharma H, Woodhead P (1994) Studies on pectin formulations for colonic drug delivery. J Control Release 30(3):225–232

Asnaghi MA, Candiani G, Fare S, Fiore GB, Petrini P, Raimondi MT, Soncini M, Mantero S (2011) Trends in biomedical engineering: focus on Regenerative Medicine. J Appl Biomater Biomech 9(2):73–86

Assifaoui A, Chambin O, Cayot P (2011) Drug release from calcium and zinc pectinate beads: impact of dissolution medium composition. Carbohydr Polym 85(2):388–393

Assifaoui A, Bouyer F, Chambin O, Cayot P (2013) Silica-coated calcium pectinate beads for colonic drug delivery. Acta Biomater 9(4):6218–6225

Assifaoui A, Lerbret A, Uyen HTD, Neiers F, Chambin O, Loupiac C, Cousin F (2015) Structural behaviour differences in low methoxy pectin solutions in the presence of divalent cations (Ca^{2+} and Zn^{2+}): a process driven by the binding mechanism of the cation with the galacturonate unit. Soft Matter 11(3):551–560

Atyabi F, Majzoob S, Iman M, Salehi M, Dorkoosh F (2005) In vitro evaluation and modification of pectinate gel beads containing trimethyl chitosan, as a multi-particulate system for delivery of water-soluble macromolecules to colon. Carbohydr Polym 61(1):39–51

Auriemma G, Cerciello A, Sansone F, Pinto A, Morello S, Aquino RP (2018) Polysaccharides based gastroretentive system to sustain piroxicam release: development and in vivo prolonged anti-inflammatory effect. Int J Biol Macromol 120:2303–2312

Bera H, Boddupalli S, Nayak AK (2015) Mucoadhesive-floating zinc-pectinate–sterculia gum interpenetrating polymer network beads encapsulating ziprasidone HCl. Carbohydr Polym 131:108–118

Caswell M, Kane M (2002) Comparison of the moisturization efficacy of two vaginal moisturizers: pectin versus polycarbophil technologies. J Cosmet Sci 53(2):81–87

Cautela MP, Moshe H, Sosnik A, Sarmento B, da Neves J (2019) Composite films for vaginal delivery of tenofovir disoproxil fumarate and emtricitabine. Eur J Pharm Biopharm 138:3–10

Chambin O, Dupuis G, Champion D, Voilley A, Pourcelot Y (2006a) Colon-specific drug delivery: influence of solution reticulation properties upon pectin beads performance. Int J Pharm 321(1-2):86–93

Chambin O, Dupuis G, Champion D, Voilley A, Pourcelot Y (2006b) Colon-specific drug delivery: influence of solution reticulation properties upon pectin beads performance. Int J Pharm 321(1-2):86–93

Charlton ST, Davis SS, Illum L (2007) Nasal administration of an angiotensin antagonist in the rat model: effect of bioadhesive formulations on the distribution of drugs to the systemic and central nervous systems. Int J Pharm 338(1-2):94–103

Chen J, Liu W, Liu C-M, Li T, Liang R-H, Luo S-J (2015) Pectin modifications: a review. Crit Rev Food Sci Nutr 55(12):1684–1698

Chen Y, Xu C, Huang R, Song J, Li D, Xia M (2018) Butyrate from pectin fermentation inhibits intestinal cholesterol absorption and attenuates atherosclerosis in apolipoprotein E-deficient mice. J Nutr Biochem 56:175–182
Chomto P, Nunthanid J (2017) Physicochemical and powder characteristics of various citrus pectins and their application for oral pharmaceutical tablets. Carbohydr Polym 174:25–31
Das S, Ng K-Y (2010) Colon-specific delivery of resveratrol: optimization of multi-particulate calcium-pectinate carrier. Int J Pharm 385(1-2):20–28
de Vries N, van der Baan S (1989) Toxic shock syndrome after nasal surgery: is prevention possible? A case report and review of the literature. Rhinology 27(2):125–128
Dhalleine C, Assifaoui A, Moulari B, Pellequer Y, Cayot P, Lamprecht A, Chambin O (2011) Zinc-pectinate beads as an in vivo self-assembling system for pulsatile drug delivery. Int J Pharm 414(1):28–34
El-Gibaly I (2002) Oral delayed-release system based on Zn-pectinate gel (ZPG) microparticles as an alternative carrier to calcium pectinate beads for colonic drug delivery. Int J Pharm 232(1-2):199–211
Elmowafy EM, Awad GAS, Mansour S, El-Shamy AE-HA (2009) Ionotropically emulsion gelled polysaccharides beads: preparation, in vitro and in vivo evaluation. Carbohydr Polym 75(1):135–142
Fan L, Cao M, Gao S, Wang W, Peng K, Tan C, Wen F, Tao S, Xie W (2012) Preparation and characterization of a quaternary ammonium derivative of pectin. Carbohydr Polym 88(2):707–712
Fernández-Hervás MJ, Fell JT (1998) Pectin/chitosan mixtures as coatings for colon-specific drug delivery: an in vitro evaluation. Int J Pharm 169(1):115–119
Giunchedi P, Conte U, Chetoni P, Saettone MF (1999) Pectin microspheres as ophthalmic carriers for piroxicam: evaluation in vitro and in vivo in albino rabbits. Eur J Pharm Sci 9(1):1–7
Glinsky VV, Raz A (2009) Modified citrus pectin anti-metastatic properties: one bullet, multiple targets. Carbohydr Res 344(14):1788–1791
Hagesaether E, Bye R, Sande SA (2008) Ex vivo mucoadhesion of different zinc-pectinate hydrogel beads. Int J Pharm 347(1-2):9–15
Hodges LA, Connolly SM, Band J, O'Mahony B, Ugurlu T, Turkoglu M, Wilson CG, Stevens HNE (2009) Scintigraphic evaluation of colon targeting pectin–HPMC tablets in healthy volunteers. Int J Pharm 370(1):144–150
Huynh UTD, Lerbret A, Neiers F, Chambin O, Assifaoui A (2016) Binding of divalent cations to polygalacturonate: a mechanism driven by the hydration water. J Phys Chem B 120(5):1021–1032
Huynh UTD, Assifaoui A, Chambin O (2017) Pellets based on polyuronates: relationship between gelation and release properties. J Food Eng 199:27–35
Illum L (2012) Nasal drug delivery - recent developments and future prospects. J Control Release 161(2):254–263
Jug M, Kosalec I, Maestrelli F, Mura P (2012) Development of low methoxy amidated pectin-based mucoadhesive patches for buccal delivery of triclosan: effect of cyclodextrin complexation. Carbohydr Polym 90(4):1794–1803
Katav T, Liu L, Traitel T, Goldbart R, Wolfson M, Kost J (2008) Modified pectin-based carrier for gene delivery: cellular barriers in gene delivery course. J Control Release 130(2):183–191
Kaur A, Kaur G (2012) Mucoadhesive buccal patches based on interpolymer complexes of chitosan–pectin for delivery of carvedilol. Saudi Pharm J 20:21–27
Kim H, Venkatesh G, Fassihi R (1998) Compactibility characterization of granular pectin for tableting operation using a compaction simulator. Int J Pharm 161(2):149–159
Kim TH, Park YH, Kim KJ, Cho CS (2003) Release of albumin from chitosan-coated pectin beads in vitro. Int J Pharm 250(2):371–383
Kohn R (1975) Ion binding on polyuronates-alginate and pectin. Pure Appl Chem 42:371–397
Lara-Espinoza C, Carvajal-Millán E, Balandrán-Quintana R, López-Franco Y, Rascón-Chu A (2018) Pectin and pectin-based composite materials: beyond food texture. Molecules 23(4):942
Li J, Mooney DJ (2016) Designing hydrogels for controlled drug delivery. Natl Rev 1:16071

Liu L, Fishman ML, Kost J, Hicks KB (2003) Pectin-based systems for colon-specific drug delivery via oral route. Biomaterials 24(19):3333–3343
Long J, Etxeberria AE, Nand AV, Bunt CR, Ray S, Seyfoddin A (2019) A 3D printed chitosan-pectin hydrogel wound dressing for lidocaine hydrochloride delivery. Mater Sci Eng C 104:109873
Maciel VBV, Yoshida CMP, Pereira S, Goycoolea FM, Franco TT (2017) Electrostatic self-assembled chitosan-pectin nano- and microparticles for insulin delivery. Molecules 22:10
Maestrelli F, Zerrouk N, Cirri M, Mennini N, Mura P (2008) Microspheres for colonic delivery of ketoprofen-hydroxypropyl-[beta]-cyclodextrin complex. Eur J Pharm Sci 34(1):1–11
Mahe J, Brachais C-H, Chambin O, Couvercelle J-P (2018) Pectin modification with a solvent free process for the formulation of drug delivery systems. Int J Appl Res Text 6(1):1–4
Maire du Poset A, Lerbret A, Zitolo A, Cousin F, Assifaoui A (2018) Design of polygalacturonate hydrogels using iron(II) as crosslinkers: a promising route to protect bioavailable iron against oxidation. Carbohydr Polym 188:276–283
Macleod GS, Fell JT, Collett JH, Sharma HL, Smith A-M (1999) Selective drug delivery to the colon using pectin:chitosan:hydroxypropyl methylcellulose film coated tablets. Int J Pharm 187 (2):251–257
Mazzitelli S, Pagano C, Giusepponi D, Nastruzzi C, Perioli L (2013) Hydrogel blends with adjustable properties as patches for transdermal delivery. Int J Pharm 454(1):47–57
Monfregola L, Bugatti V, Amodeo P, De Luca S, Vittoria V (2011a) Physical and water sorption properties of chemically modified pectin with an environmentally friendly process. Biomacromolecules 12(6):2311–2318
Monfregola L, Leone M, Vittoria V, Amodeo P, De Luca S (2011b) Chemical modification of pectin: environmental friendly process for new potential material development. Polym Chem 2(4):800–804
Moreira HR, Munarin F, Gentilini R, Visai L, Granja PL, Tanzi MC, Petrini P (2014) Injectable pectin hydrogels produced by internal gelation: pH dependence of gelling and rheological properties. Carbohydr Polym 103:339–347
Mrsny RJ (2012) Oral drug delivery research in Europe. J Control Release 161(2):247–253
Munarin F, Tanzi MC, Petrini P (2012) Advances in biomedical applications of pectin gels. Int J Biol Macromol 51(4):681–689
Mura P, Mennini N, Kosalec I, Furlanetto S, Orlandini S, Jug M (2015) Amidated pectin-based wafers for econazole buccal delivery: formulation optimization and antimicrobial efficacy estimation. Carbohydr Polym 121:231–240
Nguyen S, Hiorth M, Rykke M, Smistad G (2013) Polymer coated liposomes for dental drug delivery – interactions with parotid saliva and dental enamel. Eur J Pharm Sci 50(1):78–85
Nguyen AT-B, Winckler P, Loison P, Wache Y, Chambin O (2014) Physico-chemical state influences in vitro release profile of curcumin from pectin beads. Colloids Surf B Biointerfaces 121:290–298
Nižić L, Ugrina I, Špoljarić D, Saršon V, Kučuk MS, Pepić I, Hafner A (2019) Innovative sprayable in situ gelling fluticasone suspension: development and optimization of nasal deposition. Int J Pharm 563:445–456
Noreen A, Nazli Z-I-H, Akram J, Rasul I, Mansha A, Yaqoob N, Iqbal R, Tabasum S, Zuber M, Zia KM (2017) Pectins functionalized biomaterials; a new viable approach for biomedical applications: a review. Int J Biol Macromol 101:254–272
O'Brien AB, Philp K, Morris ER (2009) Gelation of high-methoxy pectin by enzymic de-esterification in the presence of calcium ions: a preliminary evaluation. Carbohydr Res 344(14):1818–1823
Patil H, Tiwari RV, Repka MA (2016) Recent advancements in mucoadhesive floating drug delivery systems: a mini-review. J Drug Deliv Sci Technol 31:65–71
Perera G, Barthelmes J, Bernkop-Schnürch A (2010) Novel pectin–4-aminothiophenole conjugate microparticles for colon-specific drug delivery. J Control Release 145(3):240–246

Pliszczak D, Bourgeois S, Bordes C, Valour JP, Mazoyer MA, Orecchioni AM, Nakache E, Lantéri P (2011) Improvement of an encapsulation process for the preparation of pro- and prebiotics-loaded bioadhesive microparticles by using experimental design. Eur J Pharm Sci 44(1):83–92
Prezotti FG, Cury BSF, Evangelista RC (2014) Mucoadhesive beads of gellan gum/pectin intended to controlled delivery of drugs. Carbohydr Polym 113:286–295
Ruffin E, Schmit T, Lafitte G, Dollat J-M, Chambin O (2014) The impact of whey protein preheating on the properties of emulsion gel bead. Food Chem 151:324–332
Sanzgiri YD, Maschi S, Crescenzi V, Callegaro L, Topp EM, Stella VJ (1993) Gellan-based systems for ophthalmic sustained delivery of methylprednisolone. J Control Release 26(3):195–201
Shahzad A, Khan A, Afzal Z, Umer MF, Khan J, Khan GM (2019) Formulation development and characterization of cefazolin nanoparticles-loaded cross-linked films of sodium alginate and pectin as wound dressings. Int J Biol Macromol 124:255–269
Sharma R, Ahuja M, Kaur H (2012) Thiolated pectin nanoparticles: preparation, characterization and ex vivo corneal permeation study. Carbohydr Polym 87(2):1606–1610
Sharma AK, Arya A, Sahoo PK, Majumdar DK (2016) Overview of biopolymers as carriers of antiphlogistic agents for treatment of diverse ocular inflammations. Mater Sci Eng C 67:779–791
Skjåk-Bræk G, Grasdalen H, Smidsrød O (1989) Inhomogeneous polysaccharide ionic gels. Carbohydr Polym 10(1):31–54
Sriamornsak P (1998) Investigation of pectin as a carrier for oral delivery of proteins using calcium pectinate gel beads. Int J Pharm 169(2):213–220
Sriamornsak P (1999) Effect of calcium concentration, hardening agent and drying condition on release characteristics of oral proteins from calcium pectinate gel beads. Eur J Pharm Sci 8(3):221–227
Sriamornsak P, Nunthanid J (1998) Calcium pectinate gel beads for controlled release drug delivery: I. Preparation and in vitro release studies. Int J Pharm 160(2):207–212
Sriamornsak P, Thirawong N, Puttipipatkhachorn S (2005) Emulsion gel beads of calcium pectinate capable of floating on the gastric fluid: effect of some additives, hardening agent or coating on release behavior of metronidazole. Eur J Pharm Sci 24(4):363–373
Sriamornsak P, Sungthongjeen S, Puttipipatkhachorn S (2007a) Use of pectin as a carrier for intragastric floating drug delivery: carbonate salt contained beads. Carbohydr Polym 67(3):436–445
Sriamornsak P, Sungthongjeen S, Puttipipatkhachorn S (2007b) Use of pectin as a carrier for intragastric floating drug delivery: carbonate salt contained beads. Carbohydr Polym 67(3):436–445
Stops F, Fell JT, Collett JH, Martini LG (2008) Floating dosage forms to prolong gastro-retention--the characterisation of calcium alginate beads. Int J Pharm 350(1-2):301–311
Sultatos L (2007) Routes of drug administration. In: Enna SJ, Bylund DB (eds) xPharm: the comprehensive pharmacology reference. Elsevier, New York, pp 1–3
Swarbrick J, Boylan JC (2002) Tablet formulation and tablet manufacture. In: Basel MD (ed) Encyclopedia of pharmaceutical technology. Elsevier, New York, pp 2701–2732
Thirawong N, Nunthanid J, Puttipipatkhachorn S, Sriamornsak P (2007) Mucoadhesive properties of various pectins on gastrointestinal mucosa: an in vitro evaluation using texture analyzer. Eur J Pharm Biopharm 67(1):132–140
Tian L, Scholte J, Borewicz K, van den Bogert B, Smidt H, Scheurink AJW, Gruppen H, Schols HA (2016) Effects of pectin supplementation on the fermentation patterns of different structural carbohydrates in rats. Mol Nutr Food Res 60(10):2256–2266
Torre M, Rodriguez AR, Saura-Calixto F (1995) Interactions of Fe(II), Ca(II) and Fe(III) with high dietary fibre materials: a physicochemical approach. Food Chem 54(1):23–31
Tung N-T, Pham T-M-H, Nguyen T-H, Thanh Tam P, Nguyen T-Q (2016) Pectin/HPMC dry powder coating formulations for colon specific targeting tablets of metronidazole. J Drug Deliv Sci Technol 33:19–27
Turkoglu M, Ugurlu T (2002) In vitro evaluation of pectin-HPMC compression coated 5-aminosalicylic acid tablets for colonic delivery. Eur J Pharm Biopharm 53:65–73

Valenta C (2005) The use of mucoadhesive polymers in vaginal delivery. Adv Drug Deliv Rev 57(11):1692–1712

Wakerly Z, Fell JT, Attwood D, Parkins D (1997) Studies on drug release from pectin/ethylcellulose film-coated tablets: a potential colonic delivery system. Int J Pharm 153(2):219–224

Wang S-Y, Li J, Zhou Y, Li D-Q, Du G-M (2019) Chemical cross-linking approach for prolonging diclofenac sodium release from pectin-based delivery system. Int J Biol Macromol 137:512–520

Zhang W, Xu P, Zhang H (2015) Pectin in cancer therapy: a review. Trends Food Sci Technol 44(2):258–271

The manufacturer's authorised representative in the EU is Springer Nature Customer Service Centre GmbH, Europaplatz 3, 69115 Heidelberg, Germany. If you have any concerns regarding our products, please contact ProductSafety@springernature.com

Printed and bound by CPI Group (UK) Ltd, Croydon, CR0 4YY
30/09/2026
02225503-0001